SURFACE CRYSTALLOGRAPHY

An Introduction to Low Energy Electron Diffraction

SURFACE CRYSTALLOGRAPHY

An Introduction to Low Energy Electron Diffraction

L. J. Clarke

The Cavendish Laboratory
Cambridge

A Wiley–Interscience Publication

JOHN WILEY & SONS
Chichester · New York · Brisbane · Toronto · Singapore

Library of Congress Cataloging in Publication Data:

Clarke, L. J.
Surface crystallography.

'A Wiley–Interscience publication.'
Includes index.
1. Crystallography. 2. Surface chemistry. 3. Electrons —Diffraction. I. Title.
QD921.C53 1985 548.8 84-11804

ISBN 0 471 90513 5

British Library Cataloguing in Publication Data:

Clarke, L. J.
Surface crystallography.
1. Electrons—Diffraction
I. Title
548′.83 QC793.5.E628

ISBN 0 471 90513 5

Typeset by Spire Print Services Ltd, Salisbury, Wilts
Printed by Page Bros., (Norwich) Ltd.,

To Liz

Contents

Preface

One of the major growth areas in contemporary research is the study of surfaces. ‘Surface science’ spans the traditional disciplines of chemistry and physics—on the one hand it may be applied to the problems of chemisorption and catalysis, and on the other to the properties of semiconductor surfaces, but these are only two examples from the many potential applications of this field of research. Despite the enormous importance of any insights which may be gained in surface behaviour, the study of surfaces progressed slowly until very recently due to experimental difficulties. Improvements in vacuum technology and the development of appropriate analytical techniques during the 1960s provided the impetus for a rapid expansion in the field, and as a consequence many hundreds of laboratories throughout the world are now equipped to undertake fundamental studies of surfaces on an atomic scale. The information attainable from surface studies can be divided roughly into geometrical and electronic, but the interpretation of the latter frequently requires a knowledge of the former. Low-energy electron diffraction or LEED is probably the best-known technique available for obtaining geometrical information, and its similarities to X-ray diffraction have led to its adoption as the basis for the new science of ‘surface crystallography’. Although numerous other techniques have been recently proposed which may provide information concerning the geometrical arrangement of atoms at a crystal surface, none have been studied as widely as LEED, and it is LEED which has usually provided the information by which other techniques are judged.

The number of publications which deal with LEED methods or discuss results of LEED studies now runs into thousands, and these present a bewildering collection to any newcomer to the field. The initial period of rapid development of ideas and approaches appears to have reached an end. Further developments, refinements and extensions will certainly continue to be made in the future, but in its present state LEED already provides a powerful and reliable probe of surface structures, and much work remains which may be undertaken using methods that are now available. The time seems ripe for consolidation of the position which LEED has now reached, and it is hoped that this book will provide a valuable reference for years to come, since future work will primarily build upon rather than supersede present methods.

This book is written in the hope that it will satisfy the needs of at least three groups of people: (a) undergraduates in physical and chemical sciences, to introduce an aspect of crystallography which is a vital aspect of present research; (b) researchers in any area of surface science who may use the results of LEED work, and need to be better informed of the way in which these results were obtained (this includes theoreticians as well as experimentalists); and (c) researchers who presently use or intend using LEED as a technique and require a guide to its fundamentals together with a réumé of the capabilities of the technique at present.

From a teaching point of view, the crystallography of surfaces has many advantages over traditional X-ray crystallography. Many of the basic concepts of the two crystallographies are the same (e.g. lattice types, the reciprocal lattice, space group symmetry) but the restriction to two dimensions considerably simplifies the picture, enabling the student to obtain a good understanding of the underlying principles without getting too involved in the complexities associated with three dimensional structures. Also, most research in this field has been undertaken very recently—the great majority of references in this book are to papers published within the last five to ten years. Consequently, the student can be brought very rapidly into the sphere of current research. This not only makes the study much more interesting, but, given that surface science continues to be a major growth area in contemporary research, enhances the chances that he will be able to use this knowledge directly in the future.

This books aims to cover all the essential ideas and information necessary to progress from the basics of surface crystallography to current research problems. Key references to the appropriate literature are provided so that any particular aspect may be followed up in detail as required.

As I mentioned earlier, literally thousands of papers have been published concerning the subject of LEED, and my main problem in writing this book has been to decide what should be left out. I hope that my chosen selection provides a representative overview of most aspects, without allowing any particular point to be obscured by too much detail. The references I have chosen should provide a starting-point for the reader to follow in more detail any particular line of thought or investigation. I have provided numerous examples from my own work, not in any way wishing to detract from the many alternative examples available in the literature, or from any delusions of superiority, but simply from the obvious practicality of using and discussing data with which I am most familiar.

I must extend my sincerest gratitude to all those who have helped me, both in the past, as I have developed my expertise in the subject, and more recently, in the practical aspects of putting together this book. Thank you to all who have sent me diagrams or information for inclusion: I am sorry if, for reasons just explained, if I could not incorporate everything that has been suggested to me. Responsibility for the final selection, warts and all, lies entirely upon my shoulders, as does the responsibility for any errors or misinterpretations that may have unwittingly slipped in. In particular, I would like to express my gratitude to all those with whom I have worked closely, both in the Cavendish Laboratory, Cambridge, where I began my research career, under the guidance of Professor John Pendry and Dr. Jim Wilson, and in the Université Scientifique et Médicale de Grenoble, France, where the idea for this book took shape. I would like to acknowledge the helpful discussions I had with Professor Daniel Aberdam, Dr. Robert Baudoing and Dr. Yves Gauthier during my year in France. I would like especially to thank my family for all their help and patience during the long months of toil, as the manuscript was slowly moulded into its final form.

L. J. Clarke 1984

CHAPTER 1

Surface structures and diffraction

1.1 Historical development

In 1927, working in the USA, Davisson and Germer discovered that by firing electrons with energies lying between 15 and 200 electron-volts (eV) at a crystal of nickel, angular variations in the reflected flux were produced consistent with electron diffraction.[1] Independently in Britain, G. P. Thomson and Reid used electrons with energies between 3900 and 16 500 eV to produce diffraction patterns from transmission through a thin sample of celluloid.[2] A few months later, Thomson reported the observation of transmission diffraction patterns from a thin film of platinum, using electrons with energies between 30 000 and 60 000 eV.[3] These investigations clearly demonstrated the wave-like nature of electrons, consistent with the theories of de Broglie, yet never previously observed. The difficulties associated with these early experiments should not be underestimated: a high vacuum was essential for the cleanliness of the crystal surfaces and the propagation of the electron beams, particularly in the case of the low-energy electron study, and the preparation of the samples required a degree of perfection rarely attained at the time. Davisson and Germer had been working with a polycrystalline sample of Ni for some time before an accidental overheating fortuitously annealed the surface and resulted in the coalescence of the many small crystallites into a fewer number of large crystallites, considerably aiding the observation of the diffraction phenomenon. Davisson and Thomson later received Nobel prizes for their pioneering work.

For many years, subsequent work in Britain concentrated on aspects of the high-energy electron diffraction process, and it was in the USA that the subject of low-energy electron diffraction (LEED) for surface crystallography was developed, largely through the efforts of Germer at the Bell Telephone Laboratories, where he had participated in the original observation, and Farnsworth at Brown University. A separate school of LEED studies was established in the USSR at the Leningrad Academy of Sciences. Led by Lashkarev and later by Kalashnikov, a series of important contributions were made concerning the determination of inner potentials, the effects of thermal vibrations and the use of data averaging for structural determinations. This work continued from 1932 to 1941 at which point publications ceased and the work came to an untimely end. An interesting bibliography of early work is noted in the general references to this chapter.

Diffraction from crystals using X-rays was already well established as a crystallographic technique, but X-rays are scattered relatively weakly by matter and therefore penetrate deeply into the crystal. X-Ray diffraction is therefore quite insensitive to the geometrical structure of the atoms at the surface of a crystal. By comparison, electrons interact strongly with matter and so cannot penetrate so deeply. They are also rather versatile as crystallographic probes, because it is very easy to change their energies. If E is the energy of an electron with respect to the zero of the crystal potential, then the wavelength in ångstrom units (Å) where 1 Å = 10^{-10} m is given by the equation

$$\lambda = \sqrt{\frac{150.4}{E(\mathrm{eV})}} \tag{1.1}$$

For energies between about 30 and 500 eV (1 eV = 1.9×10^{-19} J), their wavelengths are comparable to the lattice spacing of typical crystals.

At the Fermi energy itself, the path length of electrons in a perfect crystal is infinite. Energies of a few volts above the Fermi energy are sufficient to excite phonons, or collective vibrational modes of the ion cores in the crystal. More significantly, at around 10–20 eV above the Fermi energy, plasmons may be excited; these are collective modes of bound electron vibrations. These are capable of removing substantial fractions of the incident electron energy, and consequently reduce the typical penetration depth to a small number of interatomic spacings. Although the penetration depth of electrons will vary from material to material, the results of many experiments show that this depth does not vary significantly until the incident beam energy reaches 150–200 eV, above which it will steadily increase.[4] This is illustrated in Figure 1.1. The penetration depth in this context is defined as the distance at which only e^{-1} or 0.712 of the initial beam intensity continues to propagate without energy loss.

These two factors–a limited penetration depth and suitable wavelength–are the basic reasons why LEED was considered to be a suitable probe of surface geometrical structures.

Despite considerable interest in the possibilities of a surface crystallographic technique, initial development was very slow. The preparation of clean surfaces was hampered by poor vacua, and the observation of the diffraction patterns was very arduous, being made by the collection of electrons in a Faraday cup manipulated by hand. The vacuum chamber was usually constructed by glass or of brass, lacquered on the outside to provide an airtight seal. Vacuum pumps were mechanical, the base pressure limited by the vapour pressure of the oils necessary for their operation. In 1934 Ehrenberg proposed a method of displaying the electrons on a fluorescent screen by post-acceleration, thereby producing directly visible diffraction patterns.[5] Despite this notable simplification in the technique, it was not until about 1960 that suitable advances in ultra-high-vacuum (UHV) technology, for example in the manner of

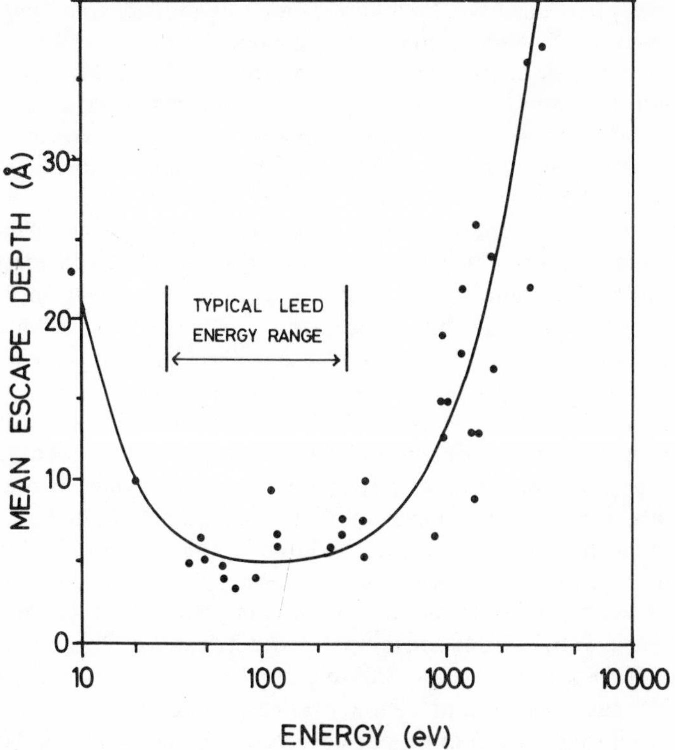

Figure 1.1. Variation in mean escape depth with increasing electron energy. (Reproduced by permission of Taylor and Francis from Rivière[4].)

construction and choice of materials for the vacuum vessels, development of contamination-free electron guns and suitable UHV pumps and progress in specimen preparation, permitted the widespread adoption of LEED as a surface structural probe.

Another significant event in the development of LEED was the introduction of Auger electron spectroscopy (AES) as a means of analysing the chemical purity of surfaces. The original observation of this phenomenon was made by Auger 1925[6] but, as with LEED, it was necessary to await the development of suitable vacuum technology and suitable electronic detection methods before it could become available as a tool for surface science. The identification of Auger transitions was first made in 1953 by Lander,[7] but an important advance was the demonstration, by Harris in 1968,[8] that Auger spectra could be enhanced considerably by electronically differentiating the detected electron energy distribution. At about the time Weber and Peria[9] showed that LEED systems employing three grid optics were suitable for use as a high-pass filter to observe Auger peaks, thus enabling LEED observations to be combined in the same piece of apparatus with a monitor of the cleanliness of the surface. Palmberg[10] took this approach one stage further by demonstrating that replacement of the middle analysis grid by a pair of grids enhanced the energy resolution of this device. Although there is considerable interest in the fine structure of Auger spectra as a means of providing information about the chemical environment or bonding mechanism of surface atoms, discussions of AES results in this book will be restricted to their role as a means of checking the cleanliness of a surface, or monitoring the fractional coverage of adsorbed species. Apart from hydrogen and helium, which do not have sufficient electrons for the process, virtually all atomic species are detectable if present in sufficient concentrations. Most can be detected in concentrations down to 0.01 monolayers using standard LEED optics as a high-pass retarding field analyser (RFA), and many can be detected down to 0.001 monolayers using a cylindrical mirror analyser (CMA) which operates as a band-pass filter.

Once confidence in experimental LEED results had been gained, stemming from better vacua and the use of AES to monitor surface contaminants, considerable efforts were made to develop suitable multiple scattering theories of electron diffraction with the intention of using LEED for surface crystallography. The first successful determinations of surface atomic geometries were made in about 1970, and were soon followed by a considerable number of investigations in different laboratories throughout the world. It took at least another five years to build up the understanding necessary to permit quantitatively data reliable to be obtained, but progress has been such that in the period of one decade details of about 100 surface structures have been reported. These investigations describe not only the structures of clean metal surfaces and adsorbed atomic and molecular species, but reconstructed surfaces of metals, semiconductors and insulators, underlayers and epitaxially grown species.

Although interest in recent years has focused on the development of LEED for three-dimensional crystallography of the surface region, a considerable body of work has been carried out observing simply the two-dimensional periodicity of crystal surfaces, investigating whether the bulk periodicity is maintained at the surface, seeing how adsorbed atoms would structure themselves on such surfaces or examining the onset and development of epitaxial growth. Chapters 1 and 6 of this book are devoted to the methods by which such information may be extracted from observations of the two-dimensional diffraction pattern.

1.2 Diffraction and the reciprocal lattice

Diffraction results from the interaction between the periodic oscillations of a wavefield and a periodic array of scattering centres. Scattering from individual centres may be very small, but if the scattered waves from successive centres are in phase, then the net result may be significant. The periodicity of the waves and scatterers will normally provide a number of different possible conditions at which strong in-phase scattering may occur, and this accounts for the production of a whole series of diffracted beams, rather than a single specular reflection.

If the array of scatterers is infinitely periodic and the incident beam has a precise wavelength, then the mathematical treatment of the process is relatively simple. When considering diffraction from surfaces, the assumption of infinite periodicity is usually well approximated in two dimensions, but clearly not in the third dimension which is truncated by the presence of the surface itself. Moreover, low energy electrons penetrate only a short distance, and so the periodicity of the structures beyond the top few layers is relatively weakly explored in the direction of the normal to the surface.

The crystal structure actually examined in a typical LEED study may therefore be considered to be a slab, infinitely periodic in two dimensions, but comprising only a limited number of layers in the third dimension (see Figure 1.2). In practice, the coherence length of the impinging electron beam will impose limits of the two-dimensional periodicity which contributes to the diffraction process and this will be discussed in detail in Chapter 2, but for the present, the assumption of infinite periodicity is a fair one.

The surface periodicity is defined in terms of a lattice. A *lattice* is defined as the simplest arrangement of points which are arranged in space with the fundamental periodicity of the crystal. These lattice points need not

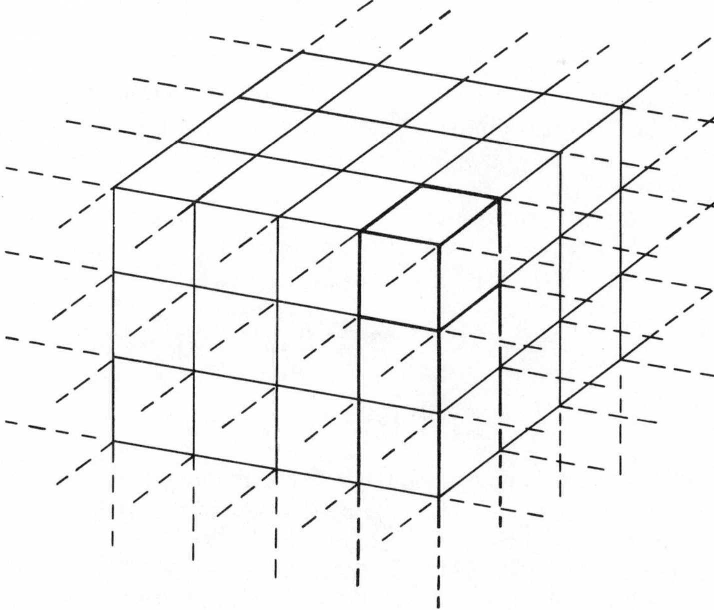

Figure 1.2. Crystal surface pictured as a two-dimentional lattice with the unit cell extended into the third dimension

correspond to actual atomic centres—the relationship between the lattice and the physical arrangement of atoms is provided by the definition of the *unit cell*, which specifies the repeat unit of the physical structure. The *primitive unit cell* is defined as the smallest repeating unit which can fully describe the crystal structure.

The real space two-dimensional lattice is specified by two *lattice vectors* $\mathbf{a}_1$ and $\mathbf{a}_2$. These define the sides of a parallelogram which form the boundaries of the unit cell.

Within its bulk, a crystal will be periodic in three dimensions and this can be described in terms of a three-dimensional lattice. Corresponding unit cells will be defined by the volume enclosed by the three lattice vectors $\mathbf{a}_1$, $\mathbf{a}_2$ and $\mathbf{a}_3$. At the surface the layers may not be equally spaced and $\mathbf{a}_3$ is ill-defined. One can think of the diffracting structure in terms of a two-dimensional lattice in which the unit cell is extended in the third dimension.

A plane wave incident on an atom or atoms within a unit cell will be scattered in all directions, but interference between waves scattered from neighbouring unit cells will restrict the net flux to those directions in which the scattered waves from all unit cells are in phase. This requires that the scattered waves from neighbouring cells differs only by an integral number of wavelengths λ. If there is only one atom per unit cell, the arrangement of atomic scatterers is the same as the lattice itself. To illustrate the effect, we may consider the scattering of a plane wave incident at an angle θ_0 with respect to a one-dimensional lattice, as shown in Figure 1.3. The in-phase condition is met for all integers n which satisfy the condition

$$a(\sin\theta_n - \sin\theta_0) = n\lambda \qquad (1.2)$$

where a is the separation distance between scatterers and λ is given by equation (1.1). This is commonly called the *Laue condition*. If the incident and emergent beams are described by unit vectors $\mathbf{s}_0$ and $\mathbf{s}'_n$ then this can be written in vector form as

$$\mathbf{a}\cdot(\mathbf{s}'_n - \mathbf{s}_0) = n\lambda \qquad (1.3)$$

or

$$\mathbf{a}\cdot\Delta\mathbf{s}_n = n\lambda$$

where

$$\Delta\mathbf{s}_n = (\mathbf{s}'_n - \mathbf{s}_0)$$

The diffracted beams are determined by $\Delta\mathbf{s}_n$ and, in the one-dimensional case, it is clear that they are given by integral multiples of the basic unit $(\lambda/|\,\mathbf{a}\,|)$. This involves the *reciprocal* of the real space lattice vector $\mathbf{a}$. We can define a *reciprocal lattice vector* $a^* = (1/a)$. In this simple example, the atoms are arranged along a single line.

A simple construction permits the diffraction beam angles to be derived from the reciprocal lattice. In describing the diffraction process, we are only

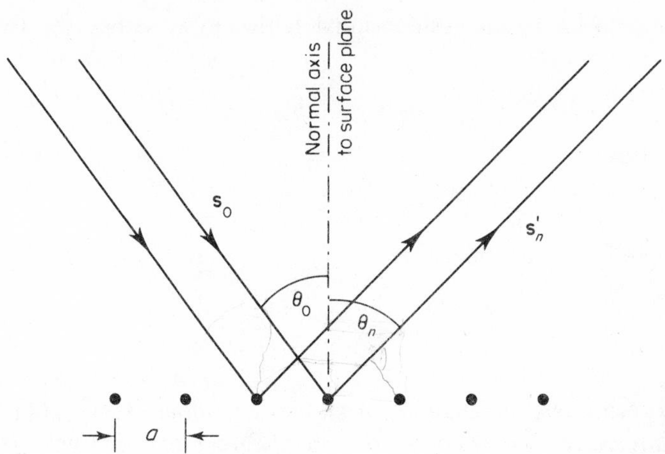

Figure 1.3. Laue diffraction from a one-dimentional array of scatterers

concerned with waves which do not lose energy, and so their wavelengths are the same on incidence and emergence. Propagating waves in reciprocal space can be represented by lines of length $(1/\lambda)$, and in this example the incident wave will have a direction at an angle θ with respect to these planes. We shall give this vector the symbol $\mathbf{k}_0$. The construction is illustrated in Figure 1.4, where, for clarity, the reciprocal lattice planes have been illustrated as lines. The wave vector $\mathbf{k}_0$ is positioned so that one end touches a lattice line, and the other end provides the centre for a circle of radius $(1/\lambda)$ or $|\mathbf{k}_0|$. Wave vectors which touch this circle have the same energy as the incident beam, and whenever this circle cuts a reciprocal lattice line the difference between the components of the resulting $\mathbf{k}$ and the incident $\mathbf{k}_0$ in a direction parallel to $\mathbf{a}$ must equal an integral number of reciprocal lattice vectors $n\mathbf{a}^*$. This satisfies the condition given in equation (1.3) for possible diffraction beams.

In two dimensions, the real-space lattice is described by two lattice vectors $\mathbf{a}_1$ and $\mathbf{a}_2$, and a corresponding reciprocal lattice can be constructed from basis vectors $\mathbf{a}_1^*$ and $\mathbf{a}_2^*$, defined by the following relations:

$$\mathbf{a}_1 \cdot \mathbf{a}_1^* = \mathbf{a}_2 \cdot \mathbf{a}_2^* = 1 \tag{1.4}$$

and

$$\mathbf{a}_1 \cdot \mathbf{a}_2^* = \mathbf{a}_1^* \cdot \mathbf{a}_2 = 0 \tag{1.5}$$

The second relation requires that $\mathbf{a}_2^*$ be perpendicular to $\mathbf{a}_1$ and $\mathbf{a}_1^*$ be perpendicular to $\mathbf{a}_2$. Any vector relating two reciprocal lattice points must take the form

$$\mathbf{g}_{hk} = h\mathbf{a}_1^* + k\mathbf{a}_2^* \tag{1.6}$$

where h and k are integers.

Diffraction from the two-dimensional lattice must satisfy the two Laue conditions

$$\bar{a}_1 \cdot \Delta\mathbf{s} = h\lambda \tag{1.7}$$

and

$$\mathbf{a}_2 \cdot \Delta\mathbf{s} = k\lambda \tag{1.8}$$

and these can be solved whenever

$$\Delta\mathbf{s} = \lambda(h\mathbf{a}_1^* + k\mathbf{a}_2^*) \tag{1.9}$$

$$= \lambda\mathbf{g}_{hk} \tag{1.10}$$

which may verified by direct substitution into equations (1.7) and (1.8). This means that *there is a direct correspondence between the observed diffraction pattern and the reciprocal lattice of the surface*.

As the lattice vector in any direction is increased, so the corresponding reciprocal lattice vector will decrease. For example, if **a** is increased indefinitely towards infinity, so the reciprocal lattice points given by $\mathbf{a}^*$ will converge towards zero, thereby forming a continuous line. In practice this means that the scattered beams in that direction could be observed at all angles and energies since the diffraction process which contributes destructive and constructive interference effects is eliminated. Equally well, we could imagine the two-dimensional plane to be a member of a three-dimensional crystal. If the separation between planes, $\mathbf{a}_3$, is increased towards infinity, the two-dimensional situation is approached. The reciprocal lattice is given by the discrete two-dimensional mesh given by equation (1.10), plus continuous lines or rods in the third dimension, aligned in a direction normal to the real-space plane.

It is then possible to envisage a geometrical construction in reciprocal space which would simulate the diffraction process. In reciprocal space, the incident beam may be represented by a vector of length $(1/\lambda)$ and direction the same as in real space. If the scattered beams do not lose energy then their wavelengths will also be λ, and they must each have a length $(1/\lambda)$ in reciprocal space. We have already seen, from equation (1.10), that the change in beam vectors on diffraction is given by the spacing between reciprocal lattice vectors about some origin ($\mathbf{a}_1^* = \mathbf{a}_2^* = 0$) in reciprocal space. If the incident beam is designated by a vector that touches a reciprocal lattice rod at one end, and this point of intersection is denoted as the origin in reciprocal space, then all possible diffracted beams are given by those vectors which have length $(1/\lambda)$ about the other end of the incident beam vector (i.e. lie on a sphere of radius $(1/\lambda)$ about that point) and meet reciprocal lattice rods. This construction, known as 'Ewald construction' is shown in Figure 1.4. For simplicity, only diffraction from a one-dimensional lattice is illustrated. The sphere which defines the locus of beam vectors which have the same energy as the incident beam is known as the 'Ewald sphere', and the intersection of this sphere with the reciprocal lattice rods defines the conditions under which

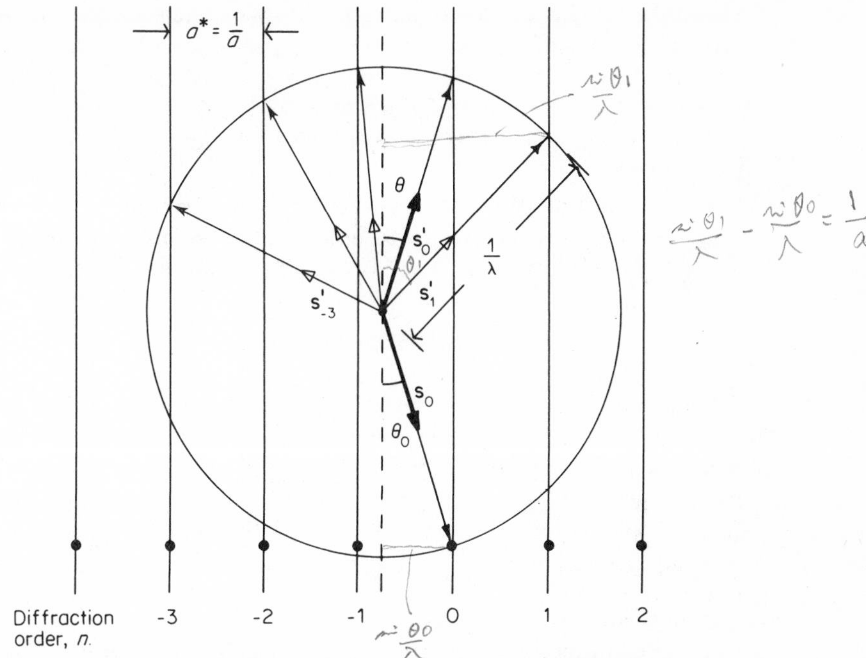

Figure 1.4. The Ewald construction for a one-dimensional system

both the diffraction condition (equation 1.10) and the conservation of beam energy are satisfied simultaneously.

As the incident energy is increased, so the sphere radius will also increase. Consequently, the number of diffraction beams increases and the angle between each diffraction beam decreases: the diffraction pattern will appear to condense towards the specular beam.

The Ewald construction provides an interesting means of visualizing the diffraction process, but is not really a very practical method for interpreting diffraction patterns. Mathematically, it is much simpler to restrict our attention to the two-dimensional reciprocal lattice. The total number of diffraction beams which can propagate at any energy can be found from the projection of the Ewald sphere on to the two-dimensional reciprocal lattice, and this forms the basis for determining the number of diffraction beams to be included in detailed computer calculations of the beam intensities. This will be described in more detail in Chapter 4.

1.3 Interpretation of diffraction patterns

1.3.1 Clean, unreconstructed surfaces

We have seen in section 1.2 that the diffraction pattern can be related to a reciprocal lattice. This enables us to derive the real-space lattice which must have produced the observed pattern. A two dimensional pattern can be described in terms of two reciprocal lattice basis vector lengths and the angle which subtends them. The relationship between the basis vectors in reciprocal and real space is given in equations (1.4) and (1.5). It is interesting to note that a simple relationship also exists between the angles which subtend these basis vectors.

If $\mathbf{a}_1$ and $\mathbf{a}_2$ in real space are separated by an angle α, and $\mathbf{a}_1^*$ is at $\pi/2$ to $\mathbf{a}_2$, then the angle between $\mathbf{a}_1$ and its complementary reciprocal vector $\mathbf{a}_1^*$ must be $(\pi/2) - \alpha$. By definition, $\mathbf{a}_1 \cdot \mathbf{a}_1^* = 1$, therefore, from the definition of the dot product

$$1 = |\mathbf{a}_1| \cdot |\mathbf{a}_1^*| \cdot \cos[(\pi/2) - \alpha]$$

so that

$$|\mathbf{a}_1| = \frac{1}{|\mathbf{a}_1^*| \cdot \sin \alpha} \tag{1.11}$$

The angle between $\mathbf{a}_1^*$ and $\mathbf{a}_2^*$, $\alpha^* = (\pi - \alpha)$ and so

$$\sin \alpha^* = \sin \alpha \tag{1.12}$$

this is illustrated in Figure 1.5.

The bulk structure of a crystal is generally well known from X-ray crystallography, and if the orientation of the surface with respect to the bulk structure is also known, then the ideal clean surface structure is easy to predict. Not infrequently, geometrical reconstruction of the surface layer

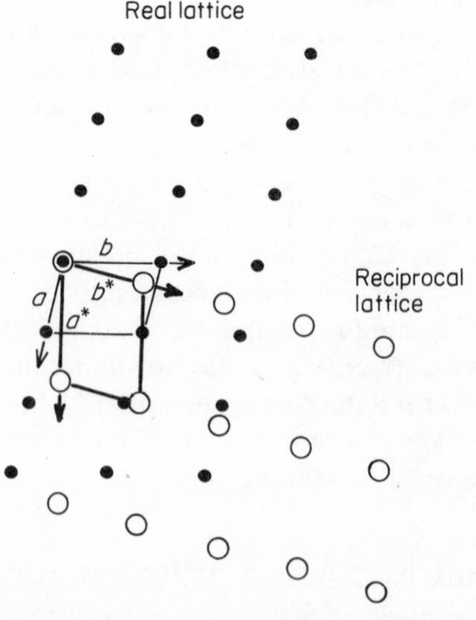

Figure 1.5. The relationship between an oblique direct lattice and its reciprocal lattice

occurs, and this will alter the diffraction pattern. It is important to verify that the diffraction pattern obtained from the clean surface does indeed correspond to the expected form. Even if the stable form of the clean surface is reconstructed from the idea bulk termination, overlayer patterns are usually defined in terms of the unit vectors of the ideal unreconstructed surface.

1.3.2 Overlayer structures

In general a surface structure may have lattice vectors $\mathbf{b}_1$ and $\mathbf{b}_2$ which differ from the substrate lattice vectors $\mathbf{a}_1$ and $\mathbf{a}_2$. They can nevertheless always be described in terms of the substrate lattice vectors as follows:

$$\begin{aligned} \mathbf{b}_1 &= m_{11}\mathbf{a}_1 + m_{12}\mathbf{a}_2 \\ \mathbf{b}_2 &= m_{21}\mathbf{a}_1 + m_{22}\mathbf{a}_2 \end{aligned} \tag{1.13}$$

which can be expressed in matrix notation:

$$\begin{pmatrix} \mathbf{b}_1 \\ \mathbf{b}_2 \end{pmatrix} = \begin{pmatrix} m_{11} & m_{12} \\ m_{21} & m_{22} \end{pmatrix} \begin{pmatrix} \mathbf{a}_1 \\ \mathbf{a}_2 \end{pmatrix} \tag{1.14}$$

or

$$\mathbf{b} = M \cdot \mathbf{a} \tag{1.15}$$

A corresponding relationship between reciprocal lattice vectors may be similarly defined:

$$\mathbf{b}^* = M^* \cdot \mathbf{a}^* \tag{1.16}$$

where

$$M^* = \begin{pmatrix} m_{11}^* & m_{12}^* \\ m_{21}^* & m_{22}^* \end{pmatrix} \tag{1.17}$$

Standard matrix algebra permits these matrices to be directly related, from which we obtain the following expression:

$$\begin{pmatrix} m_{11} & m_{12} \\ m_{21} & m_{22} \end{pmatrix} = \frac{1}{\det M^*} \begin{pmatrix} m_{22}^* & -m_{21}^* \\ -m_{12}^* & m_{11}^* \end{pmatrix} \tag{1.18}$$

where

$$\det M^* = m_{11}^* \cdot m_{22}^* - m_{21}^* \cdot m_{12}^* \tag{1.19}$$

This relationship is sufficient to permit the real lattice structure to be derived from the observed diffraction pattern, provided the appropriate reciprocal lattice vectors can be extracted from the pattern observed. If the overlayer comprises a single lattice that is simply related to the substrate lattice, then this is usually quite straightforward.

Example. The diffraction pattern shown in Figure 1.6(a) occurs very

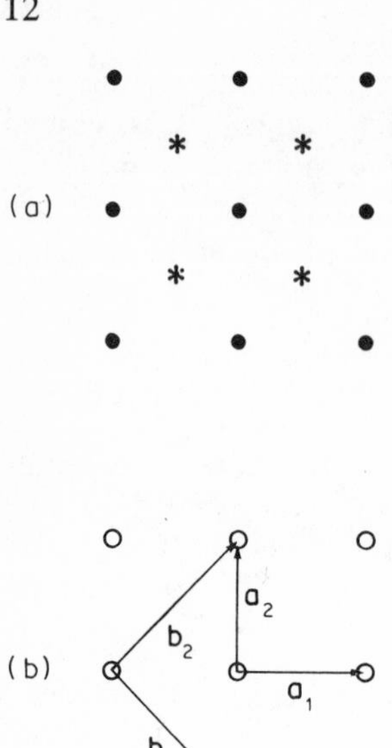

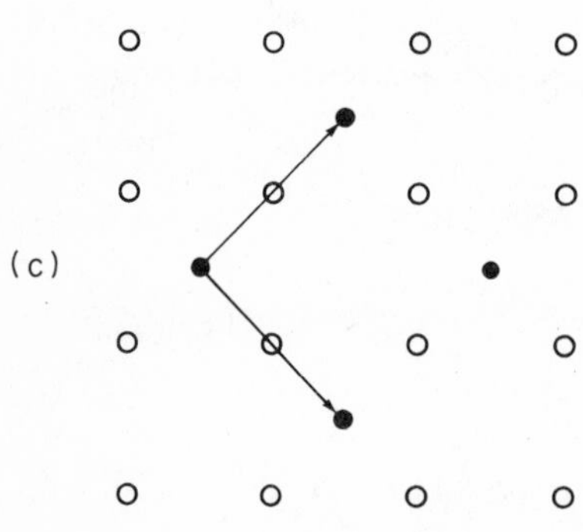

Figure 1.6. (a) Diffraction pattern: the dots represent clean surface beams, starred beams result from adsorption of an overlayer; (b) overlayer vectors derived from the diffraction pattern; (c) possible overlayer structure producing observed pattern

frequently in LEED studies. The dots indicate the arrangement of diffracted beams corresponding to a clean unreconstructed surface and the adsorption of a gas results in the appearance of additional spots, as indicated by the open circles. The diffraction pattern may be directly related to the reciprocal lattice, and the reciprocal lattice vectors $\mathbf{b}_1^*$ and $\mathbf{b}_2^*$ which describe the overlayer mesh are related to the substrate (original clean surface) vectors $\mathbf{a}_1^*$ and $\mathbf{a}_2^*$ as follows:

$$\mathbf{b}_1^* = 0.5\mathbf{a}_1^* - 0.5\mathbf{a}_2^*$$

$$\mathbf{b}_2^* = 0.5\mathbf{a}_1^* + 0.5\mathbf{a}_2^*$$

or, in matrix notation

$$\begin{pmatrix} \mathbf{b}_1^* \\ \mathbf{b}_2^* \end{pmatrix} = \begin{pmatrix} 0.5 & -0.5 \\ 0.5 & 0.5 \end{pmatrix} \begin{pmatrix} \mathbf{a}_1^* \\ \mathbf{a}_2^* \end{pmatrix}$$

from which

$$\det M^* = (0.5)(0.5) - (0.5)(-0.5) = 0.5$$

and

$$M = \frac{1}{0.5} \begin{pmatrix} 0.5 & -0.5 \\ -(-0.5) & 0.5 \end{pmatrix} = \begin{pmatrix} 1 & -1 \\ 1 & 1 \end{pmatrix} \tag{1.20}$$

but

$$\mathbf{b} = M \cdot \mathbf{a}$$

so, in real space, the overlayer structure is described by the lattice vectors

$$\mathbf{b}_1 = \mathbf{a}_1 - \mathbf{a}_2$$

$$\mathbf{b}_2 = \mathbf{a}_1 + \mathbf{a}_2$$

as illustrated in Figure 1.6(b).

It is apparent that, even though the overlayer lattice vectors have been related to the substrate lattice vectors, this only conveys information about the lattice periodicity. It does not, for example, indicate the registry of the overlayer structure (the relationship between substrate bonding sites and overlayer atoms). The most common site for adatom adsorption on such a square substrate mesh is the high-coordination site as illustrated in Figure 1.6(c), but the identification of the bonding site may only be accomplished through a detailed analysis of the diffraction beam intensities or by some complementary surface-sensitive technique.

1.3.3 Domains

Often, more than one orientation of the overlayer may exist on the surface. A region that comprises a particular overlayer structure and orientation is called a *domain*. If the *domains* are surrounded by areas of clean surface, then they are usually described as *islands*. If typical domain sizes are small compared with the *coherence length* of the incident electron beam, interference may occur between the diffracted waves from regions of different type. This interference, and the estimation of the coherence length will depend on the nature of the domains and is discussed in detail in Chapter 6. If the domains are much larger than the coherence length of the electrons but smaller than the total beam area, the resulting diffraction pattern will be a superposition of the patterns that would be expected from each domain individually. It is quite common for this condition to be met. Common forms of experimental apparatus do not usually possess the resolution necessary to distinguish whether the observed pattern has been produced from a few or many domains. We shall refer to the combined result of superposing a number of domain *types*, but of course this may be applied to any number of actual domains.

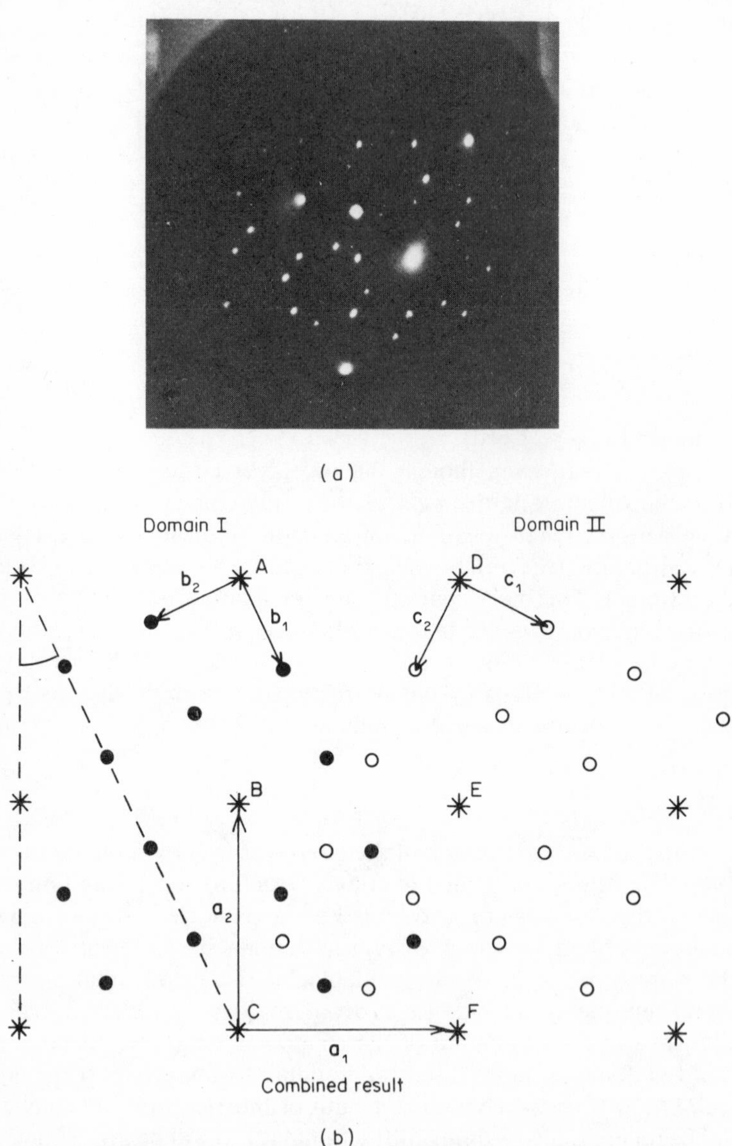

Figure 1.7. The $(\sqrt{5} \times \sqrt{5})R26^\circ$ pattern formed by the adsorption of sulphur or oxygen on Mo(001). (a) Observed diffraction pattern; (b) interpretation in terms of separate domains

As an example, consider the pattern given in Figure 1.7. This pattern has been observed with a number of different combinations of overlayer and substrate atomic species. This particular photograph is taken from an experiment in which oxygen was adsorbed on Mo(001). The first step in interpreting such a pattern is to trace out the pattern and look for lines which join substrate and overlayer spots. If we refer to Figure 1.7, it can be seen that several overlayer beams lie on the diagonal joining substrate beams A and F, and all the others lie on diagonals joining equivalently related spots, such as CD. If CF and CB represent substrate reciprocal lattice vectors $\mathbf{a}_1^*$ and $\mathbf{a}_2^*$ respectively, then one domain can be specified by the vectors $\mathbf{b}_1^*$ and $\mathbf{b}_2^*$ as illustrated. Symmetry permits the existence of an equivalent domain type with a different orientation, and this accounts for the remaining observed spots.

Sometimes, more than one domain orientation is physically possible but is not observed. Either the surface is in reality covered by only one domain type,

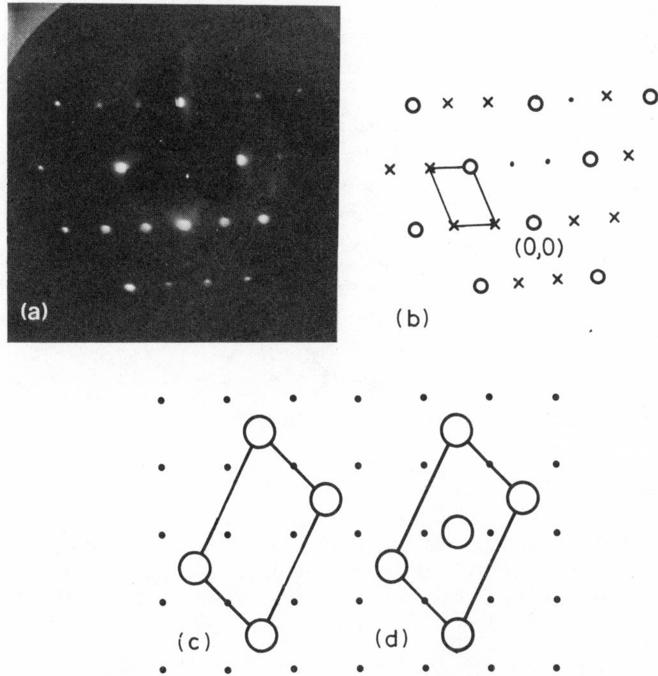

Figure 1.8. A $c(2 \times 4)$ pattern formed by the adsorption of sulphur on Mo(001): (a) observed pattern, (b) diagram of pattern: (o) are substrate beams, (x) are overlayer beams visible at this energy (·) are overlayer beams either obscured or with low intensity at this energy; (c) and (d) are possible interpretations of the adlayer structure, distinguished by the corresponding total coverage. Only one of the possible equivalent $c(2 \times 4)$ domains is present on this surface. (Reproduced by permission of North Holland Physics Publishers from Clarke[11].)

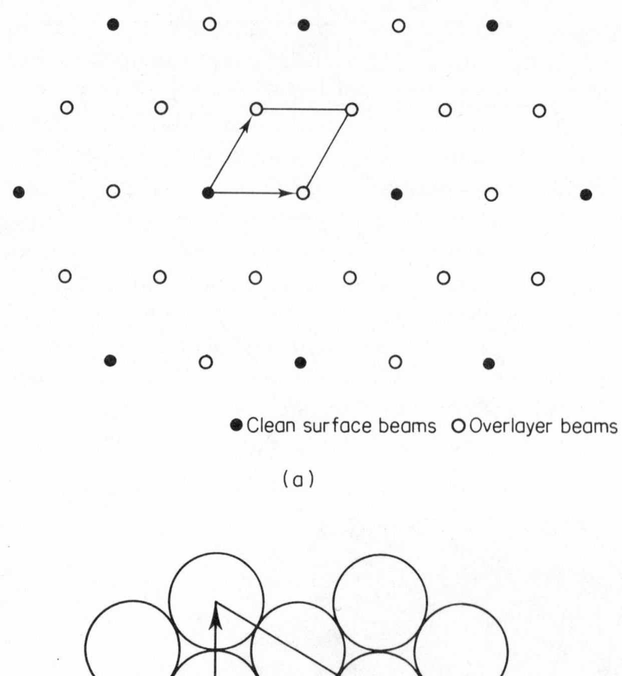

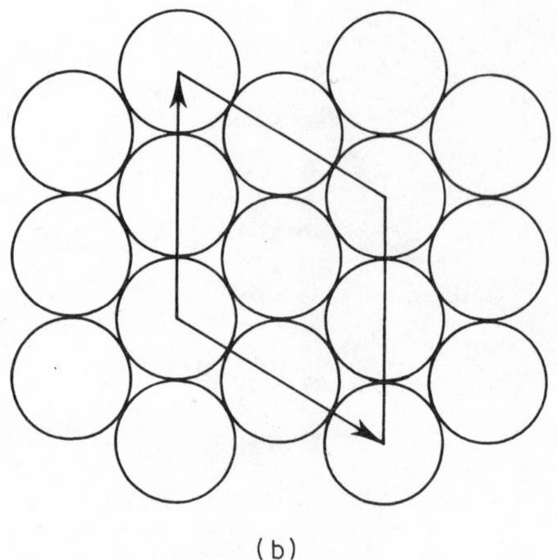

Figure 1.9. The same pattern of spots is formed on a hexagonal-symmetry surface by a (2 × 2) overlayer lattice, (a) and (b), or by three equivalent (2 × 1) domains, (c) and (d). Coverage measurements or intensity analysis is necessary to distinguish between these options

or the domains are very large. The latter possibility may be checked by moving the electron beam across the surface, to see if the other orientation appears. Statistically, the growth of such large domains is extremely unlikely on a perfectly planar surface. Growth of a single domain type may occur if the initial adsorption sites of the overlayer atoms lie along a series of parallel steps of dislocations in the surface. These steps or dislocations may be sufficiently low in density that they do not produce any noticeable change in the clean surface diffraction pattern, but may nevertheless be dense enough to

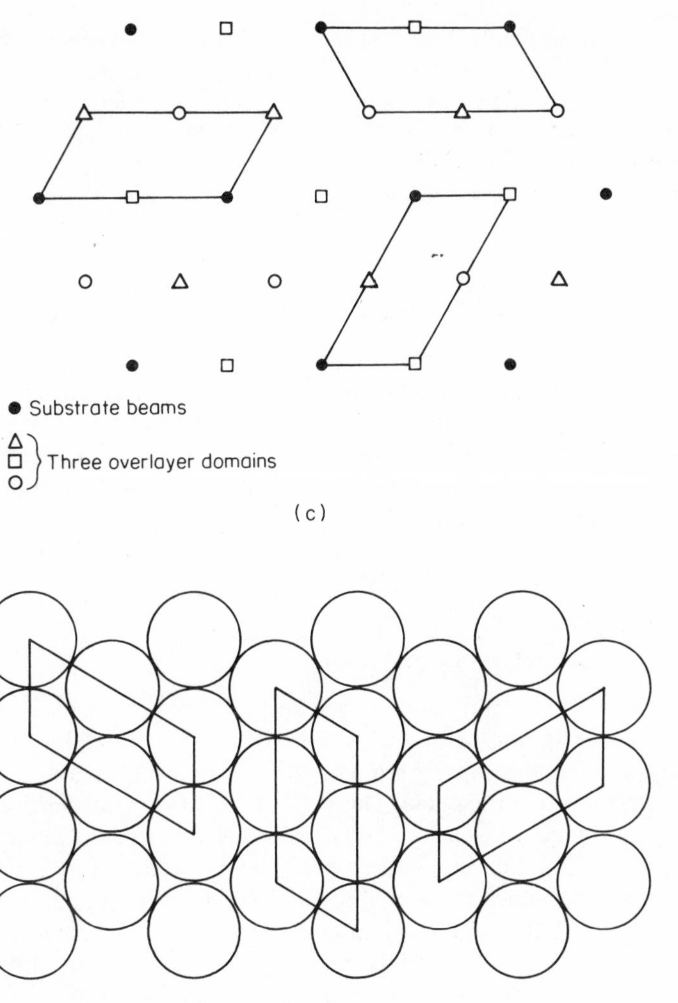

(c)

(d)

influence the overlayer structure. An example is shown in Figure 1.8, where the asymmetric pattern results from only one of two possible domain types formed when sulphur was adsorbed on an imperfect Mo(001) surface.

If the surface lattice possesses two- or fourfold symmetry, then the interpretation of the overlayer lattice is usually straightforward using the guidelines given above. However, ambiguity may remain in the interpretation of structures formed on surfaces with threefold symmetry. For example, the pattern in Figure 1.9(a) could be produced by the single '2×2' structure given in Figure 1.9(b) or by a superposition of the '2×1' domains as given in Figure 1.9(c) rotated through each of the three equivalent orientations. In the specific case of oxygen adsorbed on Re(0001), Pantel *et al*. elegantly resolved this ambiguity by preparing surfaces cut such that they possessed steps of different orientations.[12] With steps aligned parallel to one of the surface atomic rows, a single orientation of the '2 × 1' type was found to nucleate, whereas with steps aligned at a suitable angle to the atomic rows, just two of the three possible orientations were produced. The coverage of oxygen was found by AES to be identical in each case, thus confirming that the original pattern had in fact been formed by the superposition of three equivalent '2 × 1' domains.

1.4 Diffraction from successive planes of a crystal

If atoms bonded to a surface are arranged with the same lattice structure as the substrate, then the spatial appearance of the diffraction pattern will be unchanged, although the relative intensities of the individual beams may be altered as a result of the different scattering properties of the adsorbed atoms. Often, however, the overlayer will have a lattice structure which differs from that of the substrate. In order to understand how the resulting diffraction pattern will be affected, we must consider the process by which scattering occurs from successive layers of atoms.

1.4.1 Rationally related overlayers

In the case of a clean, unreconstructed surface, the periodic lattice structure of each plane is identical (see Figure 1.10(a)). If the unit cell is defined to be the same in all planes, there remains the possibility of a translation of the origin laterally within the plane. This is often called a *registry shift*. Diffraction from a single plane is insensitive to lateral translations of the plane because it is the phase difference in the scattering from successive atoms within the plane which determines the spatial form of the diffraction pattern, and this is fixed for a given lattice and incident beam energy. An electron beam incident upon the surface layer is partly scattered by each of the atom centres. Interference between scattered waves will result in a series of diffracted beams

$$\mathbf{g}_{hk} = h\mathbf{a}_1 + k\mathbf{a}_2$$

where h and k are integers.

One set of beams will propagate backwards from the plane, as described earlier in this section, but a similar set will propagate forwards into the crystal. The former set may be called the *reflected set* and the latter set of beams, plus the unscattered part of the forward propagating beam will comprise a *transmitted set*.

Because we are only concerned at the present with the directions of the diffracted beams, rather than their intensities, the unscattered fraction will be indistinguishable from the forward scattered beam corresponding to $h = k = 0$. The wavefield impinging on the second plane of atoms will comprise not one but a whole set of diffracted beams. However, each

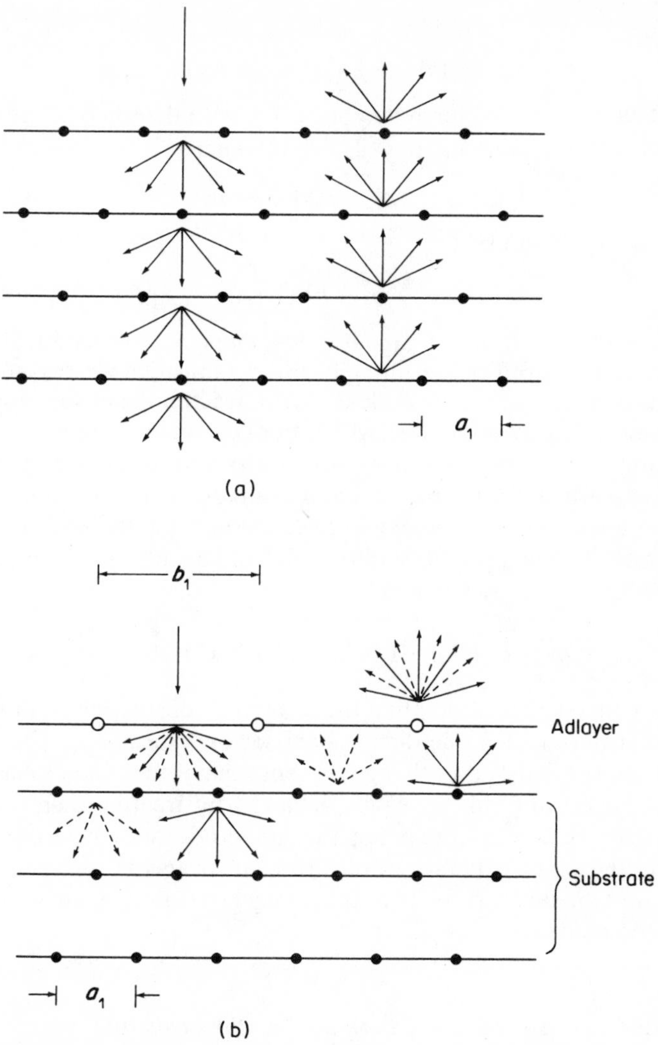

Figure 1.10. Diffraction from several layers of a crystal. (a) Surface layer with same periodicity as bulk; (b) adlayer periodicity double that of bulk: dotted lines indicate the additional ('fractional order') set of beams associated with the presence of the adlayer

diffracted beam satisfies the condition that $\Delta\mathbf{s} = \lambda\mathbf{g}_{hk}$ (see equation 1.10), and this is a valid relationship independent of whether the incident beam $\mathbf{s}_0$ or emergent beam $\mathbf{s}_{hk}$ is varied.

Consider a beam $\mathbf{s}_{hk}$ produced by the forward diffraction of an incident beam $\mathbf{s}_0$ where

$$\mathbf{s}_0 - \mathbf{s}_{hk} = \lambda(h\mathbf{a}_1^* + k\mathbf{a}_2^*) \qquad (h \text{ and } k \text{ integers}) \tag{1.21}$$

If this beam is incident upon another plane with the same lattice structure, then a new set of beams $\mathbf{s}_{nm}$ may also result where

$$\mathbf{s}_{hk} - \mathbf{s}_{nm} = \lambda(n\mathbf{a}_1^* + m\mathbf{a}_2^*) \tag{1.22}$$

substituting equation (1.21)

$$\mathbf{s}_0 - \mathbf{s}_{nm} = \lambda[(n-h)\mathbf{a}_1^* + (m-k)\mathbf{a}_2^*] \tag{1.23}$$

where $(n - h)$ and $(m - k)$ are also members of the set of all possible integers, and therefore the resulting beams $\mathbf{s}_{nm}$ must correspond to the same set of angles as the $\mathbf{s}_{hk}$ set. It is clear that diffraction from successive planes with identical lattice structures will introduce no new diffraction beams, merely intermix their intensities. Any diffracted beam $\mathbf{s}_{nm}$ will pick up contributions which result from the scattering of each and every member of the incident beam set $\mathbf{s}_{hk}$. If we could calculate in detail the way in which each possible incident beam would be diffracted from the plane then any diffracted beam $\mathbf{s}_i'$ will be given by the sum

$$\mathbf{s}_i' = m_{ia} \cdot \mathbf{s}_a + m_{ib} \cdot \mathbf{s}_b + \ldots + m_{ii} \cdot \mathbf{s}_i + \ldots + m_{iN} \cdot \mathbf{s}_N \tag{1.24}$$

where for convenience of notation the array of N diffraction beams has been relabelled in terms of a one-dimensional set of N beams $\mathbf{s}_i$. The scattering elements m_{ij} describe the change in both amplitude and phase of the diffracted beams and this is readily achieved by treating them as complex numbers (the real part containing the amplitude and the imaginary part containing the phase). If the set of diffracted beams is represented by the vector $\mathbf{S}'$ and the incident set by $\mathbf{S}$, these may be related by the square matrix M using the equation

$$\mathbf{S}' = M \cdot S \tag{1.25}$$

Consequently, *a complete description of the scattering from a plane of periodic scatterers may be given by two complex $(N \times N)$ matrices, one describing the 'reflected' set and the other the 'transmitted' set of diffracted beams.*

Consider now what would happen if the overlayer had a lattice structure which differed from the substrate. The transmitted set of beams from the surface would be given by the condition

$$\mathbf{s}_0 = \mathbf{s}_{hk} = \lambda(h\mathbf{b}_1^* + k\mathbf{b}_2^*) \tag{1.26}$$

Scattering of $\mathbf{s}_{hk}$ by a substrate layer would produce beams $\mathbf{s}_{nm}$ where

$$\mathbf{s}_{hk} - \mathbf{s}_{nm} = \lambda(n\mathbf{a}_1^* + m\mathbf{a}_2^*) \tag{1.27}$$

substituting equation (1.26)

$$\mathbf{s}_0 - \mathbf{s}_{nm} = \lambda(n\mathbf{a}_1^* + m\mathbf{a}_2^* + h\mathbf{b}_1^* + k\mathbf{b}_2^*) \tag{1.28}$$

If one of the resultant beams $\mathbf{s}_{nm}$ were scattered from a second substrate layer, then any subsequent beam $\mathbf{s}_{pq}$ would satisfy the relation

$$\mathbf{s}_0 - \mathbf{s}_{pq} = \lambda[(n-p)\mathbf{a}_1^* + (m-q)\mathbf{a}_2^* + h\mathbf{b}_1^* + k\mathbf{b}_2^*] \tag{1.29}$$

Therefore, although the overlayer would introduce additional beams into the set, propagation throughout the crystal would not increase this set any further. The total number of beams is determined purely by the set given in equation (1.28).

EXAMPLE. Consider the case where $\mathbf{b}_1 = 2\mathbf{a}_1$ and $\mathbf{b}_2 = \mathbf{a}_2$. In reciprocal space $\mathbf{b}_1^* = \mathbf{a}_1^*$ and $\mathbf{b}_2^* = \mathbf{a}_2^*/2$. The corresponding lattice vectors, from equation (1.28), reduce to

$$(n+h)\mathbf{a}_1^* + m\mathbf{a}_2^* + (k/2)\mathbf{a}_2^* \tag{1.30}$$

Whenever k is an even integer, a substrate-type set is obtained. Whenever k is odd, the beam corresponds to a member of the substrate-type set plus $\mathbf{a}_2^*/2$. Scattering from subsequent substrate layers affects n and m but not k, and so the beam sets corresponding to k even and odd are not mixed within the substrate. This has important consequences for the matrices which are used to describe scattering from atomic planes. This situation is illustrated for a one-dimensional case in Figure 1.10(b).

The total set of possible beams upon a substrate plane is given by the set

$$n\mathbf{a}_1^* + m\mathbf{a}_2^* + (d/2)\mathbf{a}_2^* \tag{1.31}$$

where n and m are integers and $d = 0$ or 1. If n and m are limited to N values each, then a complete description of the scattering by the plane must relate the intensities of all diffracted beam amplitudes and phases $\mathbf{S}'$ to all possible incident beams $\mathbf{S}$ via a complex, square matrix M as given in equation (1.25), but now the dimensions of the matrix will be $(2N \times 2N)$. The time a computer takes to calculate the complete scattering from a crystal may, as will be described in Chapter 4, increase with the cube of the dimensions of the planar scattering matrices, and therefore the calculational time could increase by a factor of eight as a result of the extra beams produced by the overlayer. More complex overlayer structures would result in even more dramatic increases in computational time. Fortunately, because the two sets of diffracted beams corresponding to k even and odd do not mix within the substrate, elements of the scattering matrices for substrate planes will be zero whenever the scattering of a diffracted beam corresponding to one value of k is related to an incident beam from the set corresponding to the other value of k, and *vice versa*. If the vectors which list all the possible beams are ordered in such a way that all beams with k odd are grouped together, the relevant matrix may be factorized as follows:

$$\begin{pmatrix} \mathbf{S}'_{k\text{ even}} \\ \mathbf{S}'_{k\text{ odd}} \end{pmatrix} = \begin{pmatrix} k\text{ even} & 0 \\ 0 & k\text{ odd} \end{pmatrix} \begin{pmatrix} \mathbf{S}_{k\text{ even}} \\ \mathbf{S}_{k\text{ odd}} \end{pmatrix} \tag{1.32}$$

If the number of beams with k even are P and the number with k odd are Q, with P greater than or equal to Q, then computer memory space may be conveniently saved by condensing the matrix into rectangular form with dimension P by $(P + Q)$, plus a considerably reduced redundant region (hatched):

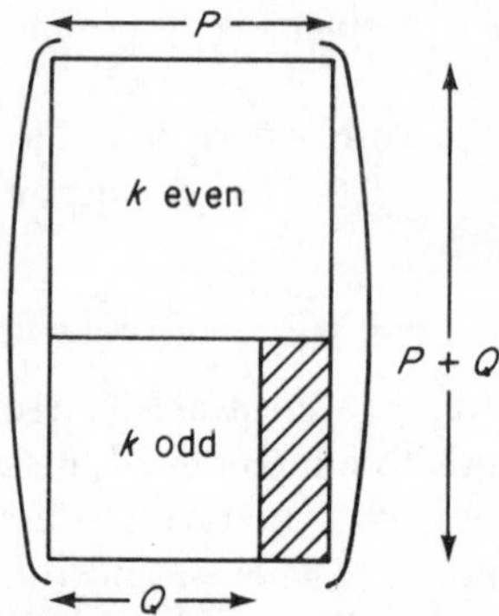

This process of reduction of the planar scattering matrices into factors has been illustrated for the simple case of a (2×1) structure, but it is clearly possible to adapt the approach for any overlayer structure of its lattice vectors are rationally related to the lattice vectors of the substrate.

1.4.2 Incommensurate overlayers

If gas atoms or molecules are weakly bonded to a metal surface, the lattice structure they adopt may be almost completely independent on the lattice structure of the substrate. This is a common occurrence whenever rare gases such as Ar, Kr or Xe are adsorbed, or molecules such as CO are bonded undissociatively to a surface. If a completely new crystal structure is formed on a surface (*epitaxial growth*) perhaps through oxidation or the deposition of a metal film, then again the lattice structure of the adsorbed layer or film may be independent of the substrate lattice. Even if the orientation of the overlayer is influenced in some way by the structure of the substrate, the lattice vectors may be *irrationally related* to the substrate lattice vectors. What will be the appearance of the resulting diffraction pattern? The overlayer will produce diffraction beams related to its own periodicity:

$$\mathbf{g}_a = h\mathbf{b}_1^* + k\mathbf{b}_2^* \tag{1.33}$$

and each of these beams incident upon a substrate will result in a new set:

$$\mathbf{g}_{total} = \mathbf{g}_a + \mathbf{g}_s = h\mathbf{b}_1^* + k\mathbf{b}_2^* + n\mathbf{a}_1^* + m\mathbf{a}_2^* \tag{1.34}$$

but now $\mathbf{b}_1^*$ and $\mathbf{b}_2^*$ are not related to $\mathbf{a}_1^*$ and $\mathbf{a}_2^*$. As noted earlier, subsequent diffraction within the crystal will not multiply further the total number of beams, since it only mixes members of the substrate set $\mathbf{g}_s$ for any given $\mathbf{g}_a$. This means that we need only concern ourselves with the possible diffraction beam mixing processes which occur between the overlayer and the substrate, since $\mathbf{g}_s$ may represent the effect of scattering from any number of substrate layers. For the present we may restrict our analysis to the case of one dimension, so that

$$\mathbf{g} = h\mathbf{b}^* + n\mathbf{a}^* \tag{1.35}$$

Generally, back-scattering from the substrate will be strong compared with the overlayer, because we are implicitly including the total back-scattering from the whole substrate crystal. If the adlayer is a weak scatterer, the largest contribution to the scattering from the overlayer will be in the forward ($h = 0$) direction. In this case a prominent set of diffraction beams will correspond to the substrate set $\mathbf{g}_s = n\mathbf{a}^*$. The back-scattered diffraction pattern from the overlayer will correspond to the set $\mathbf{g}_a = h\mathbf{b}^*$, and this will be reinforced by transmission diffraction through the overlayer of the back-scattered $n = 0$ substrate beam. If the lattice vectors of the overlayer are of similar length to those of the substrate, then this will introduce 'satellite' beams about the original substrate set, as illustrated in Figure 1.11. Each beam back-scattered from the substrate may be diffracted by the overlayer and this would produce the pattern given by the complete set given in equation (1.34). In practice, if the overlayer is a weak scatterer, intensity will be concentrated about the forward-scattered direction and scattering will be successively weaker as the diffraction orders involved increase. If we consider the satellites about the specular beam ($h = n = 0$) we

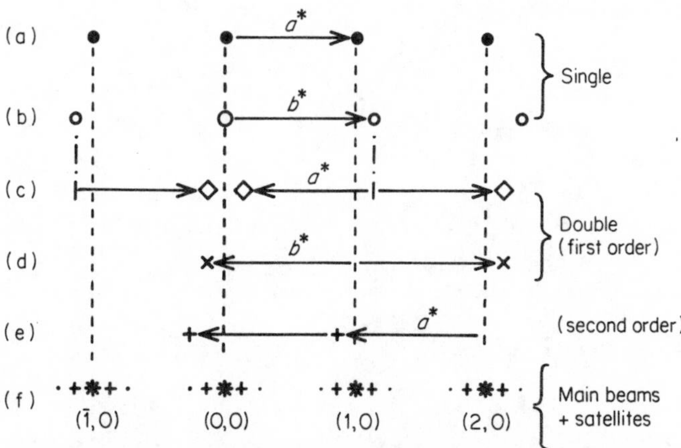

Figure 1.11. Representation, in one dimension, of the way an incommensurate adlayer, periodicity b on a surface with periodicity a, combines to cause satellite beams about the main substrate beams. Here, (a) and (b) indicate contributions from either the adlayer or bulk alone; (c) and (d) are caused by scattering from both adlayer and bulk, and (e) by double scattering from the adlayer; (f) shows the net effect of these contributions

see that the first beam is produced by the combination $h = \pm 1$, $n = \pm 1$, the second is produced by the combination $h = \pm 2$, $n = \pm 2$ and so on. Each satellite will be weaker than the previous, and after the first one or two orders may become too faint to be observed. This assumption that scattering from the overlayer is weak carries with it the assumption that multiple scattering between the layers will also be negligible, and so this interpretation is often called the *double-diffraction* model for incommensurate overlayer diffraction. This condition frequently occurs when rare-gas atoms are adsorbed on to the basal plane of graphite.[12] Close-packed layers of the adsorbate have a lattice spacing similar to that of the substrate, but find it energetically favourable to reorientate their close-packed hexagonal arrangement with respect to the graphite. Rotational symmetry permits a number of different orientations to

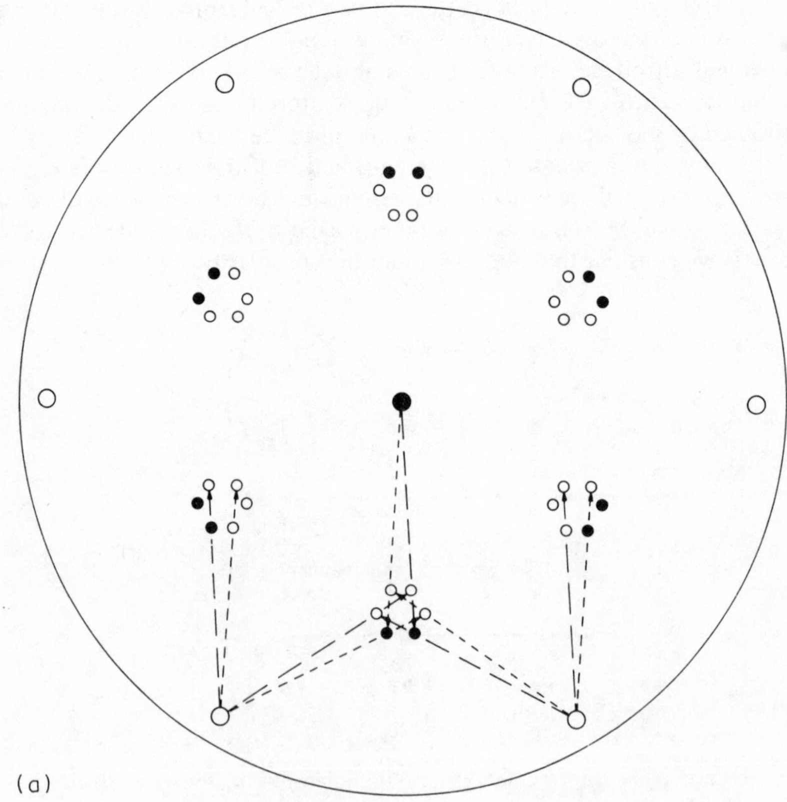

Figure 1.12 (a) Schematic LEED pattern for Ar overlayer with two oppositely rotated groups of Ar atoms having mean nearest-neighbour distance d = 3.86 Å and orientations $\theta = \pm 3.5°$. Diffraction beams represented are the specular (centre closed circle), first-order graphite (peripheral open circles), first-order Ar (12 closed circles) and graphite plus Ar (remaining 24 open circles). Some of the argon reciprocal-lattice vectors are shown as dashed lines (long dashes

occur and the net result is a diffraction pattern in which each original substrate beam is surrounded by a hexagonal arrangement of satellite beams, as shown in Figure 1.12.

This restriction on the number of beams means that a calculation of the intensities of the beams with the intention of determining further details of the bonding geometry would not necessarily involve excessive computer time. Scattering from the substrate would need to be solved for as many incident beam angles as there are satellite beams (if the electron beam is incident normally on the surface, symmetry clearly reduces this to the number of orders of satellite beams) plus one for the zeroth order or specular beam. Each satellite beam results from only one combination of substrate and overlayer scattering events and so may be treated on an individual basis. The time

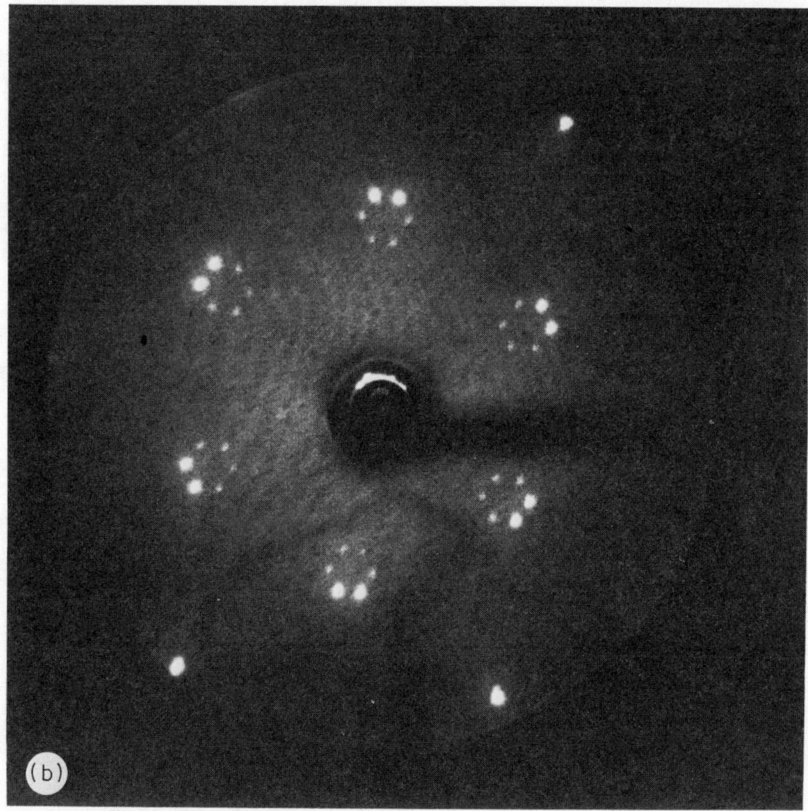

for $\theta = +3.5°$, short dashes for $\theta = -3.5°$). (b) Photograph of LEED pattern at 133 eV, 39 K and 5×10^{-7} Torr Ar pressure. Three of the six first-order graphite diffraction beams are visible at the edge of the pattern; the remaining 36 spots result from the Ar solid monolayer; $d = 3.83 \pm 0.01$ Å and $\theta = 3.55° \pm 0.15°$. (Reproduced by permission of the American Physical Society from Shaw *et al.*[15])

of calculation will increase only linearly with the number of orders of scattering considered.

1.4.3 Symmetry in diffracted beam intensities

If the direction of the incident electron beam is perpendicular (or 'normal') to the surface plane of the crystal, then symmetrically related diffraction beams will have the same intensity. If we label the beams by the indices (h, k) that correspond to the reciprocal vectors

$$\mathbf{g}_{hk} = h\mathbf{a}_1^* + k\mathbf{a}_2^* \tag{1.36}$$

then, for normal incidence on a surface with a rectangular lattice, the scattering into the beam for which $h = 1$ and $k = 0$ or $(1, 0)$ must be the same as into the beam $h = -1, k = 0$ or $(\bar{1}, 0)$, since the infinite two-dimensional periodicity of the plane makes it impossible to distinguish between the $+\mathbf{a}_1^*$ and $-\mathbf{a}_1^*$ directions. In the particular case of a square lattice, where $|\mathbf{a}_1^*| = |\mathbf{a}_2^*|$, symmetry will exist between all four first-order beams $(1, 0)$, $(0, 1)$. $(\bar{1}, 0)$ and $(0, \bar{1})$. The diffraction intensity of the $(1, 1)$ beam will differ from the $(1, 0)$ beam, but nevertheless it is clear that the normally incident electron beam will be scattered equally into each of the members of the set $(1, 1)$, $(\bar{1}, \bar{1},)$ $(\bar{1}, 1)$ and $(1, \bar{1})$. These two sets or 'stars' may be collectively described using the notation $[1, 0]$ and $[1, 1]$. Higher order stars such as $[2, 1]$ may include eight members in this special case of a square lattice, as illustrated in Figure 1.13.

Once the incident beam has been scattered by the surface plane, not one but a complete set of diffracted beams will propagate through the crystal, and only the specular beam will be incident along the normal to subsequent atomic planes. Any other beam, incident at some finite angle to the plane, will

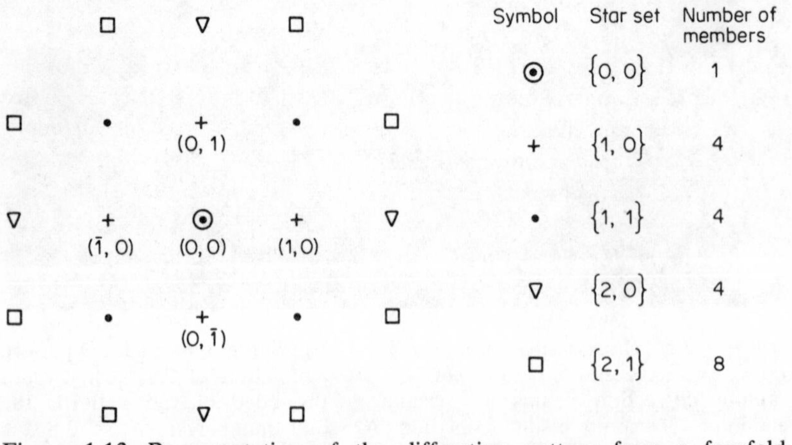

Symbol	Star set	Number of members
⊙	{0, 0}	1
+	{1, 0}	4
•	{1, 1}	4
▽	{2, 0}	4
□	{2, 1}	8

Figure 1.13. Representation of the diffraction pattern from a fourfold symmetric surface, showing the various equivalent beam sets

not produce equal amplitudes for each member of any diffracted-beam star. However, it is only necessary to take the mean value of the amplitudes and phases for each member of the resulting beam star because symmetry amongst the incident beams will automatically effect this averaging process in the final diffraction pattern. For example, in the case of a square lattice, a (1, 0) beam incident on an atomic plane may be diffracted into beams (1, 0), (0, 1), ($\bar{1}$, 0 and (0, $\bar{1}$) with amplitudes and phases given by matrix elements which we may call m_1, m_2, m_3 and m_4. The (0, 1) beam incident on the same plane would be related via matrix elements m_2, m_1, m_4 and m_3 respectively: the order would be changed, but not the values of the elements (in fact, in this case symmetry will also mean that $m_3 = m_4$). The relevant information may be encapsulated by the identification of a single matrix element

$$M = (m_1 + m_2 + m_3 + m_4)/4 \tag{1.37}$$

which relates the incident beam star [1, 0] with the corresponding diffracted beam star [1, 0]. The scattering matrix which relates the first two beam stars [0, 0] and [1, 0] which originally comprises 25 matrix elements

$$\begin{pmatrix} (0,0) \\ (1,0) \\ (0,1) \\ (1,0) \\ (0,1) \end{pmatrix} = \begin{pmatrix} m_{11} & m_{12} & m_{13} & m_{14} & m_{15} \\ m_{21} & m_{22} & m_{23} & m_{24} & m_{25} \\ m_{31} & m_{32} & m_{33} & m_{34} & m_{35} \\ m_{41} & m_{42} & m_{43} & m_{44} & m_{45} \\ m_{51} & m_{52} & m_{53} & m_{54} & m_{55} \end{pmatrix} \begin{pmatrix} (0,0) \\ (1,0) \\ (0,1) \\ (1,0) \\ (0,1) \end{pmatrix} \tag{1.38}$$

can be rewritten with only four elements

$$\begin{pmatrix} [0,0] \\ [1,0] \end{pmatrix} = \begin{pmatrix} M_a & M_b \\ M_b & M_b \end{pmatrix} \begin{pmatrix} [0,0] \\ [1,0] \end{pmatrix} \tag{1.39}$$

The precise relationship between the elements of the reduced matrix and the original full matrix can be most easily appreciated by a more specific example. Consider the fourfold symmetric diffraction pattern illustrated in Figure 1.13. If the matrix element relating the incident (0, 0) beam with the diffracted (0, 0) beam has the value a, and it is clear that no other beam in general is related by symmetry in the same way that the incident and diffracted (0, 0) beams are related, then a must be unique and irreducible. If the incident (0, 0) beam is related to the diffracted (1, 0) beam by an element with value b, then it is clear that the elements relating (0, 0) to (0, 1), ($\bar{1}$, 0) and (0, $\bar{1}$) must also take the value b. The diffraction process is time-invariant so that if we consider the process in reverse and suppose the (1, 0) beam to be the incident beam and the (0, 0) beam to be the diffracted beam, the corresponding matrix element would have the same value b as related the incident (0, 0) beam and diffracted (1, 0) beam. An incident (1, 0) beam will be related to diffracted (1, 0), (0, 1), ($\bar{1}$, 0) and (0, $\bar{1}$) beams via matrix elements d, e, f and g respectively. The resulting amplitude A' of the (1, 0) beam is the sum of the components from all the input beam amplitudes A, as modified by the matrix

$$A'(1,0) = bA(0,0) + dA(1,0) + gA(0,1) + fA(\bar{1},0) + \ldots \quad (1.40)$$

but from the symmetry of the input wavefield

$$A(1,0) = A(0,1) = A(0,\bar{1}) = A(\bar{1},0) \quad (1.41)$$

which may be collectively described by a single beam star amplitude $A\{1,0\}$. We wish to find the values β and δ from which the reduced matrix may be constructed:

$$A'\{1,0\} = \beta A\{0,0\} + \delta A\{1,0\} + \ldots \quad (1.42)$$

where $\beta = b$ and $\delta = d + e + f + g$. This is illustrated in Figure 1.14.

The general formalism for deriving the reduced matrix elements M_α can be shown[14] to take the form

$$M'_\alpha = \sum_{i,j} \frac{m_{ij}}{\sqrt{(N_j N_i)}} \quad (1.43)$$

where N_j is the number of columns of elements related by symmetry and N_i the number of rows, whilst $\Sigma_{ij} m_{ij}$ is the sum of all the elements included in the

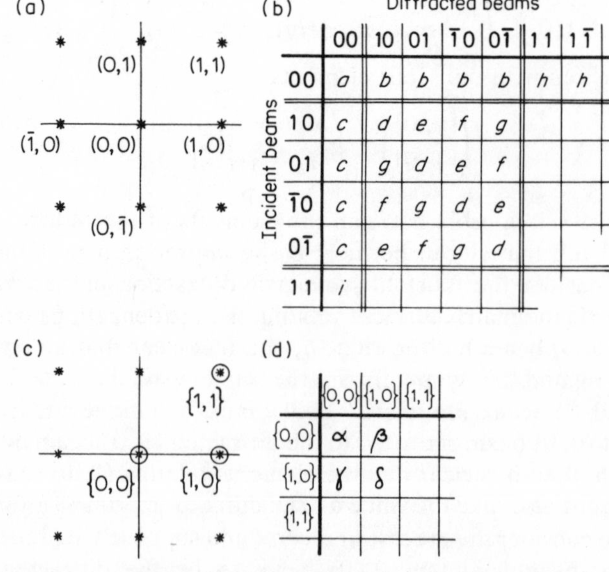

Figure 1.14. (a) Fourfold symmetric diffraction pattern; (b) representation of the scattering matrix, identical matrix elements having the same letter; (c) reduction of the original pattern to three independent beam sets; (d) representation of the reduced matrix

set or 'star'. It is easy to see by substitution that the above example is consistent with this algorithm. In fact, because each row of a 'star' contains a permutation of the elements in each of the other rows of the star (see Figure 1.14) it is more efficient computationally to scan elements one row at a time, and use the algorithm

$$M_\alpha = N_j \sum_i \frac{m_{ij}}{\sqrt{(N_i N_j)}} = \sqrt{\frac{N_j}{N_i}} \sum_i m_{ij} \tag{1.44}$$

where $\Sigma_{ij} m_{ij}$ is the sum of elements in one row of any particular star α, and to then consider only the first rows of each star, which would be rows 1, 2 and 5 in the above example if it were extended to include the {1, 1} star.

The savings in computer memory space and calculational time that can be gained from utilizing available symmetry may be very significant.[14] In the list given in Figure 1.13, the scattering matrix would be reduced in dimensions from 21 (total number of beams considered) to 5 (total number of independent beam stars).

It may have been noted that additional symmetry relationships exist even within a single row of matrix elements of a beam star. For example, referring to Figure 1.14, it is clear that diffraction of, say, an incident (1, 0) beam into the (0, 1) beam will be equivalent to scattering into the $(0, \bar{1})$ beam so that elements e and g must be identical (the reader may wish to check that this equivalence is valid for diffraction from any member of the beam star). Identification of equivalent matrix elements of this sort and incorporation within a new reduced matrix element algorithm does not save much computer time, and indeed it is probably quicker to utilize the more general but simple relationship given in equation (1.44). In special cases, where whole beam stars are equivalent, so that certain reduced matrix elements can be dropped, the exploitation of additional symmetry relations may be practical. In particular, Rundgren and Salwen[14] have shown that a $p(2 \times 2)$ overlayer results in sets of fractional order beams certain of which are equivalent and therefore need not be calculated twice. We have already noted in section 1.3.3 that the presence of a rationally related overlayer results in essentially independent sets of beams propagating through the bulk of the crystal which can be treated by separate matrix blocks. In the case of a $p(2 \times 2)$ overlayer, four such separate sets of beams (or Laue nets) exist, which may be described as $L(0, 0)$, $L(\frac{1}{2}, 0)$, $L(0, \frac{1}{2})$ and $L(\frac{1}{2}, \frac{1}{2})$. As a consequence of symmetry, the $L(\frac{1}{2}, 0)$ and $L(0, \frac{1}{2})$ nets are equivalent, so that only one of the two need be calculated. This is illustrated in Figure 1.15. In the case of a $p(2 \times 2)$ structure the number of beams which need be considered may be reduced from, for example, 69 to 13. This results in a very substantial saving in computer effort, and it is clear that exploitation of beam star symmetry is an important feature in any attempt to investigate surface structures where the number of fractional order beams is large.

Beam star symmetry is only preserved if the initial beam is incident along the normal to the crystal surface. If the incident direction is not normal, yet

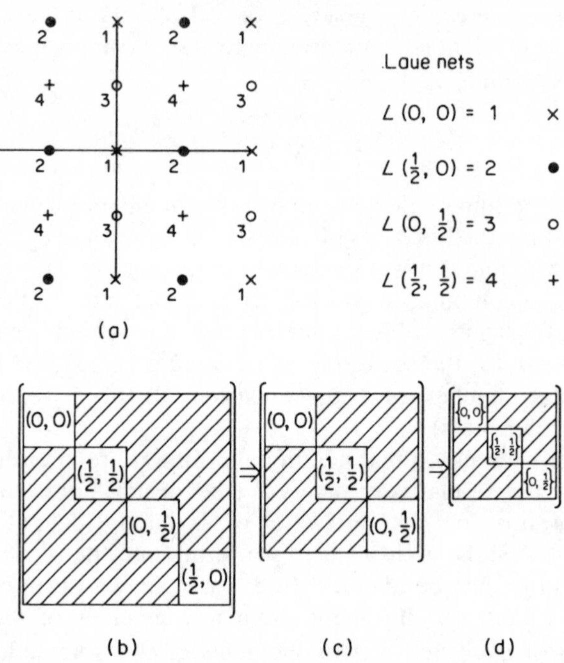

Figure 1.15. (a) A $p(2 \times 2)$ diffraction beam, which has a acattering matrix of the block form given by (b), where each block comprises only those beams in a single Laue net, can be reduced to the form (c) since the $(0,\frac{1}{2})$ and $(\frac{1}{2}, 0)$ beam sets are equivalent at normal incidence. These in turn can be reduced in size by the process given in Figure 1.14

inclined within a high-symmetry plane, fewer beams will be equivalent. Nevertheless, the number of beams may be approximately halved, which is itself a not insignificant advantage of methods which ignore beam star symmetry.[15]

Symmetry amongst beam intensities will be preserved only if scattering from successive atomic layers preserves the same degree of symmetry. Bulk crystal structure is highly ordered and frequently satisfies this condition. However, bonding of the surface layer may entail lower symmetry bonding sites, and therefore alter the symmetry properties of the resulting diffraction pattern. Consider a square lattice with one atom per unit cell. Fourfold rotational symmetry exists about the centre or any corner of the square. The diffracted beams will possess fourfold symmetry if the incident beam is normal to the surface and a normal axis can be drawn which passes through the fourfold symmetry points of the two-dimensional lattice of each layer.

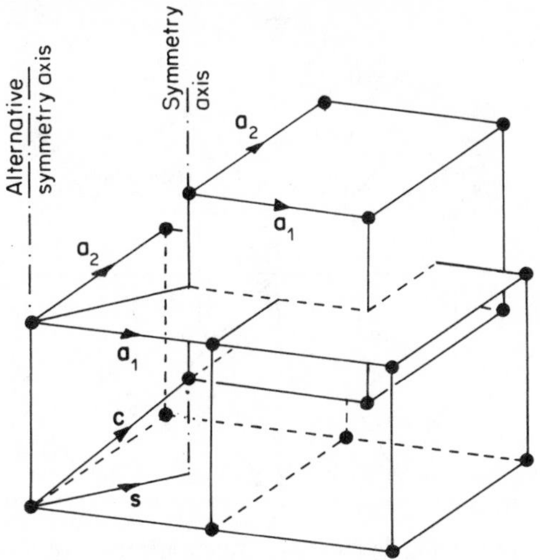

Figure 1.16. Two possible symmetry axes for the description of a (100) bcc crystal surface. The vector **s** relates the two axes

The *registry* of each layer can be defined in terms of a vector **s** which, in the present example, must be either zero or $0.5(\mathbf{a}_1 + \mathbf{a}_2)$. Of course, the normal axis could be chosen arbitrarily, but, for the sake of analysis, it would then be necessary to translate it to a symmetry axis, which would revert to the special case we are considering here.

Scattering from successive layers is dependent on its position in space, as defined by the lattice vector **c**. This may be described in terms of an interlayer spacing and a registry, that is, the components of **c** normal and parallel to the layer respectively (see Figure 1.16). In the case of a cubic lattice as described above, only two registries can occur, and the intensity analysis can be performed efficiently by considering the diffraction from successive layers as though there were two independent layer types, each possessing fourfold symmetry about a particular normal axis. This provides a more rigorous approach to the identification of situations in which beam symmetry can be exploited to reduce the size of the scattering matrices, and is the basis of the method adopted by Van Hove and Tong in their LEED computer program library (CSM: see Chapter 4).[16] Although layers with different symmetries are treated independently, calculation of the layer scattering matrices for different registries is easily effected by the systematic incorporation of different phases associated with the propagation paths of the various beams between successive layers.

In principle, diffraction beam symmetries could be calculated directly from a knowledge of the crystal structure and the incident beam direction. However, the CSM program library[16] requires the specific input of layer

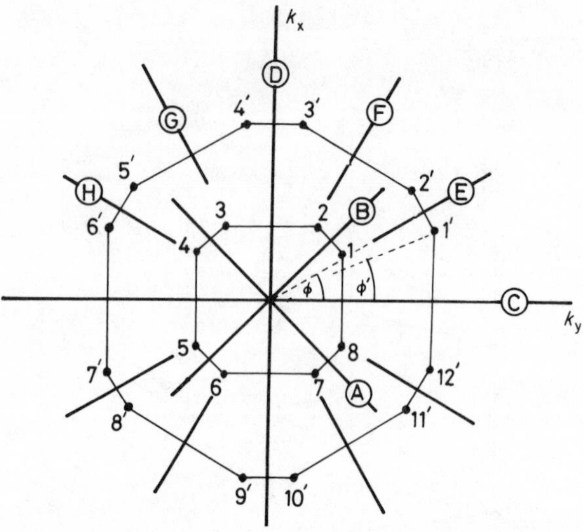

Figure 1.17. Symmetry relationships of beams (numbered here 1, 2, . . ., 8, 1′, 2′, . . ., 12′), relative to an axis 0 and mirror planes at azimuths −45°, 0°, 45° and 90° (denoted A, C, B and D respectively); and at azimuths 30°, 60°, 120° and 150° (denoted E, F, G and H respectively); ϕ and ϕ' are arbitrary angles, not confined to $< \phi \leq 45°$, $0 \leq \phi' \leq 30°$, defining the azimuth of a beam. Groups of symmetry-related beams are given code numbers as defined in Table 1.1. Reproduced by permission of Springer Verlag from Van Hove and Tong[16]

Table 1.1 Possible forms of beam symmetry

Symmetry code number	Description	Examples of groups of beams
1	Single or unsymmetrized beam	(1), (5′)
2	2 beams, twofold axis	(1, 5), (2′, 8′)
3	2 beams, mirror plane A	(2, 5), 3, 4)
4	2 beams, mirror plane B	(1, 2), (4, 7)
5	2 beams, mirror plane C	(1, 8), (3′, 10′)
6	2 beams, mirror plane D	(6, 7), (7′, 12′)
7	4 beams, fourfold axis	(1, 3, 5, 7), (2, 4, 6, 8)
8	4 beams, 2 mirror planes A, B	(1, 2, 5, 6), (3, 4, 7, 8)
9	4 beams, 2 mirror planes C, D	(1, 4, 5, 8), (2′, 5′, 8′, 11′)
10	8 beams, 4 miror planes A, B, C, D	(1, 2, 3, 4, 5, 6, 7, 8)
11	3 beams, threefold axis	(1′, 5′, 9′), (2′, 6′, 10′)
12	6 beams, threefold axis, mirror plane C	(1′, 4′, 5′, 8′, 9′, 12′)
13	6 beams, twofold axis, mirror plane D	(3′, 4′, 7′, 8′, 11′, 12′)
14	6 beams, sixfold axis	(1′, 3′, 5′, 7′, 9′, 11′)
15	12 beams, sixfold axis, 2 mirror planes C, D	(1′, 2′, 3′, 4′, . . . 11′, 12′)

Source: From p. 38 of Van Hove and Tong.[16]

registry vectors **s** and the resulting beam symmetry, which can take one of 15 possible forms (see Figure 1.17 and Table 1.1).

1.5 Surface structures and notation

1.5.1 Crystal structures and Miller indices

The requirement that a crystal should comprise a periodic array of regular units imposes clearly definable constraints on the possible structure of these unit cells. In terms of symmetry properties of the unit cells, 32 different classes of structure are possible, but of these only 11 are directly distinguishable from Laue patterns using X-ray crystallography. From these, seven basic types can be identified: triclinic, monoclinic, orthorhombic (sometimes simply called rhombic), trigonal, tetragonal, hexagonal and cubic.[17] Roughly 90 per cent of all known crystals fall into the first three categories. Fortunately, the structures of the elements mostly fall into the cubic or hexagonal classes, which are relatively simple and highly symmetric. Most LEED investigations to date have been made using crystal substrates of single-component metals or one- or two-component semiconductors or insulators. A selected list of the elements, with their crystal structural types, is given in Appendix 1 (p. 311).

The most common structural forms of the elements can be subdivided into one of the five following categories: simple cubic (sc), body-centred cubic (bcc), face-centred cubic (fcc), hexagonal close-packed (hcp) or diamond. These will be described in detail later in subsequent sections. The bulk geometrical structure of each elemental solid may thus be described by its structural type and relevant unit cell dimensions.

Although the bulk structures of the elements fall into a very limited number of types, corresponding surface structures may be continuously varied by changing the angle at which the surface is cut with respect to the basic unit cell axes. Resulting surface planes are normally defined by their *Miller indices*, which are derived as follows. Any plane related to the crystal may be described by the points or intercepts at which it cuts the three unit cell axes. These intercepts may be expressed in terms of the lengths of the corresponding lattice vectors. In practice, it is convenient to use the reciprocals of these intercepts, so that a plane parallel to a unit vector can be specified by zero rather than infinity. If the intercepts of the three axes are $|\mathbf{a}_1|$, $|\mathbf{a}_2|$ and $|\mathbf{a}_3|$, the Miller indices are found by taking the reciprocals of α, β and γ and dividing each by any common factor **c**, so as to reduce the smallest integers which have the same ratio. The plane is then described by the indices (hkl) where

$$h = 1/(\alpha c), \qquad k = 1/(\beta c) \qquad l = 1/(\gamma c)$$

These indices can always be reduced to integral form since any physical surface formed from the crystal must correpond to the discrete periodicity of the crystal lattice. The usual form of presentation is within parentheses, the indices sometimes but not universally separated by commas: (h, k, l) or simply (hkl).

The vector normal to the plane may be described in a similar manner. Square brackets are used to indicate that a vector, or direction, rather than a plane is being described: $[h, k, l]$ or $[hkl]$. For cubic lattices, the vector with the same indices as the plane will always be normal (or orthogonal) to that plane. For non-cubic lattices, this is not always so.

As an example, consider the various faces that can be obtained from a cubic lattice. Any plane formed by the vectors $\mathbf{a}_1$ and $\mathbf{a}_2$ cuts both these axes at infinity (α and $\beta = 0$) so that h and k are zero. is finite and so can be given the value 1. The Miller index of this face is therefore (001) and the surface normal is [011]. Similarly, other planes may be formed which have low values of the indices, as shown in Figure 1.18. Low-index planes are characterized by high densities of coplanar atoms, and the higher the density the more stable the face tends to be. Consequently, the great majority of LEED studies have been devoted to the structure of low-index surfaces. At the other extreme, very high index faces can usually be thought of as low-index faces with steps. Such vicinal planes are very interesting from a different point of view, and will be considered separately.

The atomic structure of a perfect crystal surface will depend on the atomic arrangement within the unit cell. Most metals have bcc, fcc or hcp bulk forms. The first two are cubic and so can be characterized by a single lattice length a, whereas the last requires two lattice lengths a and c to specify the lattice

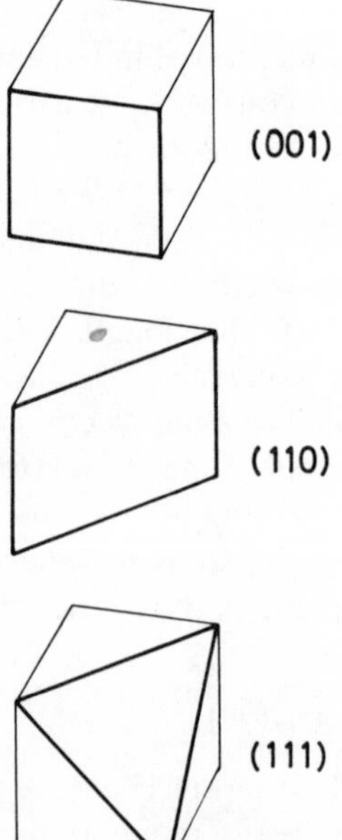

Figure 1.18. Three low-index faces of a cubic crystal unit cell, with Miller indices

vectors within the hexagonal planes and the separation between these planes respectively.

1.5.2 Cubic lattices

If a is the lattice vector length of the cubic cell, the volume of the primitive unit cell for a bcc lattice is $V = a_2/2$ and $= a^2/4$ for a fcc lattice.

The spacing between (hkl) planes (through lattice points) for cubic lattices is[18]

$$d_{hkl} = a[Q(h^2 + k^2 + l^2)]^{-1/2} \tag{1.45}$$

where for bcc: $Q = 1$ if $h + k + l$ is even, and $Q = 2$ if $h + k + l$ is odd; for fcc: $Q = 1$ if h, k and l are all odd, and $Q = 2$ if h, k and l are of mixed parity.

The density of atoms within a (hkl) plane is for bcc:

$$\sigma_{\text{bcc}} = 2[Qa^2(h^2 + k^2 + l^2)]^{-1/2} \tag{1.46}$$

for fcc:

$$\sigma_{\text{fcc}} = 4[Qa^2(h^2 + k^2 + l^2)]^{-1/2} \tag{1.47}$$

It is clear that, given the density of points in the unit cell σ, the separation between planes d_{hkl} can be obtained from the relation

$$d = V\delta \tag{1.48}$$

1.5.3 Body-centred cubic crystal surfaces

There are two atoms per unit cell in a bcc crystal. The content of the unit cell

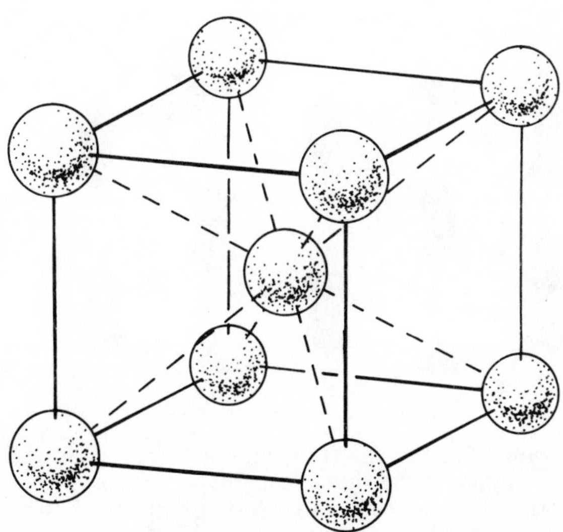

Figure 1.19. A body-centred cubic (bcc) crystal unit cell

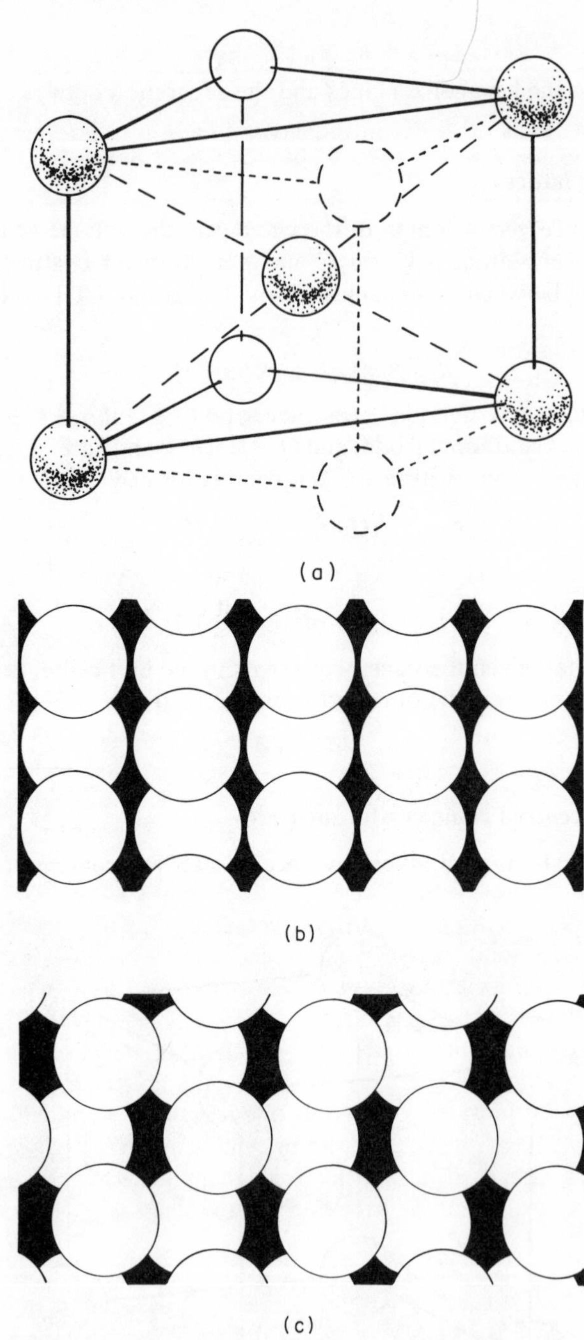

Figure 1.20. (a) The (110) face of a bcc crystal; (b) the surface, if the surface atom layer remains in the positions corresponding to a perfectly truncated bulk; (c) a possible lateral shift of the surface layer towards the threefold symmetric sites offered by the second-from-top layer

can be described by the projection given in Figure 1.19, where the circles indicate the positions of atoms in their two-dimensional projection and the small number inside each circle indicates the height of that atom above the basal (001) plane, in units of the lattice vector length a. The structure can be thought of as two interlinking simple cubic sublattices, one formed by atoms at the corners of the lattice the other by the atoms at the centre of each cell (see Figure 1.16). Within the bulk, each atom has eight nearest neighbours, their centres separated by ($\sqrt{(3)/2}$) lattice units, and six nearest neighbours, separated by one lattice unit.

The most densely packed surface is the (110), as given in Figure 1.20(a). The density is $0.707/a^2$, and the interplanar spacing is $0.707a$. It is consequently a very stable face. When compared with the bulk crystal, any surface atom has 'lost' two nearest-neighbour atoms and two next nearest neighbours. Looking at the surface from above, it can be seen that the surface atoms sit in twofold coordinated sites above their two nearest neighbours in the second layer (Figure 1.20(b)). If the surface layer were to shift laterally (or undergo a registry shift), the constituent atoms would increase their number of nearest neighbours to three, and it might be expected that this would be a more stable situation (see Figure 1.20(c)). The spatial appearance of the two-dimensional diffraction pattern would not be affected by such a shift (see section 1.3.1), but the individual beam intensities would be different in each case, and it is necessary to use LEED intensity analysis to resolve this ambiguity. In fact, all investigations to date, on Na,[19] Fe,[20] Mo[21] and W,[22,23] have concluded that the twofold coordination site is maintained. It seems that the surface atoms find it energetically preferable to maintain twofold symmetry throughout their neighbours than to mix twofold symmetry amongst their neighbours within the surface layer with threefold symmetry amongst nearest neighbours in the second layer, despite any possible energy gains associated with increasing the number of nearest neighbours.

The (001) surface is the second most densely packed bcc surface. The density is $0.5/a^2$ and the interplanar spacing, if the bulk structure is maintained, is $0.5a$. When compared with the bulk crystal, any surface atom has lost four nearest neighbours but only one next nearest neighbour. LEED intensity analyses have shown that the interplanar spacing between surface and second layer is generally contracted from the typical bulk crystal value, by an extent varying from near zero for Fe,[24,25] 7 per cent for W[26–32] and 12 per cent for Mo.[15,33]

The third most densely packed surface with a bcc crystal is not the (111) but the (210) which has a density $0.408/a^2$. By comparison the density of the (111) surface is only $0.289/a^2$. These are illustrated in Figures 1.21(a) and 1.21(b). The relatively low density of the (111) bcc surface makes it rather unstable, and liable to reconstruct into regions or facets that comprise more densely packed structures. An example of this is described in section 1.5.9.

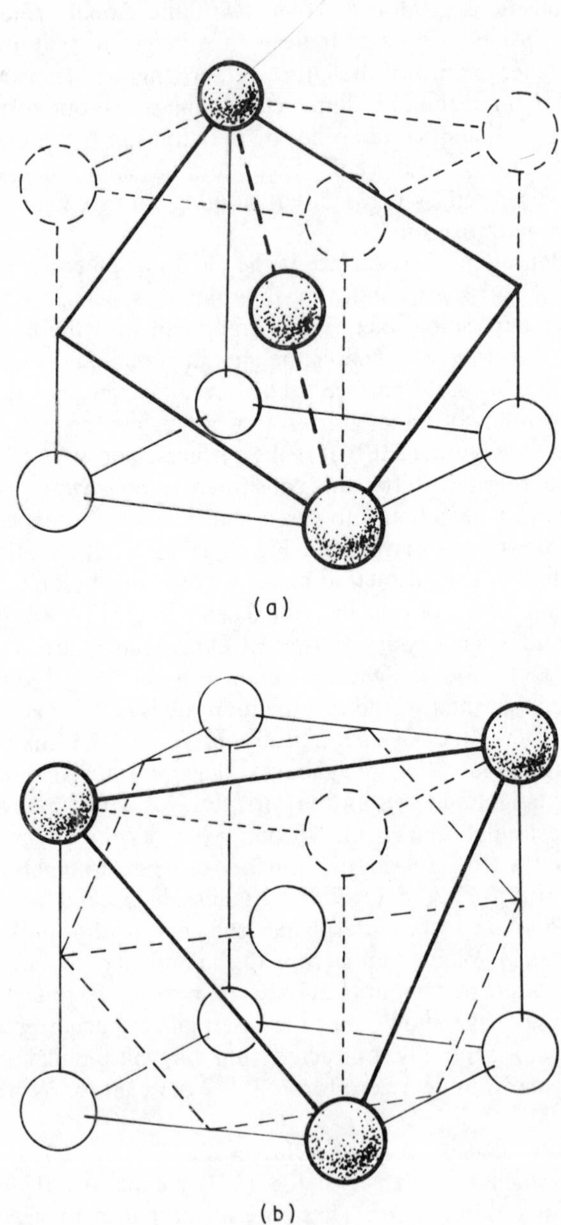

Figure 1.21. (a) The (210), and (b) (111) surfaces of a bcc crystal

1.5.4 Face-centred cubic crystal surfaces

An fcc unit cell comprises four atoms. The projection of atoms within this cell is given in Figure 1.22(a). Within the bulk, each atom has 12 nearest neighbours, the centre of each separated by a distance $a/\sqrt{2}$. Each atom has six next-neighbours, with a separation distance a.

The most densely packed fcc surface is (111), the primitive unit cell of which has a density $2.31/a^2$, and an interplanar separation distance $0.58a$. The plane view of the surface, Figure 1.22(b), shows that this face has a close-packed hexagonal structure. The second layer provides atoms beneath every other hollow, but translational symmetry means that there are two possible registries that could occur with this structure. If the registry of the top layer is called A and the registry of the second layer is B, we see that atoms within the third layer may either sit immediately below atoms in the surface layer in registry A again, or else sit beneath the remaining unoccupied hollows in registry C. In the case of an fcc(111) surface, it is this latter option which occurs, and the stacking sequence of layers may be described as ABCABCAB. . . . If the third layer had occupied sites with registry A, then the stacking sequence would have been set at ABABAB . . . and the structure would have turned into hcp(0001). The structural energies of fcc and hcp arrangements are very similar, and the truncation of the sequence at the surface could conceivably result in a change in the stacking order. In fact, no such change has ever been observed, despite the analysis by LEED of many surfaces, both fcc(111) (Al, Ni, Cu, Ag, Ir, Pt, Au and a high-temperature phase of Co) and hcp(0001) (Be, Ti, Zn, Cd, Sc and a low-temperature phase of Co): see Reference 34 for further discussions and literature references of these analyses. One exception may be Co(0001) which appears to have a surface transformation from hcp to fcc at a temperature below the critical temperature of the bulk,[35] this could be observed because the temperatures involved are relatively easy to attain. It would seem likely that any crystal which may change its bulk structure as a function of temperature could possess a different structure in the bulk and surface in a narrow temperature range just below the bulk critical temperature, although for most materials this may not be easy to realize in practice.

The second most densely packed fcc surface is the (001), with a density $2.0/a^2$ and an interplanar separation distance $0.5a$. Surface atoms lose 4 of their 12 nearest neighbours and in the case of certain heavy elements (Ir, Pt and Au) the surface layer reorders into a close-packed hexagonal structure. This reduces coordination between the surface and second layer, but increases the number of nearest neighbours within the top layer from four to six.

The (110) surface of a fcc crystal has a density $1.4/a^2$ and an interplanar spacing $0.35a$. A certain lack of stability has made the study of fcc (110) surfaces rather difficult; some, for example Pt(110), reconstruct to a (2×1) structure.[36] A recent LEED analysis has indicated that the surface of Ni(110)

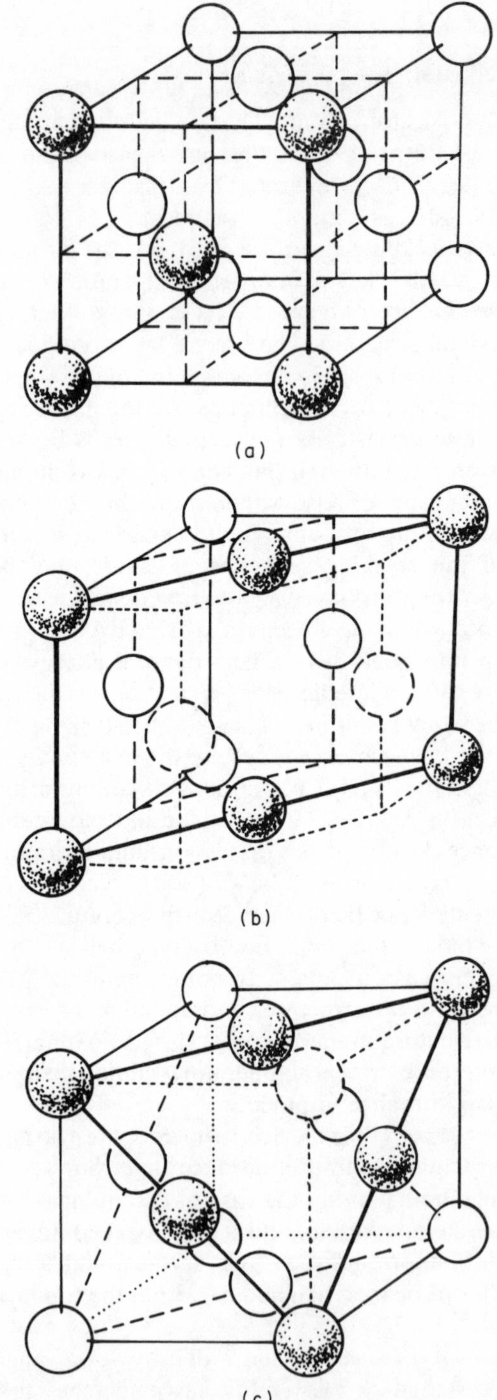

Figure 1.22. A face-centred cubic (fcc) crystal, showing (a) the (100), (b) the (110) and (c) the (111) surfaces

is relatively well ordered with a surface interlayer spacing contracted by about 7 per cent from the typical bulk value.[37] The (110) surface of Al appears to be contracted by up to 15 per cent,[38] which makes this the largest interplanar displacement yet discovered, comparable with the contraction of certain bcc (001) surfaces.

1.5.5 Hexagonal close-packed crystal surfaces

Any surface plane of an hcp crystal may be indexed with three Miller indices (*hkl*) as with the cubic crystals. The two vectors within the hexagonal plane are separated by angle of 120° rather than the 90° of the cubic lattice (see Figure 1.22). However, it is common practice to index hcp surfaces by a set of four indices (*hklm*) where h, k and l refer to the reciprocals of the intercepts of the plane in question with respect to three vectors within the hexagonal plane, mutually separated by 120°. A consequence is that there is redundant information in the four-index notation, and in practice a surface which can be described as (*hkl*) would be indexed $(h\ k\ h + k\ l)$ in the four-index form.

If the lattice is defined in terms of the three indices (*hkl*), a is the lattice vector length within the plane, c the separation distance between close-packed planes and r the axial ratio c/a (see Figure 1.22), then the volume of the primitive unit cell is

$$V = (\sqrt{(3)}/2)a^2c \tag{1.49}$$

The spacing between (*hkl*) planes (through lattice points) is

$$d_{\text{hcp}} = c\sqrt{(3)}\,[4r^2(h^2 + hk + k^2) + 3l^2]^{-1/2} \tag{1.50}$$

and the density of atoms within a surface plane (*hkl*) is

$$\sigma_{\text{hcp}} = 2/a^2[4r^2(h^2 + hk + k^2) + 3l^2]^{1/2} \tag{1.51}$$

The stacking sequence has been noted above as the feature which distinguishes an hcp(0001) from an fcc(111) surface. In fact, this assumes that the axial ratio c/a corresponds to the theoretical value $\sqrt{(8/3)} = 1.633$ for close packing throughout the structure. In fact, a further distinction may occur due to departures from this ideal ratio. Zinc, for example, has $r = 1.85$ which is much greater than the ideal. Mg and Co with $r = 1.62$ are quite close, whilst Be, with $r = 1.58$ is rather smaller. A number of metals transform easily from hcp to fcc, and Co has been studied by LEED both in its low-temperature hcp phase and its high-temperature fcc phase. Bulk phase changes between bcc and either fcc or hcp may also occur with certain metals, but the much greater differences in structure mean that the specimen may recrystalize into many discretely ordered polycrystals, rendering it unacceptable for LEED analysis. For this reason Fe, which is bcc at room temperature but hcp at high temperatures, must not be heated above 700 °C during cleaning.

The basal plane (0001) provides the most stable and commonly studied hcp surface. If a is the lattice vector length within the close-packed plane, then the density is $1.15/a^2$ and the interlayer separation distance is given by c. If r ($= c/a$) does not correspond to the ideal, then one cannot generalize about areas or interlayer spacings of other faces. Equations (1.46) and (1.47) above provide direct means of deriving these values for any plane (hkl). The structure of a number of low-index hcp planes are illustrated in Figure 1.23.

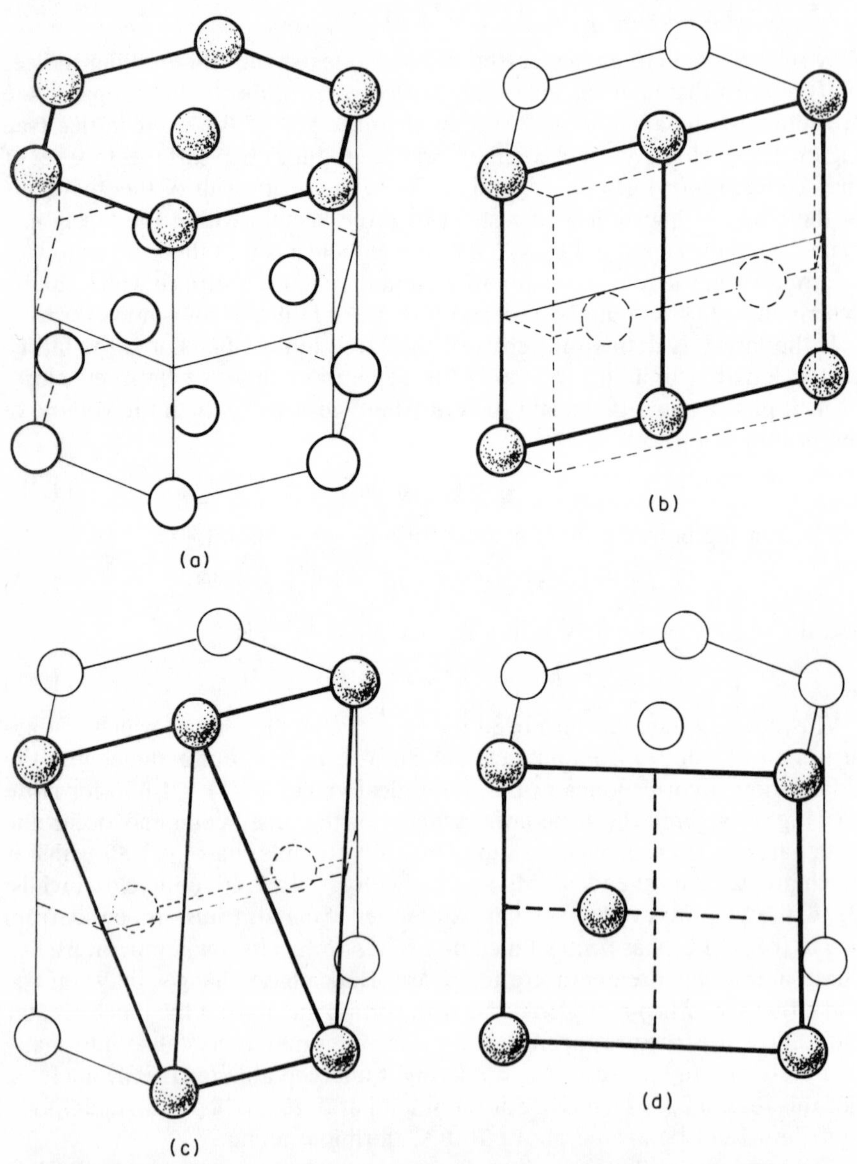

Figure 1.23. A hexagonal close-packed (hcp) crystal unit cell, illustrating (a) the (0001), (b) the $(10\bar{1}0)$, (c) the $(10\bar{1}2)$ and (d) the $(11\bar{2}0)$ surface planes

The only reported observation of a phase transition of an hcp (0001) surface is that by Bridge *et al.*[35] who noted the transition of the surface layer from hcp to bcc below the critical temperature of the bulk. The Ti(10$\bar{1}$1) surface has also been found to transform to a bcc phase, in this case at about 80 °C below the bulk critical temperature—a result indicated by ultraviolet photoemission (UPS), but not as yet substantiated by LEED.[39]

A particularly interesting plane is the (10$\bar{1}$2), because successive planes are separated alternately by short and long distances, and it is not clear whether a clean metal cut with this surface will have its surface layer separated from the second layer by the shorter or longer of these two separations. In a study of Co(10$\bar{1}$2), Prior *et al.*[40] used a kinematic LEED intensity analysis to show that the double-layer termination is the more likely of the two possibilities.

1.5.6 Diamond structure surfaces

The diamond structure is characteristic of strongly directional covalent bonding. The positions of the atoms are most easily represented by their projection as given in Figure 1.24. It has the form of two intersecting fcc lattices, one comprising the atoms at heights 0 and $\frac{1}{2}$, the other comprising the atoms at heights $\frac{1}{4}$ and $\frac{3}{4}$. The cubic cell contains eight atoms. Carbon (diamond!), Si, Ge and Sn (grey tin) may all take this structure. Each atom is tetragonally bonded to its neighbours. Despite the well-known cohesive strength of bulk diamond, a LEED and AES investigation of diamond surfaces[4] has shown that the (001) surface reconstructs to give a structure with (2 × 1) periodicity, and that any diamond surface could actually transform into the hcp graphite structure if subjected to sustained heating above 1,000 °C.

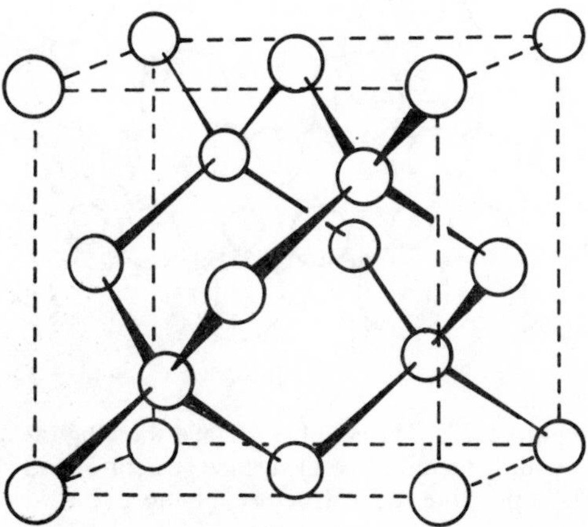

Figure 1.24. A diamond structure unit cell

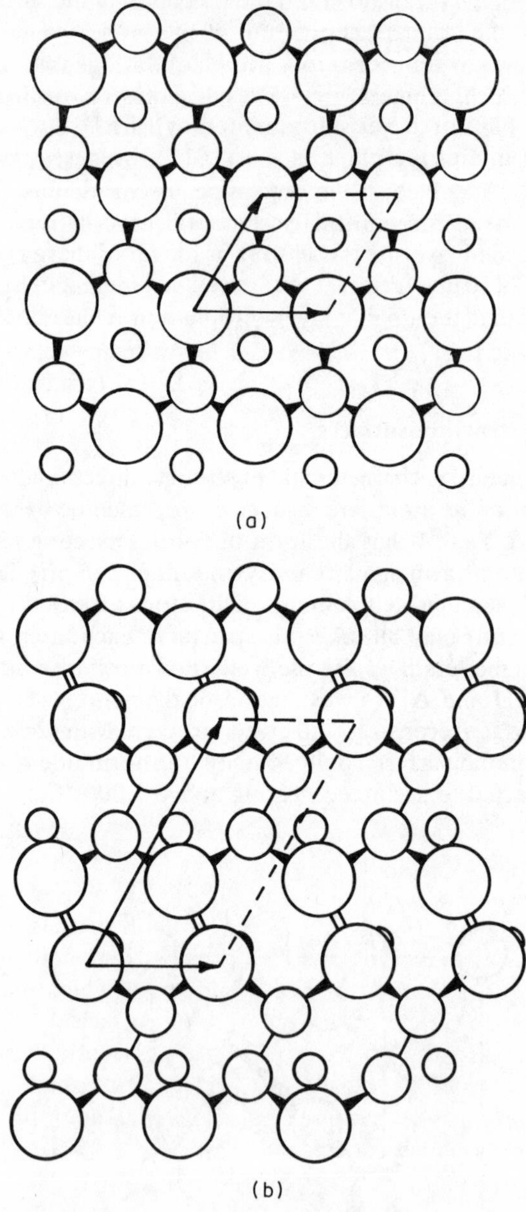

Figure 1.25. The Si(111) surface showing (a) the undistorted (1 × 1) surface structure, and (b) a possible reconstructure, giving a (2 × 1) structure

Obvious applications in the field of microelectronics have made Si and Ge the subject of many surface studies. The directionality of the bonding is closely linked to the electronic structure of the material, and the presence of the surface causes perturbations which may result in geometrical reconstruction. Compared with metals, these materials tend to be rather brittle in certain directions, and a flat surface can be obtained by cleavage (pressing a knife-edge against the crystal so that a crack propagates between weakly bonded planes). The cleavage plane is often the densest one since this requires the fewest bonds to be broken between planes. In the ideal diamond structure the densest plane is the (111).

In practice, if a Si(111) or Ge(111) plane is produced by cleavage under vacuum at room temperature, the resulting structure has (2×1) periodicity as noted above for diamond, so that twice as many beams appear in the diffraction pattern as would be expected from the ideal termination. A possible interpretation, suggested by Lander *et al.*[42] is given in Figure 1.25. An alternative, and simpler, interpretation is that of Haneman[43] in which alternate rows of surface atoms move outwards and inwards respectively to produce a buckled surface. The former model assumes that the reconstruction involves predominantly lateral shifts, the latter assumes vertical or normal shifts. This structure, however, is only metastable. If annealed at roughly 500 K the (2×1) structure irreversibly transforms to a (7×7) structure. The transformation temperature is quite sensitive to experimental uncertainties such as the terrace width of the cleavage steps and in practice may occur between 470 and 660 K.

The ideal unreconstructed Si(100) surface has a square unit cell, length $a_0/\sqrt{2}$, where a_0 is the characteristic length of the cubic unit cell. The bulk interlayer spacing is $a_0/2\sqrt{2}$. For Si, $a_0 = 5.341$ Å, so this spacing is 1.92 Å. In practice, the surface is also reconstructed into a (2×1) structure. A variety of possible interpretations of the reconstruction have been made, and several tested by LEED analysis. None, however, have met with sufficient success to provide an unambiguous solution. Recent theories of the reconstruction, combined with evidence from LEED analyses and helium ion-scattering experiments indicate that subsurface strains exist which may cause reconstruction in three to five layers adjacent to the surface.[44]

Semiconductor surfaces are complex, and are currently the subject of many independent investigations. Kahn[45] has recently published a review of progress to date, and this may provide a useful starting-point for the assessment of any future developments in this field.

1.5.7 The surface structures of some common binary compounds

Departure from single elements opens up an enormous number of possible new structures. Some such as NaCl are very simple, but others may be much more complex. Given the bulk structure, it is possible to use the methods described above to describe planes, but the actual atomic structure of those planes can only be determined from a detailed consideration of the configuration of atoms within the unit cell. It would not be appropriate in this book to enter into discussion of all the different structures examined to date. However, it is worth noting that numerous important surface studies of binary compounds fall into a few major categories: NaCl, CsCl, zinc-blende and wurzite. The first three can be derived from the simple cubic, bcc and

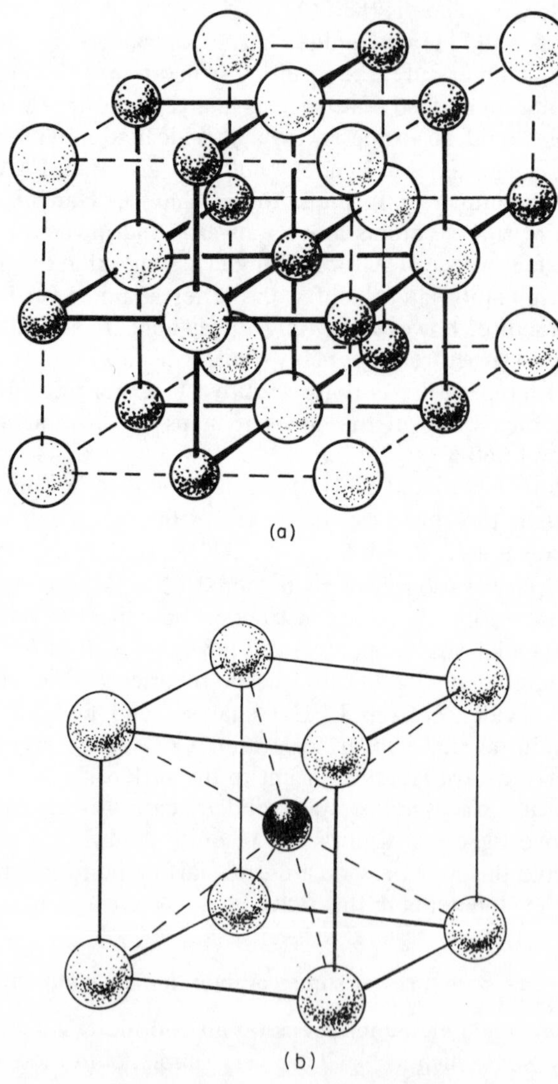

Figure 1.26. Cubic crystals comprising two atoms per unit cell: (a) NaCl; (b) CsCl

Table 1.2 Binary compounds possessing simple crystal structures at room temperature

General type	NaCl	Zinc-blende	Wurzite	CsCl
Monoxides	Mg, Ca, Ti, V, Mn, Fe, Co, Ni, Sr, Zr, Cd, Ba, Eu, Th, Pa, U, Np, Pu, Am		Be, Zn	
Sulphides	Mg, Ca, Mn, Sr, Ba, Pb, La, Ce, Pr, Nd, Sm, Eu, Tb, Ho, Th, U, Pu	Be, Zn, Cd, Hg	Mn, Zn, Cd	
Selenides	Pb, Mg, Ca, Sr, Ba	Be, Zn, Hg	Zn, Cd	
Tellurides	Sn, Pb, Ca, Sr, Ba	Be, Zn, Cd, Hg	Mg, Zn	
Metallic halides	Metals with coordination number (CN) = 6	CN = 4	CN = 4	CN = 8
Alkali halides	All except those with CsCl structure			CsCl, CsBr, CsI
Carbides	Ti, V, Zr, Nb, Hf, Ta			
Nitrides		B	Al, Ga, In	
Phosphides	La, Sm, Th, U, Zr	B, Al, Ga, In		
Arsenides		B, Al, Ga, In		
Sb		Al, Ga, In		

Alloys with CsCl structure: CuZn (ρ-brass), AgMg, BeCu, FeAl, CoAl, NiAl, LiAl, LiHg, MgTl, LiTl, SrTl.

diamond structures respectively by the alteration of elemental types at successive atomic sites.

NaCl has a cubic structure in which each of the two elemental species occupy alternate sites, as given in Figure 1.26(a). It is clear from Table 1.2 that a wide variety of compounds possess this structure. LiF,[46] MgO[47] and NiO[48,49] have all been examined using LEED intensity analysis and positive evidence found for a rumpling of the surface, the cationic species moving inwards and the anionic species outwards by a few per cent of the ideal interlayer spacing. In addition, the mean position of the surface layer may also shift slightly towards the bulk. EuO has also been examined using LEED,[50] but the surface layer was found to be disordered after cleavage, possibly due to surface-enhanced covalency of the cationic species, or to attempts to form an antiferromagnetic surface layer.

CsCl has a structure equivalent to the bcc structure, in which the centred atom has one elemental type and the corner atom has the other, as illustrated in Figure 1.26(b). This is commonly adopted by metallic alloys and metal halides for which the metal has a coordination number 8.

The zinc-blende or cubic zinc-sulphide structure is formed when alternate sites in a diamond structure are occupied by each of the two component species. We noted earlier that the diamond structure has the form of two intersecting fcc lattices, and the zinc-blende structure is formed when, for example, Zn atoms are placed in one of the fcc lattices and S atoms in the other. The space lattice is fcc and each atom has four equally spaced neighbours of the complementary type. Numerous materials possess this structure, but of particular interest is the semiconductor GaAs. Other related materials with this structure are given in Table 1.2. Such materials display two distinct types of surface, which differ widely in structural properties. One is the non-polar type of surface which terminates ideally in an equal number of the two elemental species (and consequently neutralize each other's residual charge). The other is polar, comprising entirely one of the lattice constituents. (A specially cut stepped surface may of course lie between these extremes.) Cleavage always occurs along the non-polar surface with the lowest index: the (110) surface (see Figure 1.27). This is in contrast with pure Si or Ge which cleave along the (111) planes. Although the covalent bonding energy of a (110) surface is higher than a (111), the additional electrostatic forces which apply to the (111) polar surfaces make this energetically less favourable. It is conventional to define the cationic face as (111) and the anionic face as $(\bar{1}\bar{1}\bar{1})$. It has been found that, in several cases, the (111) face reconstructs to form a structure with (2×2) periodicity and the $(\bar{1}\bar{1}\bar{1})$ face reconstructs to produce a (3×3) structure.[40]

All clean, non-polar surfaces of these materials produce (1×1) LEED patterns. This does not preclude the possibility of some reconstruction, since the unit cell contains two types of atoms which may shift with respect to each other without affecting the overall periodicity. The GaAs(110) surface has been the subject of numerous LEED studies; see reference 53 for a discussion of this and other covalently bonded semiconductor (110) surfaces. A recent study of GaAs(110) using multiple-scattering theory for LEED intensity analysis by Masud[54] confirms an earlier finding that reconstruction occurs with bonds rotating so that the anionic species (As) move outwards and the cationic species (Ga) move inwards, together with a slight shift in the mean surface layer towards the bulk.

Finally, we mention the wurzite structure. This is also based on tetrahedral bonding, but alternate groups are aligned to form a hexagonal rather than a cubic space lattice. CdS and ZnO are important materials which have this structure. The (0001) and $(000\bar{1})$ surfaces of such materials are polar. In this direction, the structure comprises double layers (e.g. Cd–S) stacked in such a

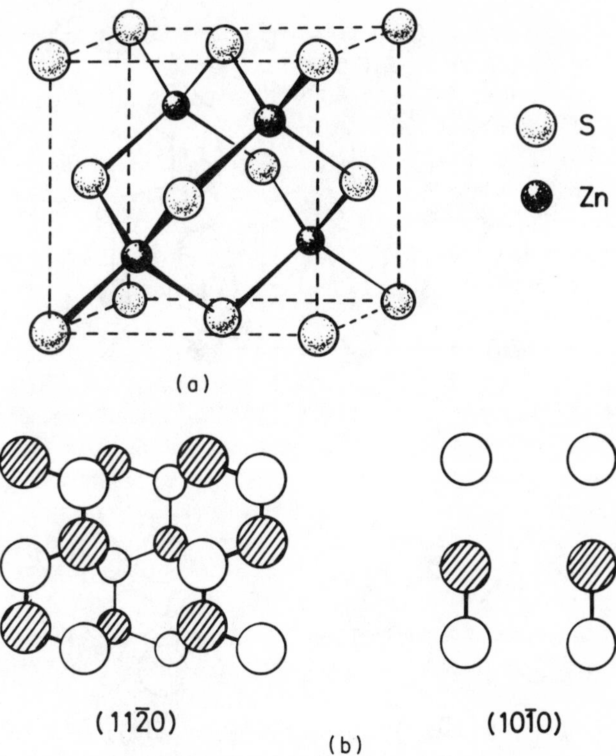

Figure 1.27. The zinc-blende structure: (a) unit cell, (b) surface arrangements of the (11$\bar{2}$0) and (10$\bar{1}$0) faces

way that every other double layer is identical except for translation. Adjacent double layers are related by a 60° rotation about the surface normal in addition to translation. Uncertainty in the interpretation of even these basal planes continued for many years because they almost always exhibit sixfold rather than threefold rotational symmetry in their LEED patterns. This, however, can be attributed to the presence of single height steps, revealing terraces rotated through 60° with respect to each other, the superposition of patterns from adjacent terraces combining to produce the apparently sixfold symmetric pattern.[55] The lowest energy non-polar surfaces can be considered to be the sides of the hexagonal prism, normal to the basal plane and rotated 30° with respect to each other. In practice two of these prismatic faces, the (10$\bar{1}$0) and (11$\bar{2}$0), are nearly equivalent in energy, but the former is the usual cleavage plane. These are illustrated in Figure 1.28.

1.5.8 Overlayer notation

The lattice vectors $\mathbf{b}_1$ and $\mathbf{b}_2$ of an overlayer can always be described in terms of the substrate lattice vectors $\mathbf{a}_1$ and $\mathbf{a}_2$. Advantages of using this method of

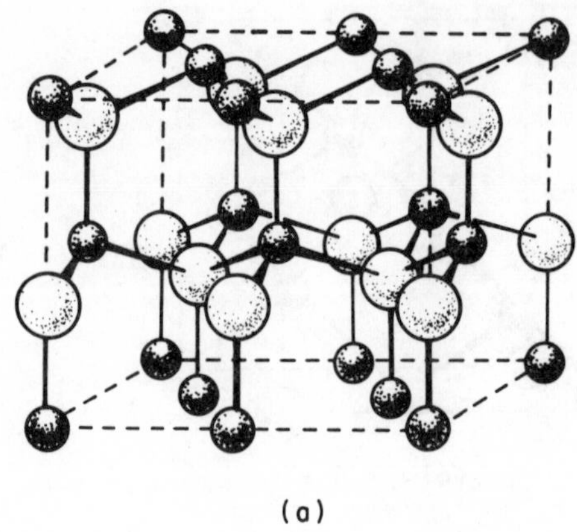

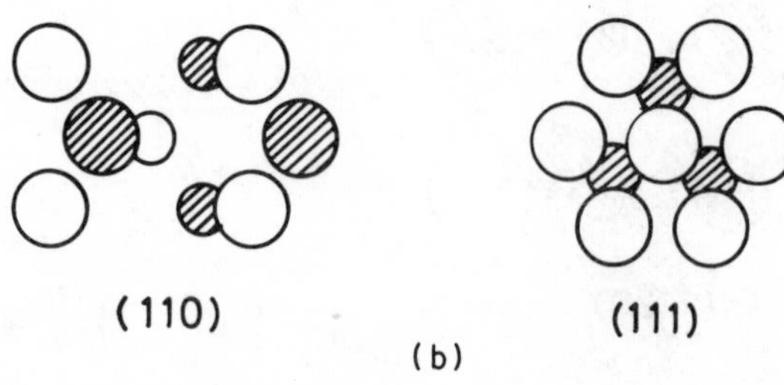

Figure 1.28. The wurzite structure: (a) unit cell, (b) surface arrangements of the (110) and (111) faces

notation are that it is always applicable, and, as we have seen in section 1.3, the resulting matrix elements can be transformed to or from reciprocal space, enabling the real-space lattice and the resulting diffraction pattern to be rapidly identified. A disadvantage is that matrices are not convenient to write, and it is not always straightforward to recognize special features such as the existence of centred symmetry structures. Certain simple structures occur so frequently that a shorthand notation is preferable. The commonest nomenclature is that proposed by Wood,[56] as described in the following paragraph.

If the lengths $|b_1| = m\,|\mathbf{a}_1|$ and $|\mathbf{b}_2| = n\,|\mathbf{a}_2|$ and the angle between $\mathbf{b}_1$ and $\mathbf{b}_2$ is the same as that between $\mathbf{a}_1$ and $\mathbf{a}_2$, then the overlayer may be described as $p(m \times n)$ where p indicates that the overlayer cell is primitive. If the overlayer lattice vectors are subtended by the same angle as the substrate vectors, but the whole lattice is rotated through an angle α with respect to $\mathbf{a}_1$, the lattice may be described as $(m \times n)R\alpha°$. Examples are given in Figure 1.29.

If the sides of the unit cell are greater than the substrate unit cell by an even number of substrate lattice spacings, then it is possible for an additional atom

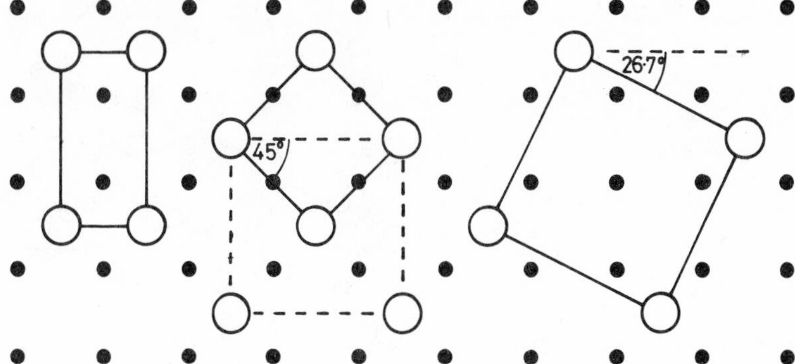

Figure 1.29. Three common adlayer structures: (a) $p(2 \times 1)$; (b) $(\sqrt{2} \times \sqrt{2})R45°$ or $c(2 \times 2)$, and (c) $(\sqrt{5} \times \sqrt{5})R26.7°$

to be placed at the centre of the cell, with a bonding symmetry identical to that of the corner atoms. If there are no other atoms within the unit cell, or if the positions of any additional atoms relative to the centre atom are identical to their positions relative to the corner atoms, then a special symmetry condition is produced. Referring to Figure 1.29, we see that the overylayer cell can be defined in either of two equivalent ways, which are indistinguishable from each other. The unit cell is called 'centred', and in Wood's notation, takes the form $c(m \times n)$. The translational symmetry properties of this type of structure result in destructive interference which causes alternate beams to disappear (see Figure 1.30(a)). For comparison, the pattern resulting from a $p(2 \times 2)$ structure is given in Figure 1.30(b).

In the case of a $c(2 \times 2)$ structure, this argument is rather trivial, since it is easy to interpret the pattern in terms of the primitive unit cell used previously to illustrate matrix notation (see Figure 1.6), which may also be written in Wood's notation as $(\sqrt{2} \times \sqrt{2})R45°$. However, it provides a useful means of interpreting more complex structures. For example, the $c(4 \times 2)$ structure occurs quite frequently. The primitive unit cell is oblique and cannot be accommodated within the standard format of Wood's notation (see Figure 1.30(c)). Nevertheless, the overlayer lattice is easily identifiable from the diffraction pattern, due to its simple relationship to the pattern which would be expected from a $p(4 \times 2)$ overlayer (see Figures 1.30(c) and (d)).

An important point to note is that a structure can only be described as centred if the corner and centred atoms are indistinguishable, in other words, if a lattice based on a corner atom has an identical unit cell to a lattice based on a centre atom. For example, when sulphur is adsorbed on to a Mo(001) [11] or W(001)[57] surface, there is strong evidence that the structure formed at saturation is a $p(2 \times 1)$ with a second atom at the centre of the cell, as shown in Figure 1.31. This may not be termed $c(2 \times 1)$ since the centred atom has a different bonding site and so is distinguishable from a corner atom.

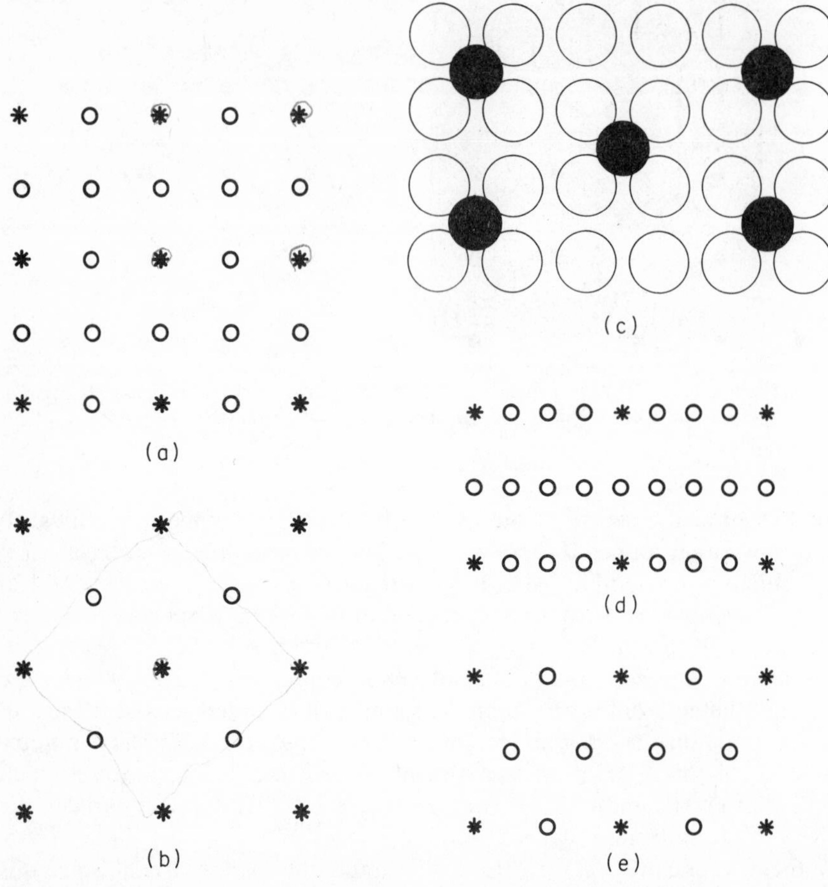

Figure 1.30. (a) $p(2 \times 2)$, and (b) $c(2 \times 2)$ diffraction beam patterns, showing how the centred arrangements causes alternate beams to be absent; (c) a $c(2 \times 4)$ adlayer: the corresponding pattern can be related to the $p(2 \times 4)$ pattern (d) by removing alternate fractional-order beams as in (e)

Destructive interference cannot result in the systematic disappearance of alternate beams from the $p(2 \times 1)$ pattern.

The complete description of a surface structure formed by the adsorption of an atomic or molecular species E on a (hkl) surface of a substrate M is thus

$$M(hkl) - {}^{p}_{c}(m \times n)E$$

or

$$M(hkl) - {}^{p}_{c}(m \times n)R\alpha^{\circ} - E$$

with either p or c used as appropriate.

Wood's notation has the advantage of simplicity, but can be ambiguous when applied to non-square lattices. It is conventional with rectangular

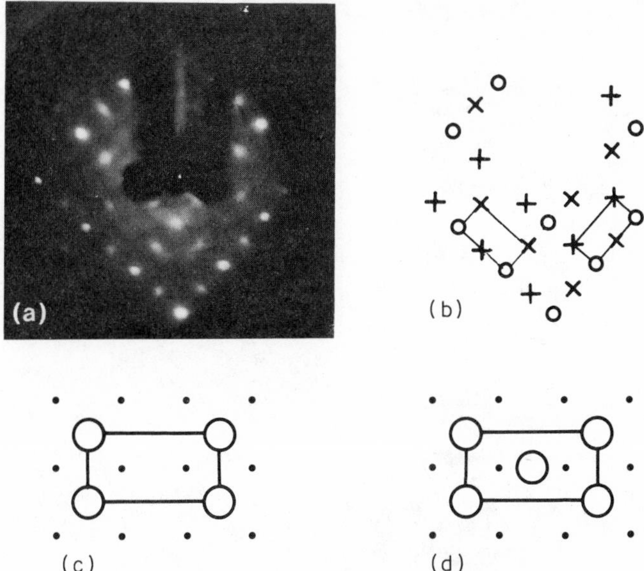

Figure 1.31. A $p(2 \times 1) + p(1 \times 2)$ pattern, as formed by sulphur on Mo(001): (a) observed pattern; (b) diagram of pattern; (c) possible adlayer arrangement, corresponding to a half-monolayer coverage; (d) possible adlayer arrangement, corresponding to a full monolayer coverage. (Reproduced by permission of North Holland Physics Publishing from Clarke[11])

lattices to order the lattice vectors starting with the shorter of the two real space axes. Centred rectangular meshes, as obtained with bcc(110) surfaces may often be described in terms of the non-primitive rectangular ideal surface mesh rather than the oblique primitive mesh. This is of considerable benefit when complex adsorbed structures are to be described since they are frequently based on the rectangular mesh. The non-primitive mesh may be given the notation $(1 \times 1)p$[58] or $p'(1 \times 1)$.[59] The latter notation is preferable because it is more consistent with Wood's original notation and less likely to cause confusion if carbon or phosphorus superstructures are to be described. Any overlayer with rectangular symmetry will take the form $p'(m \times n)$ or $c'(m \times n)$, where m refers to the periodicity along the shorter of the two real axis directions. Examples are given in Figure 1.32.

1.5.9 Non-planar surfaces: steps and facets

The orientation of a flat crystal surface is described by the relationship between the bulk crystal axes and the macroscopic plane which, for example, would reflect optical light. Indeed, one of the best ways of determining the orientation is to combine X-ray diffraction, which gives the bulk crystal orientation with the reflection of laser light, which detects the macroscopic surface plane.

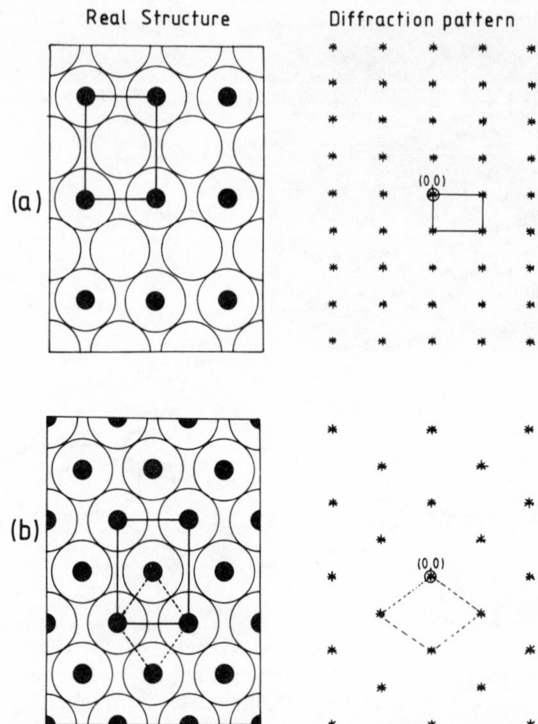

Figure 1.32. Nomenclature for adsorbates on bcc(110) surface: (a) $p'(1 \times 1)$ based on rectangular mesh with corresponding diffraction pattern; (b) clean surface ($p(1 \times 1)$ in the usual Wood notation) which would be designated $c'(1 \times 1)$ in the present notation leading to alternate 'missing' spots in the diffraction pattern. (Reproduced by permission of North Holland Physics Publishing from Clarke and Morales de la Garza[32])

It is always possible to describe the macroscopic plane orientation by means of the Miller indices (see section 1.5.1). However, a description of this plane does not necessarily inform us of the actual surface structure on an atomic scale.

In general the stability of a crystallographic plane will increase with the density of atoms within that plane, as we have demonstrated earlier in this chapter. If the crystal is cut at some inclination to one of the very stable faces, then the surface may break up into a series of facets of 'hill-and-valley' structures, which maintain the macroscopic orientation, but on an atomic scale involve the juxtaposition of planes with low Miller indices and high atomic densities. Alternatively, the same macroscopic plane may be reproduced as closely as possible subject to the discrete nature of the crystal lattice by a series of parallel terraces separated by monoatomic steps. A wide

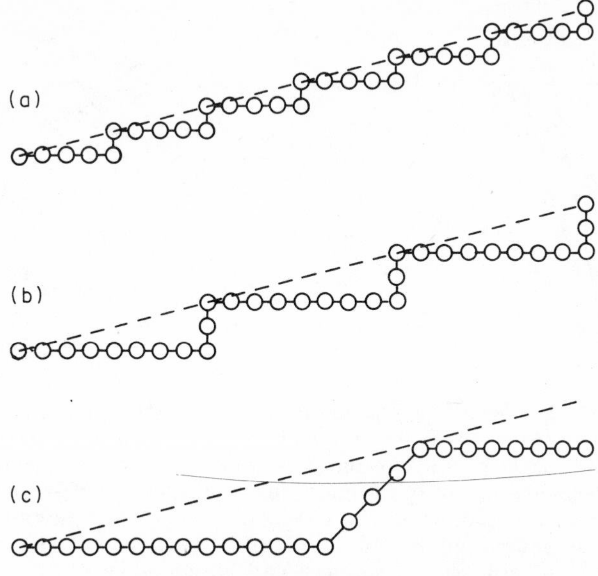

Figure 1.33. (a) Monoatomic regular step structure; (b) regular diatomic step structure; (c) facets

variety of alternatives between these extremes are also possible, and a number of possibilities are illustrated in Figure 1.33 for the simple situation of a one-dimensional surface.

The thermodynamically most stable surface structure is obtained when the total surface free energy is a minimum. A graphical method of determining the equilibrium structure is the Wulff construction or γ-plot.[60] In a γ-plot, the surface free energy is plotted as a function of crystallographic orientation. The more stable low-index faces define cusps in the diagram, as shown in Figure 1.34. In three dimensions, the polygon defined by the planes normal to, yet touching, the cusps is similar to the equilibrium shape of the crystallite. Planes with orientation in the vicinity of the low-index planes (i.e. 'vicinal' surfaces) are represented by portions of the γ-plot near the cusps. The surface energy which results from the faceting of such a vicinal surface has been shown by Herring[61] to be given by the projection of the facet plane intersection point (C in Figure 1.34) upon the normal vector to the original vicinal plane (OM)*. OM defines the surface of a sphere which passes through the origin and the three cusps whose planes have positive projections on n. If, on the other hand we consider that the surface does not form into large-facet planes but rather into a regular series of low-index terraces and monoatomic steps, then its surface energy will be given by the surface free energy of the low-index terrace face γ, plus some energy contribution β associated with each step. The step contribution will be a function of the step density; if the γ-plot in the vicinity of a cusp is spherical (as would be

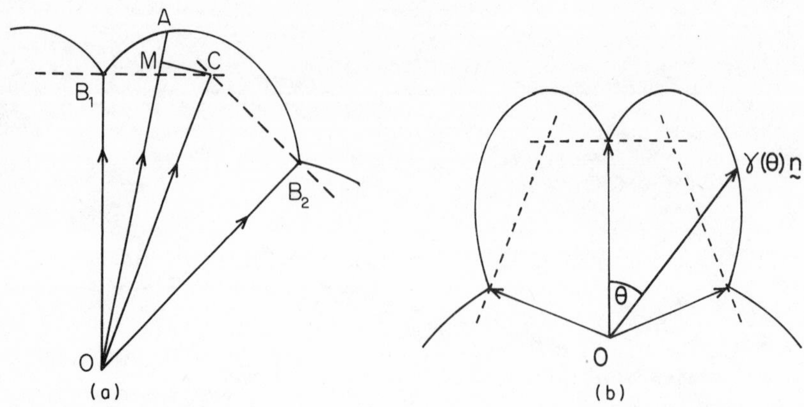

Figure 1.34. The Wulff construction for predicting the equilibrium shape of a crystal. (a) γ-plot showing the surface free energy as a function of crystallographic orientation described by the surface normal **n**. The dashed lines indicate surface plane orientations of minimum energy. (b) Section of γ-plot illustrating the reduction in surface energy for direction n_s when the surface assumes a hill and valley structure comprising facets of orientation n_1 and n_2. (Reproduced by permission of Springer-Verlag from Wagner[71])

predicted by a pairwise interaction model), it can be shown[61] that

$$\gamma(\theta) = \cos\theta\left(\gamma_0 + \frac{\beta}{d}\tan\theta\right) \tag{1.52}$$

where d is the step height and θ the inclination angle of the face with respect to the low-index plane. This defines a circle passing through the origin and cusp (a sphere in three dimensions) of radius

$$R = \frac{1}{2}\sqrt{\left[\gamma_0^2 + \left(\frac{\beta}{d}\right)^2\right]} \tag{1.53}$$

If this radius is less than the radius of the circle defined by the locus of OMs (Figure 1.34) then the monoatomic step structure will be more stable than large facets, and vice versa.

If there are impurities present, significant changes may occur in the step behaviour. If the impurity atoms preferentially bond at the steps, then the free energy associated with the steps, β, will be reduced. On the other hand, if the impurities bond on to the low-index terraces then it can be shown[62] that the cusp will become more pronounced—this is because the surface free energy of the low-index face itself will be effectively lowered. In the case of edge adsorption, the face energy at the cusp is unaffected (at the cusp there are no steps, so there is no impurity-induced energy reduction) but β is less, so the curvature of the γ-plot is increased about the cusp. In either case, the result could be that faceting preferentially occurs.[63] It is not always necessary that the original surface have particularly high Miller indices: a (111) surface on a bcc material is relatively open and could be considered as a very high density stepped surface. It is found that small densities of oxygen on Mo(111) may cause it to facet into relatively large subplanes of {110} and {211} orientation.[64–66]

Even on a perfectly clean surface the above analysis is not quite complete. In general, steps on a vicinal surface will tend to repel each other. This may arise from a number of causes: for example, due to the change in configurational entropy associated with a step edge,[67] or due to potential interactions involving either long-range strain fields[68] or pairwise interactions.[69] Energetically, this will add higher power terms to the tan θ term in equation (1.52). This may result in a fairly regular spacing between steps so that the local surface plane (defined by the locus of successive edge site atoms) is maintained as close as possible to the macroscopic surface plane.

The way in which LEED may be used to ascertain which of the various possible atomic arrangements occurs is discussed in Chapter 6. At this stage we shall only concern ourselves in ways of describing the structure mathematically.

Faceted surfaces are not normally given a detailed atomic description, since the sizes of the facets tend to be large and ill-defined. The general topology may be found from reflection high-energy electron diffraction (RHEED) or perhaps electron microscopy, but it is not usually possible to specify the exact number of atoms which lie in any facet face.

Vicinal surface, on the other hand, can often be given a precise description. Since inter-step repulsive interactions tend to order the steps into a regular array, vicinal surfaces may often be described in terms of terraces with low Miller indices (h, k, l) or simply (hkl) of width m atoms, and steps with edge planes $(h'k'l')$ of atomic height n. Using this as a basis, Lang *et al*[70]. suggested the following nomenclature:

$$E(S) - [m(hkl) \times n(h'k'l')] \tag{1.54}$$

where E is the chemical notation for the substrate and the S is included simply to designate the fact that it is a stepped surface. If $n = 1$ it is usually omitted. If the step edge lies along a low-index plane so that there are no kinks, then m is straightforward to determine.

Ambiguity may arise in describing the orientation of the step, since this depends precisely on which atoms in the terrace are assumed to form part of the step surface. For example, the monoatomic step in a fcc(111) surface as illustrated in Figure 1.35 may be defined as (111) or (110) depending whether rows AA′ and BB′ or AA′ and CC′ are chosen in the definition. Lang *et al.*[70] proposed that the step be always defined by the plane that makes the smallest angle with the terrace normal (in this case the (111) plane defined by AA′–BB′). In the above definition, it is usual to include the row of atoms which lie in the terrace below an edge row in evaluating the value of m. Further ambiguity may also arise if the step edge corresponds to a higher

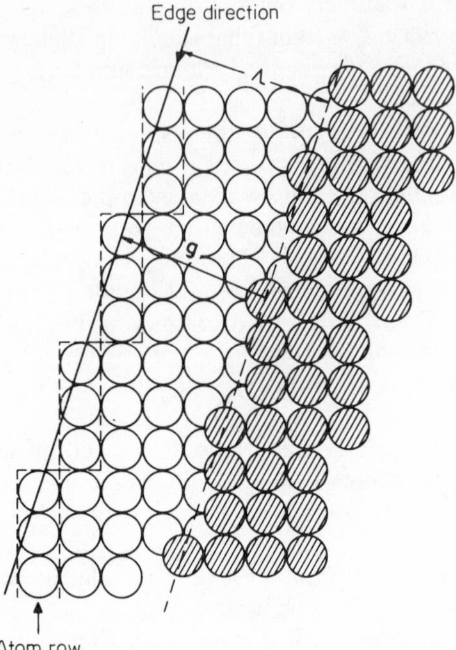

Figure 1.35. Top view of a stepped surface indicating the definitions of atom rows on the terrace; $4\frac{2}{3}$ atom rows are accomodated on the lower terrace (open circles). The edges are formed by smooth segments of close-packed atoms and periodically spaced kinks. The vector **g** connects atoms in equivalent positions on neighbouring terraces. (Reproduced by permission of Springer-Verlag from Wagner[71])

index plane so that it includes kinks (steps within the step plane itself). In this case, planes drawn through equivalent atoms in successive step edges will not lie parallel to atomic rows, so that the distance between edges is not strictly an integral number of rows. Wagner[71] has suggested that all edge atoms between successive equivalent edge sites belong to one atom row. In Figure 1.36 each egde row comprises three atoms and, depending exactly how one counts the atom rows across the terrace, one arrives at a mean number which is a fraction, in this case $4\frac{2}{3}$. To be precise, one could use the separation between successive equivalent edge planes which, in a two-dimensional projection on to the low-index terrace plane is $\sqrt{(58)}/2(= 3.81)$. This is approximately one row less than the previous estimate because the previous scheme includes terrace atoms under the next step edge.

It is clear that this nomenclature, although quite suitable for simple stepped (non-kinked) surfaces, becomes increasingly unsatisfactory as the surface

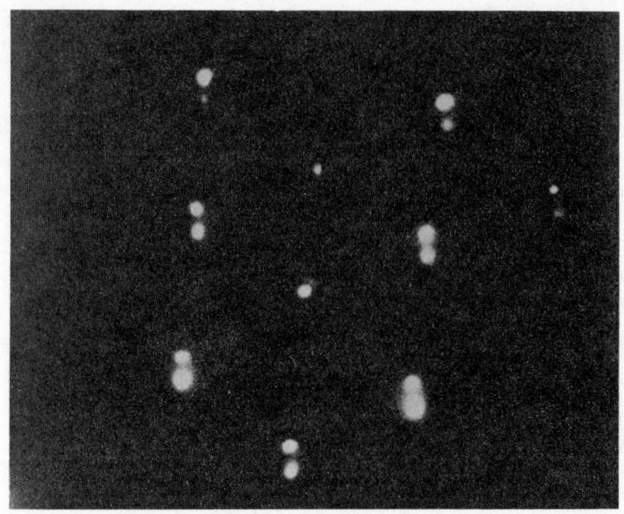

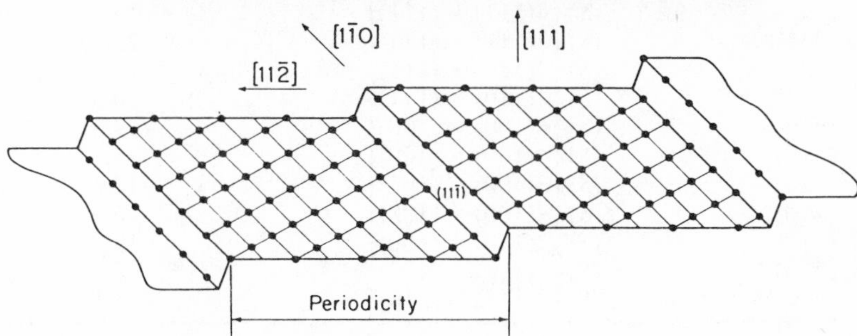

Figure 1.36. Diffraction pattern and schematic representation of a stepped surface that was cut 6.5° from the (111) crystal face. The notation to identify this surface is Pt*S*) − [9(111) × (111)]. (Reproduced by permission of North Holland Physics Publishing from Baron *et al.*[63])

becomes more and more complex. In Table 1.3 a list of high-index planes on an fcc crystal and corresponding stepped surface designations are given.[72] For less complex surfaces, Nicholas[18] has compiled an atlas of ball models which provide direct visualisation together with associated data on bonding configurations, and this can be a valuable source book for many LEED studies.

Mindful of the limitations of the Lang *et al.*[70] stepped surface nomenclature, Van Hove and Somorjai[73] have recently devised an improved system for vicinal surfaces which is not only more general, but also provides a direct mathematical link between the Miller indices of the macroscopic surface orientation and the component low-index planes which make up the terraces and steps. The Miller indices are decomposed vectorially into low-index components or 'microfacets' which form a terrace, step or low-index com-

Table 1.3 Miller indices, stepped surface designations and angles between the macroscopic surface and terrace planes for fcc crystals. (Reprinted with permission from Castner and Somorjai, *Chem. Res.*, **79**, 233 (1979). Copyright 1979 American Chemical Society)

Miller index	Stepped surface designation	Angle between the macroscopic surface and terrace (degrees)
(544)	(*S*)-[9(111) × (100)]	6.2
(755)	(*S*)-[6(111) × (100)]	9.5
(533)	(*S*)-[4(111) × (100)]	14.4
(211)	(*S*)-[3(111) × (100)]	19.5
(311)	(*S*)-[2(111) × (100)]	29.5
(311)	(*S*)-[2(100) × (111)]	25.2
(511)	(*S*)-[3(100) × (111)]	15.8
(711)	(*S*)-[4(100) × (111)]	11.4
(665)	(*S*)-[12(111) × (111)]	4.8
(997)	(*S*)-[9(111) × (111)]	6.5
(332)	(*S*)-[6(111) × (111)]	10.0
(221)	(*S*)-[4(111) × (111)]	15.8
(331)	(*S*)-[3(111) × (111)]	22.0
(331)	(*S*)-[2(110) × (111)]	13.3
(771)	(*S*)-[4(110) × (111)]	5.8
(610)	(*S*)-[6(100) × (100)]	9.5
(410)	(*S*)-[4(100) × (100)]	14.0
(310)	(*S*)-[3(100) × (100)]	18.4
(210)	(*S*)-[2(100) × (100)]	26.6
(210)	(*S*)-[2(110) × (100)]	18.4
(430)	(*S*)-[4(110) × (100)]	8.1
(10, 8, 7)	(*S*)-[7(111) × (310)]	8.5

ponent of a kink. For example, it is possible to decompose a Miller index (h, k, l) which has been ordered for convenience such that $h > k > l$ on to the three lowest index planes of a cubic structure as follows:

$$(h, k, l) = l(111) + (k - l)(110) + (h - k)(100) \qquad (1.55)$$

In general, it is possible to describe the stepped surface of any cubic structure in terms of three non-equivalent low-index microfacets

$$(h_i, k_i, l_i) \qquad \text{where } i = 1, 2, 3$$

A coefficient a^i which comes from the vector decomposition may be identified with each of the microfacets. The number of unit cells of each microfacet which occur in each complete surface unit cell is simply related to the vector decomposition coefficients a^i, but to make the notation as informative as conveniently possible, this number of microfacet unit cells is included as subscripts n_i^{uc} to the coefficients a^i. The general form of the description is then given as

$$E(S) - [a_n^1{}_{uc}(h_1, k_1, l_1) + a_n^2{}_{uc}(h_2, k_2, l_2) + a_n^3{}_{uc}(h_3, k_3, l_3)] \qquad (1.56)$$

where E represents the substrate material and S denotes a stepped surface as with the notation described earlier in this section.

The area of the unit cell of an arbitrary surface (h, k, l) is given by the volume of the bulk unit cell divided by the interplanar spacing if the bulk unit cell is always normalized to unity. The interplanar spacing depends whether the structure is sc, bcc, fcc, hcp or some other. Using the rules given earlier in the chapter for the determination of this interplanar spacing, Van Hove and Somorjai have shown that, when the surface (h, k, l) of a cubic crystal structure described by a vector $\mathbf{u}_0$ is decomposed into three microfacets as follows:

$$\mathbf{u}_0 = a^1\mathbf{u}_1 + a^2\mathbf{u}_2 + a^3\mathbf{u}_3 \qquad (1.57)$$

then the ratio of the number of unit cells in each microfacet is given by

$$n_0 : n_1 : n_2 : n_3 = p_0 : a^1p_1 : a^2p_2 : a^3p_3 \qquad (1.58)$$

where $p = 1$ for sc lattices and bcc lattices when $h_i + k_i + l_i =$ odd; $p = 2$ for bcc lattices when $h_i + k_i + l_i =$ even, and fcc lattices when h_i, k_i and l_i are not all odd; $p = 4$ for fcc lattices when h_i, k_i and l_i are all odd.

In particular, for the decomposition into (100), (110) and (111) microfacets as given in equation (1.56), the relative number of unit cells for each of the three cubic structural types is given as

$$n_{hkl} : n_{111} : n_{110} : n_{100} = \begin{cases} p_{hkl}^{sc} : l : k - l : h - k & \text{for sc} \\ p_{hkl}^{fcc} : 4l : 2(k - l) : 2(h - k) & \text{for fcc} \\ p_{hkl}^{bcc} : l : 2(k - l) : h - k & \text{for bcc} \end{cases}$$

However, it is perhaps more useful to define the fcc and bcc surfaces in terms of the three most dense faces which actually occur. For fcc, the $(11\bar{1})$ is more pertinent than the (110) and equation (1.56) may be decomposed further using the relationship

$$(110) = (111) + (11\bar{1}) \qquad (1.59)$$

giving

$$(h, k, l) = (k + l)(111) + (k + l)(11\bar{1}) + (h - k)(100) \qquad (1.60)$$

and the ratio

$$n_{hkl} : n_{111} : n_{11\bar{1}} : n_{100} = p_{hkl}^{fcc} : 2(k + l) : 2(k - l) : 2(h - k) \qquad (1.61)$$

similarly for bcc, decomposing the open (111) face in terms of (110) and (001) planes gives

$$(hkl) = k(110) + (h - k)(100) + l(001) \qquad (1.62)$$

so that the ratio

$$n_{hkl} : n_{110} : n_{100} : n_{001} = p_{hkl}^{bcc} : 2k : h - k : l \qquad (1.63)$$

If the surface is non-kinked then one of the coefficients will be zero—the surface can be described simply in terms of two microfacet faces. As an example of a simply stepped surface, the fcc (775) surface becomes

$$\text{fcc}(S) + [5_5(111) + 2_1(100)]$$

whilst fcc (10 8 7), which is kinked as well as stepped, may be described in one of several ways depending on the way in which it is decomposed. For example

$$\text{fcc}(S) - [7_{14}(111) + 1_1(310)]$$
$$\text{fcc}(S) - [7_{14}(111) + 1_1(110) + 2_2(100)]$$
$$\text{fcc}(S) - [(15/2)_{15}(111) + (1/2)_1(11\bar{1}) + 2_2(100)]$$

The first form is analogous to the original Lang *et al.*[70] notation, where the terrace is seen to be the (111) plane, and steps by the kinked (310) face. The step plane may be visualized as a series of (110) and (100) microfacets, or alternatively as (111) and (100) microfacets. This is illustrated in Figure 1.36. The terrace microfacet, non-kink step edge atoms and kink atoms are identified respectively by their decreasing numbers of unit cells n_i^{uc}.

The precise mathematical description of the terrace plane in terms of the surface plane permits the angle of inclination to be derived simply, and this is a result which is most valuable for the interpretation of LEED patterns from stepped surfaces. This and other useful data has been given an analytical form by Van Hove and Somorjai,[73] who have tabulated values for a large number of possible stepped surfaces on an fcc crystal. Data for bcc surfaces could be calculated similarly. Other structures such as diamond, zinc-blende or sodium chloride are based on the fcc structure and so the above analysis may be adapted appropriately. Non-cubic crystals such as hcp or wurzite would need a more complex analysis since the basic assumption used in the analysis is that a plane with Miller indices (*hkl*) is orthogonal to the vector [*hkl*], an assumption which does not necessarily hold for non-cubic lattices.

1.6 Labelling of diffraction beams and the incident beam direction

A variety of different notations have been used to label LEED beams. Early experimental papers frequently labelled beams (*hk*) with an orientation based simply upon the visual appearance of the pattern, for example setting according to a right-handed coordinate system. This is not satisfactory if comparisons are to be made with calculated beam intensities: it is clearly necessary to define the orientation of the beam indices with respect to the direction of the incident electron beam, since although diffraction beams may be rotationally symmetric in position about the specular (00) beam, their intensities will in general differ from each other. This applies whenever the incident beam direction is not normal to the surface plane.

Marcus and Jona[74] proposed a scheme which is now almost universally adopted whenever comparisons with theory are involved. In this scheme, the surface basis vectors are defined such that the projection of the incident beam

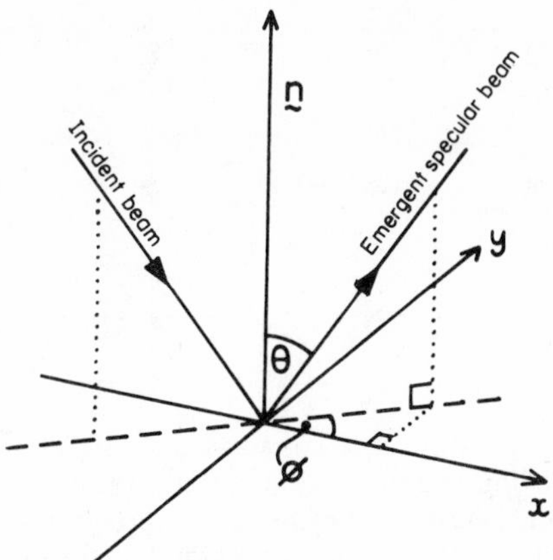

Figure 1.37. Notation for the polar (θ) and azimuthal (ϕ) beam angles upon a crystal surface, normal n and in-plane axes x and y

direction is a positively directed vector with respect to both basis vectors. Defining a left-handed coordinate system with x- and y-axes within the surface plane and the z-axis into the surface, and choosing one of the basis vectors (a_1, conventionally chosen to be the longer of the two) to be along the x-axis, then the angle subtended by a_1 and the projection of the incident beam direction within the surface plane is called the azimuthal angle ϕ, whilst the angle between the incident beam direction and the surface normal is called the polar angle ϕ. This is illustrated in Figure 1.37. The choice of a left-handed coordinate system does not affect the beam or angle notation, but is recommended because it is the system most frequently used in multiple scattering theories and computer program libraries.

The polar and azimuthal angles are generally designated in degrees. A convenient nomenclature to describe the incident beam direction defined by θ and ϕ is to write them within square brackets thus: $[\theta, \phi]$. Such a scheme, as proposed by Clarke,[15] is succinct, and should not lead to any confusion with beam direction vectors due to the context of their use.

There are two schemes commonly used to determine the values h and k of the diffraction beam indices. The more widely used of the two defines the reciprocal lattice unit cell as that obtained from the unreconstructed substrate lattice. If the surface structure has a unit cell greater than that of the substrate, then the resulting diffraction pattern will contain beams for which h and k possess fractional values. These indices are given by the reciprocal lattice matrix elements m^* from equation (1.18), and integral multiples of these values. An alternative scheme defines beams in terms of the reciprocal lattice produced by the real-surface structure or coincidence net. This has the advantage that, for simple commensurate superstructures, all beam indices

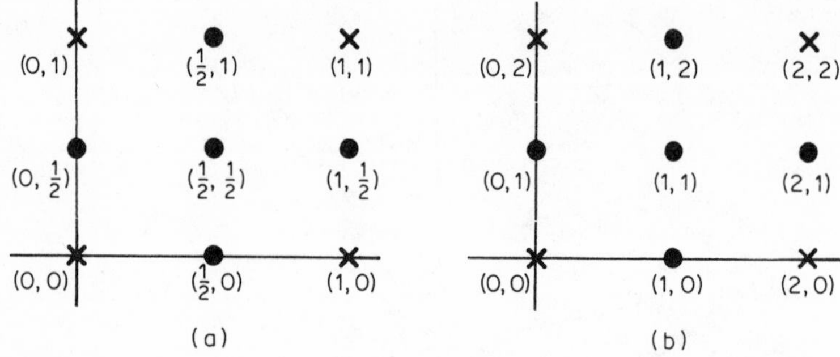

Figure 1.38. Labelling of a diffraction pattern: (a) conventional method, with respect to the substrate beams (x), overlayer beams (·) assume fractional-order values; (b) alternative method, involving a relabelling of the substrate beams, such that the overlayer unit reciprocal lattice cell has integral-order indices

have integral values. This is marginally more convenient for computational purposes, and is used in their computer programs by Rundgren and Salwen,[75] but is not convenient experimentally since it is common for several different surface structures to appear as a function of adsorbate coverage, the beams having different labels for each structure. The former is therefore the recommended method for normal purposes, and examples are given in Figure 1.38.

1.7 Space-group symmetry

A two-dimensional lattice may be classified by its symmetry properties. The distinguishing features are whether or not the basis vectors are of equal length, what is the angle subtended by them and whether there is an atom centred in the unit cell. The resulting types are known as Bravais lattices, four of which have primitive lattices (square, rectangular, hexagonal and oblique) and one type with a centred rectangular lattice. These are illustrated in Figure 1.39, from which it is clear that the rectangular lattice is the only one to which a centred atom could be added and thereby form a lattice with new symmetry properties.

In order to discuss symmetry properties of two-dimensional structures it is first necessary to note that there are two fundamental types of symmetry operators—*point symmetries* which refer to the rotational symmetry which may exist about some point or points in the unit cell, and *line symmetries*, which indicate the existence of lines or planes about which the cell has *mirror-reflection* symmetry or *glide-reflection* symmetry (one in which both reflection and translation along the mirror plane must be combined to regenerate the initial geometrical configuration of the unit cell). These fundamental symmetry properties are illustrated in Figure 1.40.

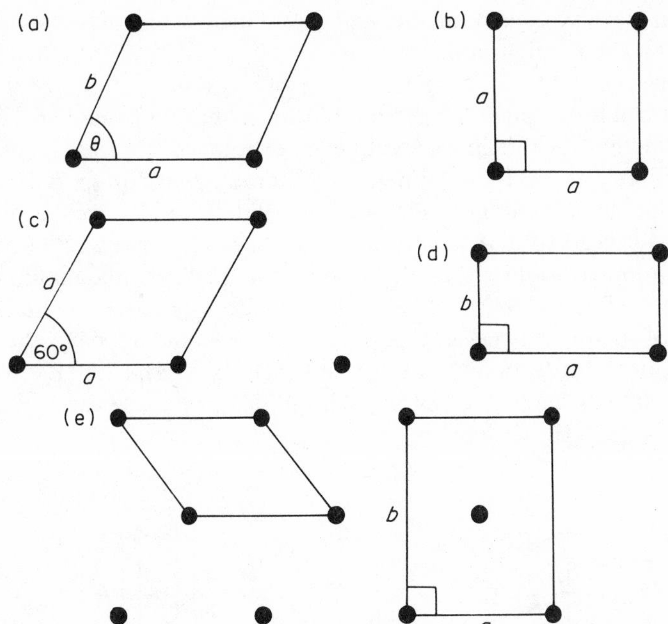

Figure 1.39. The five two-dimentional Bravais lattice unit cells: (a) oblique ($a \neq b$, θ arbitrary (not 60°, 30°); (b) square ($a = b$, $\theta = 90°$); (c) hexagonal ($a = b$, $\theta = 60°$); (d) rectangular $a \neq b$, $\theta = 90°$); (e) centred rectangular ($a \neq b$, $\theta = 90°$, with centred atom)

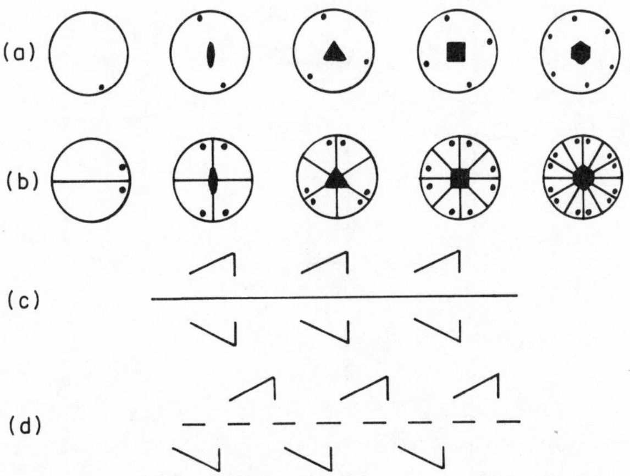

Figure 1.40. Rotational and line symmetries: (a) the five rotational symmetry operations without mirror-planes; (b) the five rotational symmetry operations with mirror-planes; (c) mirror reflection about a mirror-plane; (d) glide reflection (translation plus mirror reflection) about a glide plane

The complete symmetry of the unit cell depends not only on the lattice type, but on the configuration of atoms within. It is found that there are 17 possible types of unit cell, or *space groups*, when the symmetry of the atoms within the cell are included. The standard description of these space groups includes whether the lattice is primitive (p) or centred (c), what is the highest point symmetry to be found (one-, two-, three-, four- or sixfold), and the nature of the line symmetries, mirror (m) or glide (g) reflections, if they exist. It is taken for granted that a centred lattice necessarily involves the existence of alternate mirror- and glide-reflection symmetry lines. Such alternation of mirror- and glide-reflection lines will also occur parallel to the diagonal direction of square lattices, since a square lattice would appear to be centred if the diagonals were defined as the unit cell axes. The letter g, denoting glide-reflection symmetry, is not included to describe these cases. Only glide

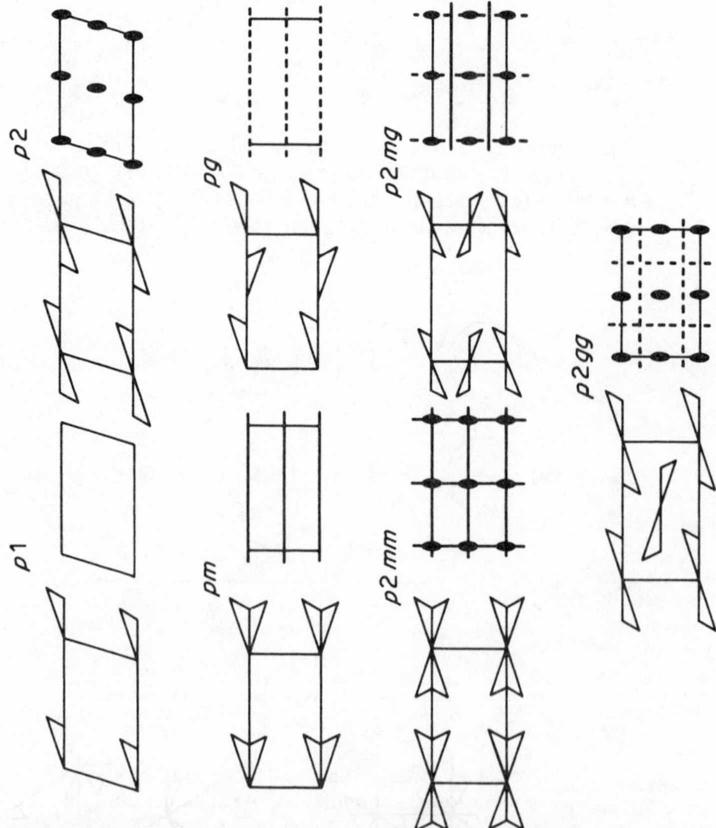

Figure 1.41. Unit cell of the 17 plane groups. (Reproduced by permission of Plenum Press from Ladd and Palmer *Structure Determination by X-ray Crytallography*, 1977)

planes parallel to the axes of square lattices are included in space-group symbols. Even then, it is found that a complete description is more than strictly necessary to identify the space group, and so for 11 of the full space-group symbols there is a shortened version. In Figure 1.41 the 17 space

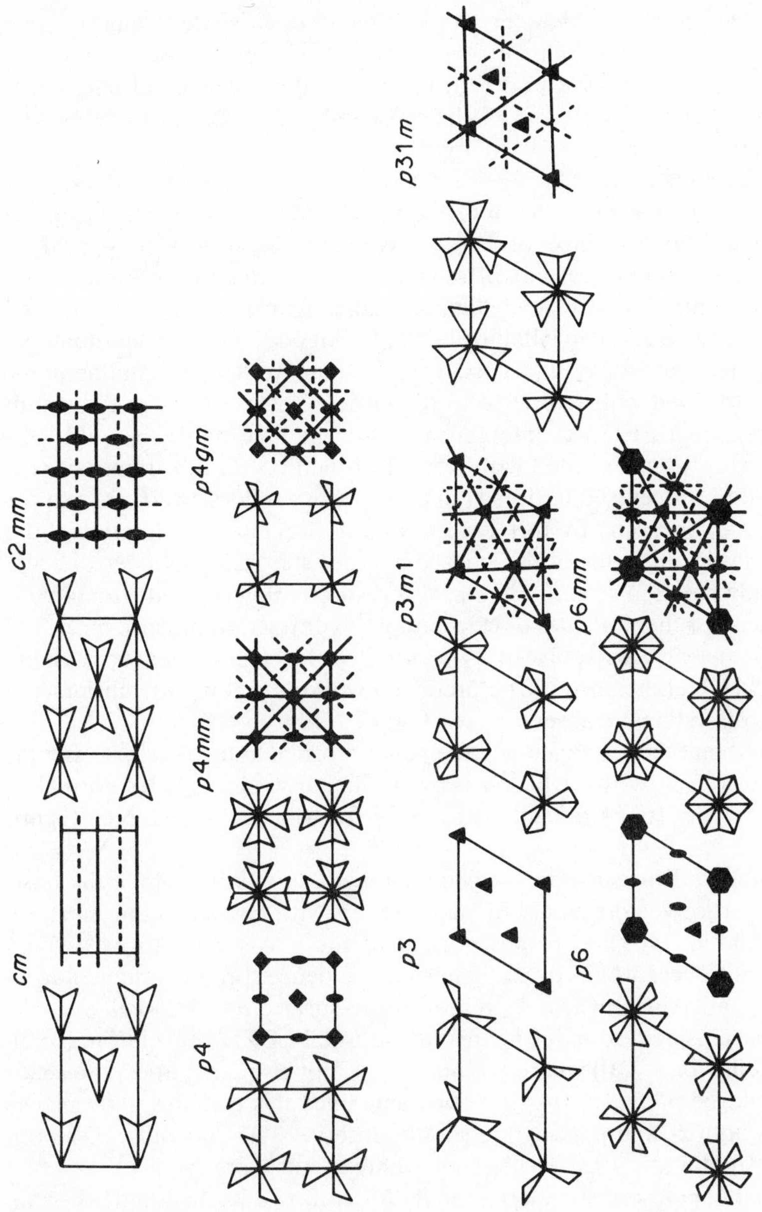

groups are illustrated with full and short symbols. The convention for illustrating the space groups is taken from the *International Tables for X-ray Crystallography*,[76] as is the order in which they are presented. The right-hand diagrams give the space groups proper, that is, as groups of symmetry operators distributed in space. On the left are the complete set of equivalent positions that any single element, marked by an open circle, would assume after application of the symmetry group. Conversely, a configuration of atomic elements within the unit cell, including all those general equivalent positions corresponding to one of the left-hand diagrams, must possess the space-group symmetry given to its right.

From the point of view of LEED analysis, perhaps the most interesting information to be gained from an examination of two-dimensional groups is the fact that the existence of certain symmetry operators may lead to systemmatic absences of certain diffraction beams. Such absences occur when the lattice is centred or when glide lines exist orthogonal to one of the unit cell axes. Care must be taken in labelling beams, to identify those which should be absent. Usually the indices (h, k) occurring in standard X-ray crystallographic literature are given with reference to the lattice vectors of the relevant unit cell, whereas in LEED the interest may, for example, lie with a $c(2 \times 2)$ overlayer with beams labelled with respect to the (1×1) substrate unit cell. The condition for centred lattices is that only those reflections (h, k) exit for which $(h + k)$ is even. In this case h and k are integers describing the reciprocal lattice in terms of the surface and not substrate basis vectors. We have already noted in section 1.5 that this restriction on centred structures is trivial, since, for the purposes of calculating the diffraction pattern, a centred rectangular net may be replaced by a primitive lozenge-shaped or rhombic cell from which all beams may be predicted without needing to consider the possibility of systematic absences (see Figure 1.30).

It is found that the existence of glide-planes normal to the y-axis restrict $(h, 0)$ reflections to those for which h is even. Similarly, glide planes normal to the x-axis restrict $(0, k)$ reflections to those for which k is even (see Figure 1.42).

For example, if the surface unit cell were a $p(2 \times 2)$ lattice, then the basis reciprocal lattice vectors would be half-integers with respect to the substrate reciprocal lattice. If glide planes normal to the y-axis were to exist, then systematic absences would occur at normal incidence for odd values of h' in the $(h', 0)$ reflections, where h' refers to the surface $p(2 \times 2)$ cell vectors, which means, using h to index beams by the usual LEED convention, for all beams labelled $(h + \frac{1}{2}, 0)$ where h is an integer. For a square lattice, the space group would be $p4gm$, or $p4g$ in short, and since this contains glide planes normal to both x- and y-axes, one should observe systematic absences both for $(h + \frac{1}{2},0)$ and also $(0,k + \frac{1}{2})$ beams, where h and k are integers.

This has been observed experimentally for carbon adsorbed under certain conditions on Ni(001), producing a Ni(001)–$p(2 \times 2)$C–$p4g$ structure, as shown in Figure 1.40.[77] In order to observe this effect it is necessary for the

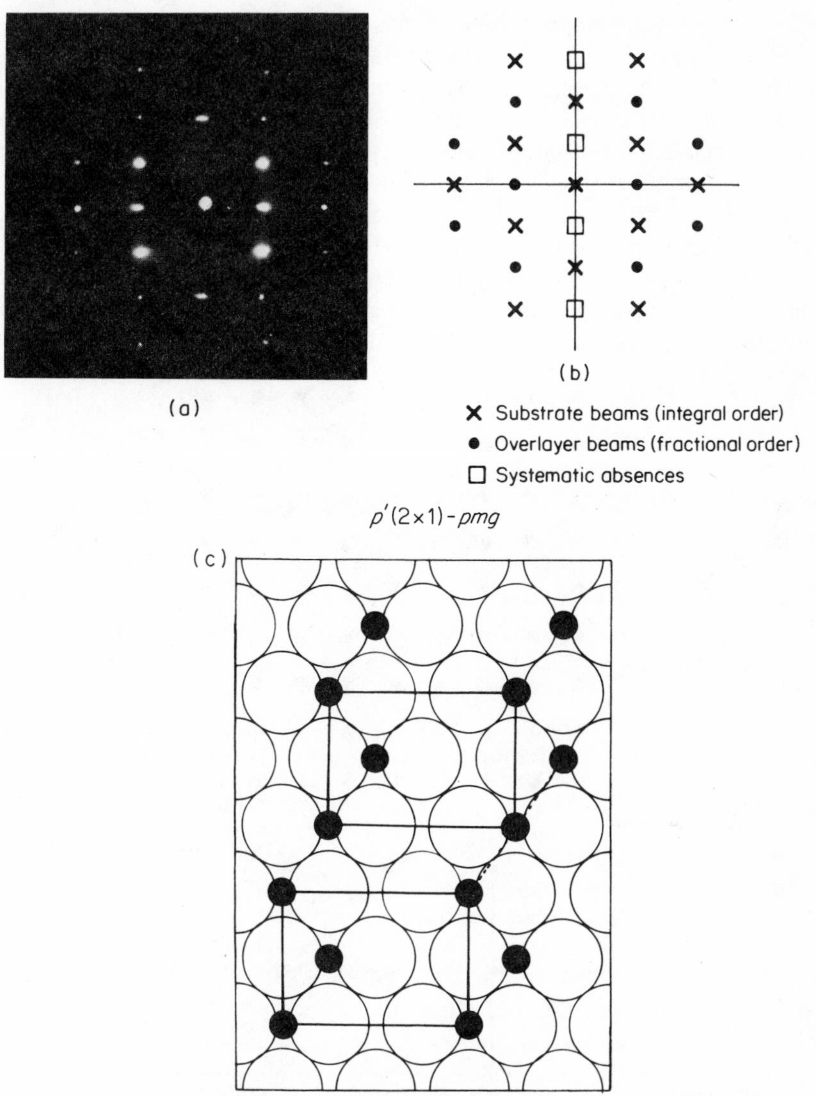

Figure 1.42. The $p'(2 \times 1)$–*pmg* structure formed by Cl on Mo(110): (a) observed pattern; (b) diagram of pattern; (c) two domains of a possible adlayer structure producing the observed pattern. (Reproduced by permission of North Holland Physics Publishing from Clarke and Morales de la Garza[59])

electron beam to be incident along a direction normal to the surface. If not, the symmetry is broken and the beams will assume finite intensities. An incident beam direction off-normal to the surface plane, yet normal to a glide line will not break the symmetry conditions for glide lines of that particular orientation, and characteristic absences will occur for beams in that direction, but not at 90° to it.

The effect of systematic absences has also been observed with the clean W(001) surface, which, when cooled, reconstructs to form a $c(2 \times 2)$ structure.[78] In this case, glide planes normal to the axes of the primitive surface cell, that is, at 45° to the substrate (1×1) net. In this case, the structure is $p2mg$ (or pmg) which means that for any one orientation of the crystal systematic absences would occur in only one of the $(h', 0)$ or $(0, k')$ directions. (In this case, h' and k' are along axes diagonal to the surface cell, so that the condition h' odd implies absences for $(h + \frac{1}{2}, h + \frac{1}{2})$ beams in the usual LEED notation.) The possibility of domains oriented $\pi/2$ to each other would normally mean that the effect would be obliterated by the

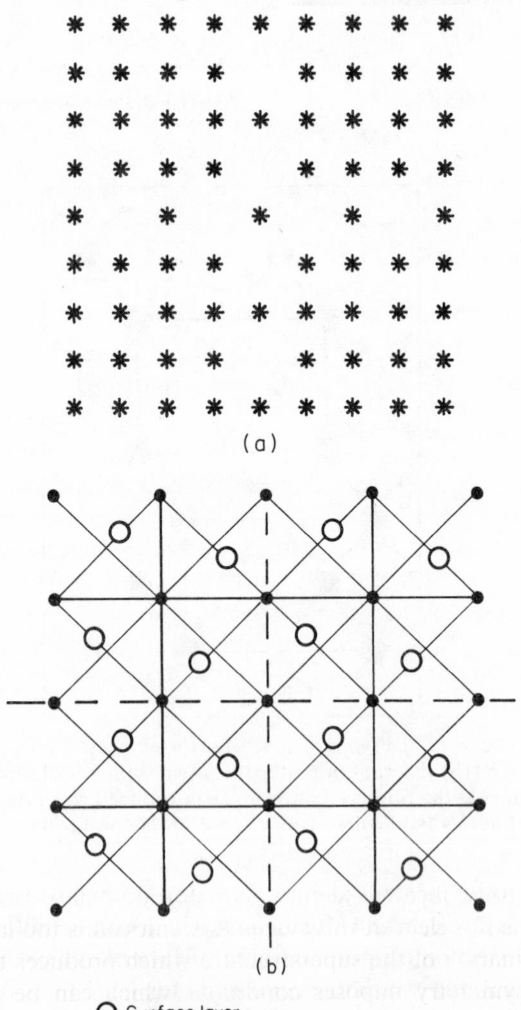

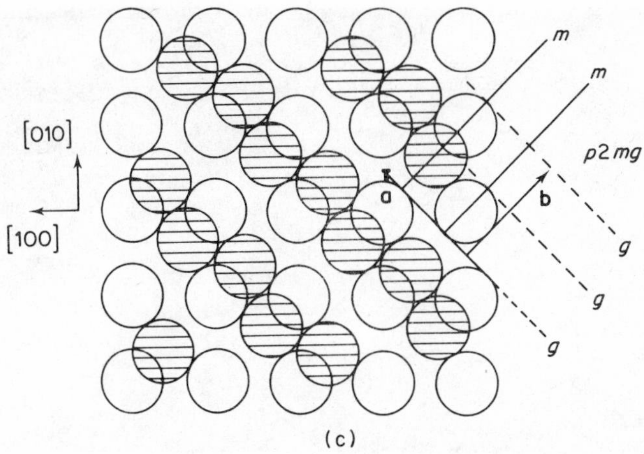

Figure 1.43. The $p4g$ superstructure formed by the adsorption of carbon on Ni(100). (a) Diagram of the diffraction pattern. (b) Diagram of the reconstructed Ni surface causing the observed pattern. The carbon atoms occupy symmetric sites that must be determined by intensity angles. (c) The clean-surface reconstruction of W(001). A top-layer rearrangement ball model for a $(\sqrt{2} \times \sqrt{2})R45°$ domain having $p2mg$ symmetry. For a single-layer rearrangement model, this structure appears to be unique. The primitive unit mesh is defined by **a** and **b** and its symmetry elements are indicated: g = glide plane, m = mirror-plane. (Reproduced by permission of the American Physical Society from Debe and King[79])

superposition of beam intensities from the two domains. However, some slight asymmetry in the surface, probably a low density of steps, led to different contributions from the two possible domain orientations and enabled this effect to be observed. Because this interpretation assumes a twofold glide-plane structure on a substrate with fourfold symmetry, the effect of the systematic absences is rather obscured, and some controversy remains in the interpretation of this particular surface structure.[79] Suggested models for the Ni(001)–$p(2 \times 2)$C–($p4g$) and W(001)–(pmg) structures are illustrated in Figure 1.43.

The diffraction pattern given in Figure 1.44 produced by the presence of carbon on Mo(110) is interesting in that systematic absences occur along the axes of the non-primitive rectangular unit cell.[59] Using the dashed symbol to describe this cell, as suggested in section 1.5.7, this structure may be conveniently described as $p'(12 \times 8)$C–($p2gg$). Whether this pattern is due purely to an overlayer structure or an adsorbate-induced substrate reconstruction is not clear at this stage. The unit cell is too large to permit a unique determination of the superstructure which produces this pattern, but the observed symmetry imposes conditions which can be used to reduce considerably the number of possibilities. Another complex overlayer structure for which the existence of glide-plane symmetries has considerably

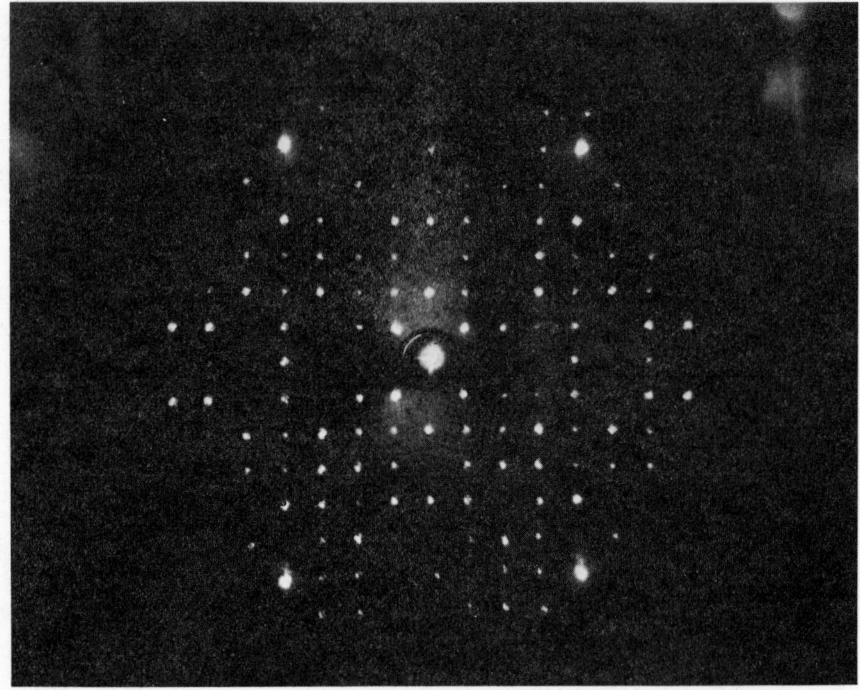

Figure 1.44. The $p'(4 \times 6)$–$p2gg$ structure formed by the adsorption of Cl on Mo(110) after heating the sample to 1200 K. (Reproduced by permission of North Holland Physics Publishing from Clarke and Morales de la Garza[59])

aided the interpretation is that obtained by the adsorption of naphthalene on Pt(111).[80] The restrictions imposed by glide-plane symmetry reduce to manageable proportions the diversity of models which would otherwise need to be considered.

CHAPTER 2

Experimental techniques

2.1 Basic elements of the standard LEED apparatus

The essential requirements for a LEED experiment are an electron source or gun, a specimen manipulator and a means of displaying or recording the diffracted beam intensities. These must be housed in a vacuum chamber within which the environment is suitable both for cleaning the specimen and also for the controlled introduction of gases or the deposition of adsorbate films. The technology of UHV systems has been developed and refined in recent years to an extent where virtually all essential items may be purchased commercially together with the necessary electronics. This in itself represents a big step forward from these *ad hoc* glass or lacquered-brass vacuum components that had to be constructed in the early days of LEED. UHV systems were becoming commerically available in the early 1960s and this paved the way for the recent considerable progress in the development of surface analytical techniques.

Details of pumps, materials, pressure gauges and much other useful information can be found in manufacturers' specifications or in one of the general references given at the end of this book.

2.1.1 The vacuum chamber

The vacuum chamber must be capable of maintaining a vacuum so low in pressure that contamination of the sample remains negligible for the duration of the experiment. In principle, if the information could be extracted instantaneously, then the upper limit would be 10^{-3} Torr, since above this pressure the mean free path of the electrons would be insufficient to permit their detection in a diffraction chamber of typical size. However, if we assume that gas atoms will adsorb with a sticking probability of unity, then at 10^{-3} Torr the surface would become covered with a monolayer of adsorbed gas in 1 msec. At 10^{-6} Torr a monolayer would form in 1 sec, and at 10^{-10} Torr, a clean surface would maintain an adsorbed impurity concentration of less than 10 per cent for up to 15 min.

A vacuum pressure level of 10^{-9} Torr is therefore the minimum acceptable level for detailed surface studies, whereas 10^{-11} Torr or better would be excellent for most work. Modern systems equipped for rapid data acquisition are clearly less critically dependent on the vacuum level than older manual systems for which data acquisition may take as long as 1 hour per diffraction beam.

Modern chambers are constructed of non-magnetic stainless steel or Mu metal (a ferrous alloy with high magnetic permeability, see section 2.5). Vacuum seals between flanges are regularly made using copper gaskets pinched between steel knife-edges although annealed rings of pure gold wire or indium strip squashed between mirror-polished flats are still sometimes used. Chambers need to be bakeable in large removable ovens to about 250 °C to accelerate the outgassing or desorption of gases from the inner walls, but temperatures much in excess of this or baking times greater than about 24 hours may well be counter-productive, since excessive heating may cause leaks at the gaskets.

2.1.2 The display system

The intensity of the electron beams diffracted from the crystal surface may be measured directly, for example by collecting an electron beam in a Faraday cup, or indirectly by causing the electrons to impinge on a screen coated with phosphorescent material and measuring the brightness of the resulting spots of light. The latter method has the advantage of producing a pattern that can be viewed by eye. This property is comparatively rare amongst modern surface analytical techniques, and has undoubtedly contributed to the popularity and usefulness of LEED. Even if the diffraction beam intensities are ultimately to be measured with a Faraday cup, a phosphorescent screen is almost always included, to provide this visual aspect.

The commonest form of LEED apparatus comprises a hemispherical phosphorescent screen with a fixed electron gun aligned along the central axis of the screen. The crystal specimen is positioned at the centre of radius of the hemispherical screen. With this arrangement, diffraction beams emerging from the surface of the crystal travel radially towards the screen. This is the basis for most commercially available LEED systems and may be described, for convenience, as the 'standard LEED system'. This is illustrated in Figure 2.1.

The diffraction pattern is usually viewed from behind the crystal, so the observer can see only a projection of the pattern that forms on the screen. Consequently the screen is not constructed to form a complete hemisphere. Usually, the angle subtended by the screen at the crystal is set at 60° about the central axis.

The screen is metallic and coated in a phosphorescent material. It is biased to 4–6 kV, so that approaching electrons are accelerated sharply on to the surface coating. Electrons which impinge on the screen create a glow proportional to the beam intensity. The metallic nature of the screen is necessary to permit the incident electrons to be conducted away, avoiding the undesirable effects of charging. Just before the screen is a series of two, three or four hemispherical meshes or grids.

The grid nearest the sample (the first grid) is normally earthed so that the

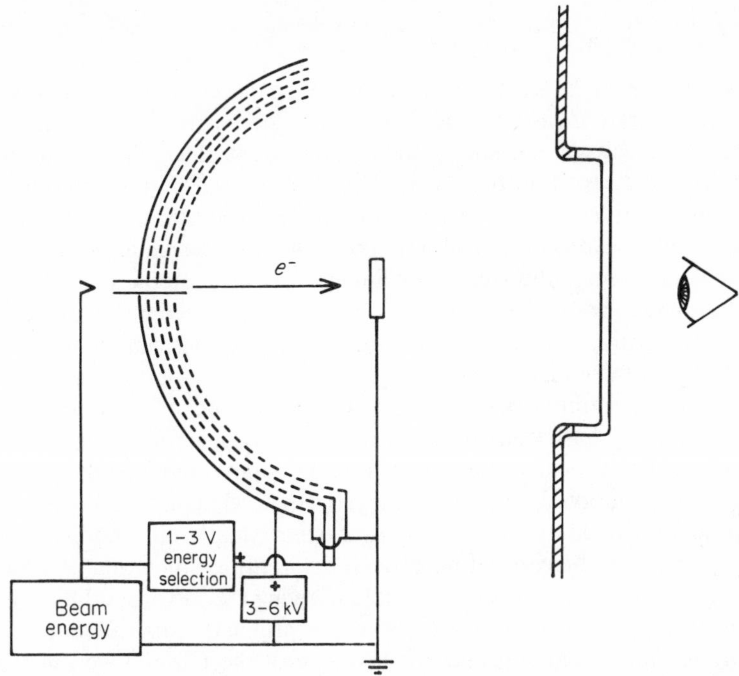

Figure 2.1. Schematic diagram of a typical ('standard') LEED system

diffracted beams follow linear trajectories in field-free space (assuming that the sample is similarly at earth potential). Because the screen does not subtend an angle of 90° about the normal to the sample surface, not all emergent beams will appear on the screen. If the sample is biased to about 150 V, with the electron gun potential similarly shifted, the resulting electron beam trajectories in the vicinity of the sample will be curved away from the surface, and may permit a grazing-emergence beam to be observed which would otherwise not be visible. The next grid away from the sample (the second grid) is maintained at a potential a few volts below the incident electron beam. It thus acts as an energy-selection grid: electrons which have lost more than a few volts, and cannot therefore be part of the 'elastic' diffraction process, have insufficient energy to pass this grid, so do not reach the phosphorescent screen.

In a four-grid optics system, the second and third grids are linked together to improve their energy-selection characteristics, and the fourth grid is earthed, to isolate the phosphorescent screen potential from the energy-selection potential. For the purposes of LEED, two grids are sufficient, but the four-grid system provides more accurate energy selection, which is an advantage when using the system for AES.

2.1.3 Measurement of intensities

As noted in section 2.1.2, a Faraday cup may be placed in front of the screen and grids for direct measurement of the diffracted beam currents. This is the most accurate way of measuring beam intensities, since the beams travel directly from the crystal into the cup. A grid system within the cup filters out inelastically scattered electrons, and a 'channeltron' (a compact electron multiplier) may be attached to allow measurements to be made using very low incident beam currents. A disadvantage of this method is its poor manoeuvrability, associated with the fact that the mechanical movements must operate within the vacuum chamber. To track a moving beam by hand can be quite time-consuming—even a small amount of mechanical slack may hinder accurate positioning. Automatic beam tracking, using servomechanisms, has been successfully applied to a Faraday cup system.[81] However, even when automated, this approach is not rapid, and therefore has not been widely adopted for LEED analysis. In principle, a Faraday cup could be used in conjunction with a display system simply for the purpose of providing a means of calibrating observed intensities in terms of absolute beam currents. In some cases it is possible to swing the cup around to face into the gun, to provide a direct measure of the primary beam current. This is probably the most common and effective use of the Faraday cup in typical LEED practice.

A telephotometer or 'spot photometer' is an optical brightness meter that is commonly used to make measurements of beams as displayed on a fluorescent or phosphorescent screen. Within the device, a partially reflecting mirror, or, more commonly now, a mirror with a small hole in it, is placed at 45° to the incident light direction. This allows the observer to view the pattern through an eyepiece and visually align the photometer with the beam to be measured. Once the beam coincides with the hole, the light passes through the mirror and its intensity is measured. The acceptance angle is narrow—typically $\frac{1}{2}°$ or $\frac{1}{4}°$, but this is necessary to restrict the measurement to the beam region only, avoiding unnecessary areas of background. The photometer is necessarily situated outside the vacuum chamber, so, in the standard LEED system, its separation distance from the phosphorescent screen may be typically 30–50 cm (see Figure 2.2).

A spot photometer is quite bulky, but it can be mounted on a precision manipulator so that tracking beams across the screen is likely to be much easier than with a Faraday cup. Although visual observation is usually sufficient to direct the photometer at any particular beam, accurate alignment can only be checked by scanning the immediate vicinity of that direction to find the position giving the maximum output signal. By hand, this may take several seconds per energy value, and to record a complete intensity–energy curve over a range of 250 eV may take up to 45 min. Again, servomechanisms may be used to automate this process, but the time saved may not be

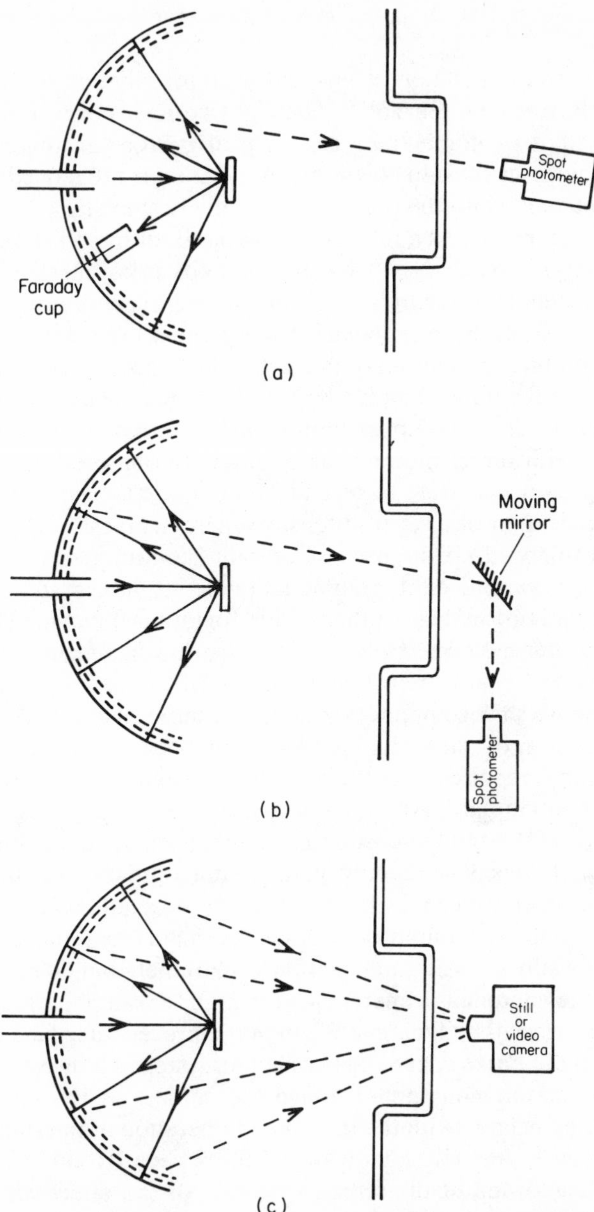

Figure 2.2. Various ways of recording diffraction beam intensities using the standard LEED system. (a) Faraday cup or spot photometer; (b) mirror deflection with a fixed spot photometer; (c) photographically, using a still or video camera

substantial, due to the bulk of the photometer and the slowness of its manipulation.

An imaginative adaptation of this approach for more rapid data collection has been described by Berndt.[82] The light from a beam spot is directed towards the spot photometer by a system of mirrors. In this case the spot photometer remains fixed in position and the mirrors are moved to reflect the light from the beam into the photometer as the beam varies. The mirrors can be moved relatively quickly, as they scan around for the position yielding the maximum output signal. Nevertheless, methods using mechanical scanning are generally unable to record a complete intensity–energy curve in less than several minutes. Only one beam can be measured at a time so the measurement of a complete set of diffraction beams for a given primary beam direction may take about half an hour. Unless the vacuum is exceptionally well maintained, this is an upper limit if surface contamination effects are to be avoided. With direct mechanical beam-measuring methods as described above, it is usually necessary to reclean and prepare the surface several times during the collection of a set of intensity–energy curves.

The need to speed up the process of data acquisition has resulted in the development of various photographic and video-camera systems for indirect data storage and processing—either on photographic film, video tape or some form of computer memory. Video systems mark a significant advance in the ease and speed of data retrieval, and are described separately in section 2.3. In this section we shall complete our review of more established methods with a brief description of photographic techniques.

Although this requires two steps—intensity data are not available until after the photographs have been analysed, and there is the additional problem to check that the intensities of the negatives lie in the linear range of the emulsion, it does have the advantage that a 'hard copy' of the diffraction patterns is obtained which may be subjected to repeated analyses if necessary. If analysis of the photographs is made by hand then the eventual time required to obtain intensity information is extremely long—many hours are required. If the microdensitometer is automated to scan photographs and give digital output as used by Stair *et al.*,[83] then the process may be a little faster.

Apart from the Faraday cup data-collection system, all the various methods of intensity measurement noted above are subject to the restrictions and uncertainties associated with the design and operation of the standard LEED apparatus. These are: (i) shadowing of the screen by the sample and its holder; (ii) shadowing of the screen by the grids (an effect which increases towards the edges of the screen due to the fact that it is a projection of the screen which is being observed); (iii) distortion of the pattern and beam shapes (also a result of the projection process); and (iv) uncertainties associated with possible defects in the phosphorescent coating and any non-linearity of the optical brightness with incident electron flux.

Shadowing of the screen by the sample holder can be reduced by careful design, but significant reduction in the bulk of the holder generally requires a reduction in rigidity and a loss of manoeuvrability (frequently, there is little or no control over the azimuthal angle of the crystal). Many types of sample holder have been used, but ultimately they must involve a compromise between the incorporation of suitable tilting, rotating, heating and mounting mechanisms and the minimization of the shadowing effect.

Typical modern grids are made of nickel-plated tungsten wire, and have a transparency of about 80 per cent per grid. This means that a beam of light incident in a direction normal to the surface of the grids will have its intensity reduced by $(0.8)^4 \times 100$ per cent = 41 per cent when transmitted through a series of four consecutive grids. If the incident beam is inclined at an angle to the surface of any grid, the intensity will be reduced yet further, as a result of the finite dimensions of the grid mesh. In a standard curved screen system in which intensities are measured by a telephotometer or vidicon camera, compensation for the increasing attenuation of the light towards the edges of the screen should be made. This is most easily achieved after the experimental intensity data has been digitized and stored in the computer. The angle of emergence of each beam can be calculated as a function of the incident angle, crystal type and beam energy, from which an appropriate renormalization procedure may be implemented. The modifications to the intensity–voltage curves resulting from this effect are not usually strong enough to cause any significant changes to the structural conclusions that result from subsequent analysis, but they could interfere with the estimates of the surface vibrational amplitudes. It is also important to include this effect if information is being extracted from the relative intensities of all the beams at individual energies.

2.1.4 Electron guns

The requirements of an electron gun will vary with the type of system used and the application intended—does it need to be able to operate at very low energies, or very high? Is energy or angular resolution going to be a critical factor? Is it necessary to have high currents for visual observation directly from a fluorescent screen, or is a channel-plate electron multiplier available to boost a weak beam current? Electron guns have been used extensively in commercial display systems—TVs and oscilloscopes, for example—as well as in electron microscopes, so that their operational characteristics have been carefully examined and well documented. LEED involves back reflection, so that in many cases the diffracted beams pass back near the gun. Consequently, only electrostatic deflection is normally permissible. Stray fields will adversely affect the observation of the diffraction pattern, and it is much simpler to shield electrostatic fields than electromagnetic ones.

The general form of an electron gun is illustrated in Figure 2.3. Electrons emitted from the cathode must be accelerated to an anode which is at a positive potential with respect to the cathode. Electrons near the cathode have low kinetic energies compared to the total potential drop and so will tend to follow field lines. However, as they continuously accelerate towards the anode, their kinetic energy may become so much greater than the rate of change of potential that they can deviate substantially from the directions of the field lines, thereby passing through the anode aperture. For a good emission system it is desirable that all electrons emitted from the cathode pass through the anode, since electrons impinging might lead to secondary emission and a background of non-focused electrons. This is achieved with the aid of an auxiliary electrode placed between cathode and anode, at, or slightly negatively biased to, the cathode potential. This tends to focus the field lines towards the anode aperture in which direction the emitted electrons will initially pass. Because the anode is substantially positive with respect to the cathode (and thus also to the auxiliary electrode) strong deviation of the field lines only occurs quite near the anode in a region where electrons are appjoaching their maximum kinetic energy. The auxiliary electrode is described variously as the 'Wehnalt' or, by its analogy with electric valves, the 'grid'.

In general, electrons emitted from the cathode follow paths which converge towards the central axis, reach a cross-over point where the beam has minimum radius and thereafter begin to diverge. The anode takes the form of a relatively long cylinder with an aperture at the far end. This aperture limits

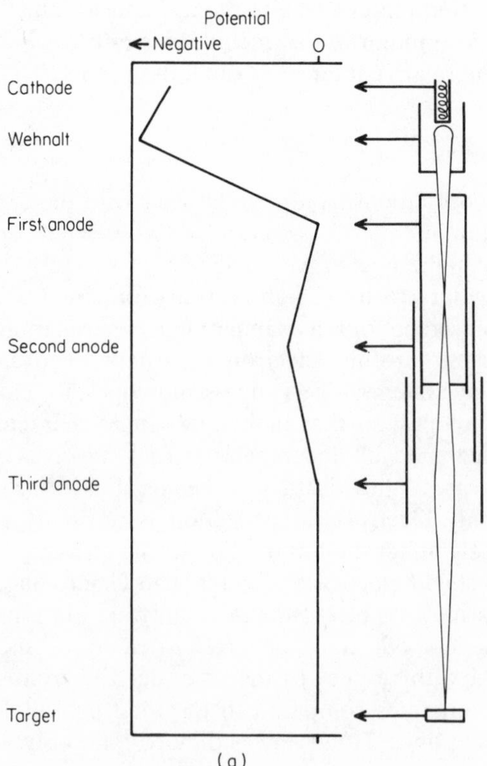

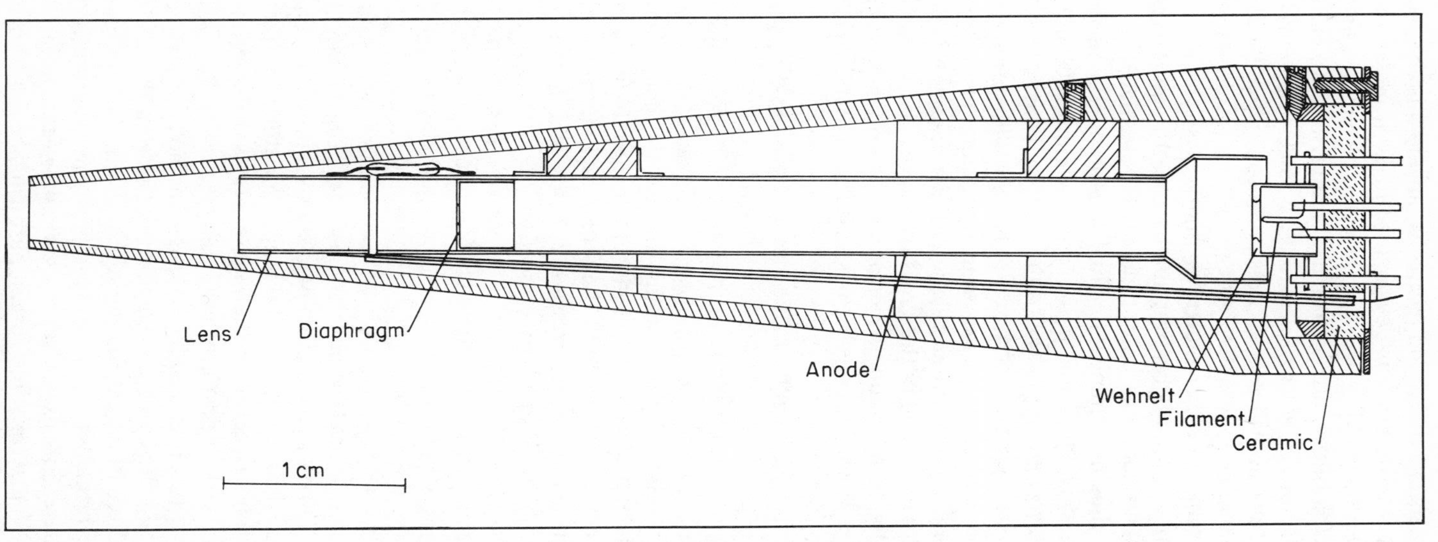

(b)

Figure 2.3. Electron guns for LEED: (a) schematic layout; (b) scale drawing of the de Bersuder mini electron gun. (Reproduced by permission of the American Institute of Physics from de Bersuder.[85])

the cone of divergent electrons, and electrons outside this cone are collected by the anode, but the geometry is designed to make it highly probable that any secondary electrons will also be collected. This divergent cone of electrons is now refocused by an electron lens comprising two more elements, a second anode, or 'focus', which is negatively biased with respect to the first anode, and a third anode which is at the same potential as the first anode.

In general, the sample will be at earth potential, and in standard LEED systems it is desirable that the electrons finally pass through a field-free regions. Putting the third (and also first) anode at earth potential ensures this, in which case the cathode must be at a negative potential. The value of the cathode potential thus determines the final kinetic energy of the electrons once the filament work function has been taken into account. The final focused beam is essentially the image of the electron beam cross-over point.

For thermionically ejected electron beams, the dimensions of the final beam diameter are invariably greater than the minimum theorectical limit. This is because the electrons are ejected with finite energies. If the electrons were emitted with zero velocity the cross-over size, and hence final image should, theoretically, be invariant as the cathode potential (V) is changed. As a result of finite electron ejection velocities, it is found that the cross-over size actually varies approximately as $1/\sqrt{V}$.[84]

In practice, most LEED guns make use of this basic configuration of electrodes, but the precise geometry will depend on the LEED system in which it is to be used. For example, in the 'standard' LEED system, the gun geometry is similar to that of figure 2.3(a). The anodes take the form of three concentric cylinders, the third protruding through the hemispherical screen and meshes of detection optics so that in both shields the emergent electron beams from the electrostatic fields on the screen and meshes and, to a lesser extent, shields diffracted beams from the fields within the gun. In the apparatus designed by de Bersuder (see section 2.3.1) the gun moves within the radius of the screen and meshes and so must be kept small. In this case, the third anode, being earthed, is allowed to form a complete, conical shield,[85] as shown in Figure 2.3(b). Other designs of compact electron guns for LEED experiments have been described by Chutjian[86] and Cowell.[87]

Laydevant's high-voltage apparatus (see section 2.3.4) works on a different principle—a 5 kV gun of the type normally used for Auger analysis is suitable for low-voltage (0–50 eV) studies, whilst a 25 kV gun of the type used for high-energy electron diffraction (HEED) is appropriate for diffraction studies of 0–250 eV electrons. Because of the high energies of the electrons in this system it is not necessary to post-accelerate the electrons on to the fluorescent screen. The gun may be optimized in its design and no additional precautions are required to shield the emergent electron beam from the screen.

UHV is an ideal environment for the operation of electron guns, but the filament must nevertheless be able to withstand numerous periods at atmospheric pressure, or exposure to various gases during adsorption experiments. However, if for some reason performance is adversely affected, it is desirable that reactivation be achieved fairly rapidly. Electrons are emitted by heating the cathodes either directly by passing a current through it, or indirectly by means of an independent heater, or else by electron bombardment. Perhaps the simplest thermionic cathode is a tungsten wire bent in the shape of a hairpin. A current of a few amps will lead to a typical operating temperature in excess of 2000 K. Emission occurs from the tip which may have a surface area about 10^{-10} m^2 or less, i.e. a diameter less than 0.1 mm. Alternatively, a flat or concave cathode of tungsten or, occasionally, tantalum may be heated indirectly by a spiral of tungsten. By operating well below the melting-point, the evaporation rate of the tungsten may be kept to negligible levels.

The bright light of the glowing filament may be considered a problem in the 'standard' LEED system where the viewing position directly faces the gun; the cathode in such systems may be mounted off-axis, additional deflection electrodes being necessary to bring the emitted beam back on-axis. However, the primary use of off-axis guns is to avoid emitting the products of evaporation towards the sample. This is more of a problem for 'matrix' cathodes, as described below, than pure tungsten cathodes, but could nevertheless occur, for example, when a gun is used after gas adsorption experiments. Lanthanum hexaboride is an efficient electron emitter, and is well suited to prolonged exposures to atmospheric pressure when cold.

A commonly used 'matrix' cathode is made from tungsten powder which has been sintered and impregnated with barium aluminate. These cathodes are robust and readily machinable, and are generally made in the form of a cylinder with a closed end. A spiral tungsten heater is enclosed within, and emission occurs from the planar end of the cylinder. Operational temperatures are below 1500 K and light emission is low. However, a new cathode may evaporate barium and although this may reduce to negligible quantities after about 100 hours it is another good reason for off-axis mounting of the cathode.

Reactivation after a cathode has been subjected to atmospheric pressure may be achieved quite readily with a plain tungsten cathode by a brief period of outgassing at a higher than normal temperature. Reactivation will occur anyway after a few hours under normal operation conditions. The barium-impregnated cathodes are slightly more variable in their behaviour after exposure to high pressures.

Stability of the emission current may be affected by the mode of operation. At one extreme, emission may be temperature limited. This is commonly the case for tungsten hairpin cathodes. All available electrons are extracted by the applied field so that the emission is insensitive to further increases in the anode voltage. The emission current will increase exponentially with cathode temperature. Flat equipotential cathodes often operate under space-charge conditions in which heating is sufficient to lead to a build-up of electron charge, or a space-charge cloud in front of the cathode. Fluctuations in cathode temperature have little effect, but the emission current (I) will increase with the anode–cathode potential (V). In a simple configuration this emission current increase may be given by the equation $I \propto V^{2/3}$. Usually the presence of a space charge is considered to be an undesirable effect, leading to additional beam spreading and defocusing due to collisions between the beam and electrons within the cloud. In particular, space-charge effects may build up for an otherwise temperature-limited cathode when operating at very low voltages, but reduction of the gun-sample distance may help to lessen this effect. Clearly the high-voltage LEED system of Laydevant (see section 2.3.4) has the additional advantage that even at very low incident energies the gun operation may remain optimized. At high voltages, space-charge effects remain negligible—at least in the vicinity of the cathode. In a conventional LEED gun a possible means of space-charge reduction is to decrease the emission current, particularly if a channel plate is available to detect diffraction patterns formed by a very low primary beam current.

Ultimately one obtains an electron beam incident on the sample with a certain energy and angular spread. Typically, the diameter of the incident beam is 0.5 mm at about 100 eV, varying from about 1.0 mm at 15 eV down to 0.25 mm at 500 eV. With a 20 kV source, the beam may be as little as 500 Å across, making it possible to examine quite small crystallites.[88] The energy spread for normal filament voltages and temperatures may be 0.5 eV for pure tungsten cathodes and less for barium-impregnated cathodes, but this can be decreased by operating at lower temperatures.

The emission current will increase with electron energy unless steps are taken to control it. There are various possible ways in which the beam current can be controlled. The simplest is to limit the emission current directly—but, because of changing gun characteristics with energy, this does not guarantee a constant beam current at all energies. More usually the beam current itself is monitored as part of a feedback circuit that then adjusts the grid voltage.

2.1.5 Sample holders and manipulators

In many diffractometers the relative positions of the electron gun and detection systems are fixed, so the incident electron beam angle can only be varied by turning the crystal itself. The range of angles through which the crystal may be turned is limited by the design of the sample holder. No unique design of sample holder can be identified as ideally suitable for LEED work, because the design is constrained by practical considerations such as the method of heating, and the complexity permissible. The physical size of the holder is roughly related to its complexity. In certain designs of diffractometer, the diffraction pattern is observed from the same side of the crystal as the incident beam, in which case the physical size of the holder may be irrelevant. However, in many cases, such as the standard LEED system, the diffraction pattern is viewed from behind the sample holder so that any increase in its size increases the extent to which it obscures or shadows part of the diffraction pattern from the observer. If this is the case, the size of the sample holder must be restricted, and its design will be a compromise between size and versatility.

The crystal can be heated either by passing an electric current through the mounting strip (resistive heating) or by electron bombardment, from a separately heated filament. For resistive heating, currents of 50–100 A may be required, whereas for electron bombardment the current can be small, but voltages of about 10 000 V are necessary. If the former method of heating is adopted, thick current-carrying wires must be used. The outer sheath of coaxial cables is sometimes suitable for this purpose. Commonly available electrical feed-throughs have upper limits to their current-carrying capacity of not more than about 40 A. If that is the case, then up to three pairs will be necessary to cary the current into the vacuum chamber and out again.

The wiring for electron bombardment is less bulky, but has the dangers and insulation problems associated with very high voltages.

The crystal temperature can usually be measured by attaching a thermocouple comprising platinum and platinum/rhenium wires. For successful electron bombardment, the crystal, rather than the filament is biased to high voltage, since this will cause the emitted electrons to travel preferentially towards the crystal rather than the entirety of the crystal holder. However, this makes it necessary for the thermocouple also to be biased to high voltages. The voltage-measuring circuit for the thermocouples will thus need to be electrically isolated.

A hot-wire pyrometer may be used to measure the temperature if it is sufficiently high to cause the crystal to glow. Hot-wire pyrometers work on the principle of matching the colour of a reference wire to that of the sample This is done visually using a telescopic sight trained on to the crystal. Correction factors must be used to compensate for absorption through the chamber window and the emissivity of the sample, but provided these are included the method is sufficiently accurate for monitoring the temperature when cleaning the crystal. This is important if the sample must be heated to near its melting-point and the heating current obstructs the normal operation of the thermocouple. A pyrometer is also useful to check the operating temperature of an electron-bombardment filament.

It is always frustrating if the vacuum chamber must be brought back up to atmospheric pressure to carry out a simple repair such as the replacement of a burned-out electron-bombardment filament. It can be even more frustrating if the crystal itself is allowed to melt! Accurate temperature measurement and the means for cross-checking values are therefore worth taking great care over.

A final detail of the sample holder worthy of consideration is the means of attaching the crystal to the holder. If resistive heating is adopted, a high current must be passed through a metal slit, and a suitable method is that sketched in Figure 2.4. A slot is spark-eroded from the sample and this provides a channel through which the heating strip may pass. The heating strip may be made of any suitable material, but there are advantages in using

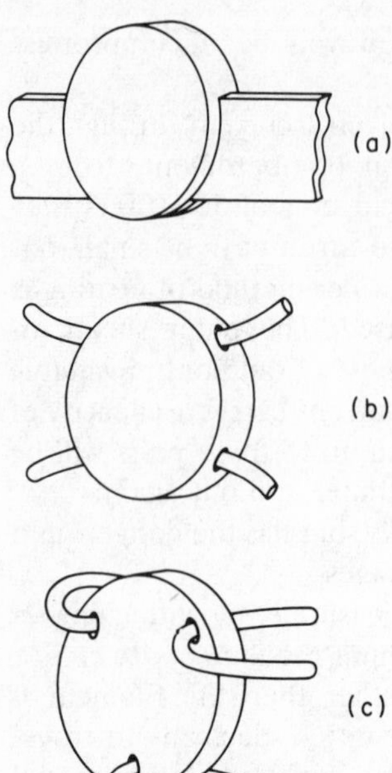

Figure 2.4. Various common ways of mounting crystals for LEED work: (a) slotted specimen on a metal strip; (b) holes spark-eroded parallel to surface; (c) holes spark-eroded perpendicular to surface

the same material for the strip as the crystal wherever possible, since this reduces the potential for contamination. If the dimensions are chosen suitably, the temperature of the strip may be about 100 °C hotter than the sample—if the crystal must be heated to near its melting-point during the cleaning process, this temperature difference will provide a safety zone, ensuring that, in the case of accidental overheating the strip will act as a fuse and melt first, thus protecting the crystal. This method may be successfully used in the case of molybdenum, for example.

If electron-bombardment heating is chosen, it will be desirable to minimize the bulk of the mounting wires to reduce the transfer of heat away from the sample. Wires threaded through spark-eroded holes either parallel to, or perpendicular to, the crystal surface may be used, as shown in Figure 2.4. Spot-welding wires to the crystal for the purposes of mounting is not recommended since this causes damage to the crystal in the locality of the weld, and such welds have a habit of breaking during extensive heating cycles. However, spot-welding may be the only means of making a good contact to the thermocouple.

Thermal expansion of the crystal mount during heating may alter the crystal orientation, and this possibility must be borne in mind when monitoring the effects of temperature on diffraction beam intensities. Because of this, the mounting wires should not be too thin, and a high-melting-point material such as tungsten may be preferable.

A variety of designs of sample holder are commercially available, so detailed descriptions will not be given here. For LEED analysis it is desirable

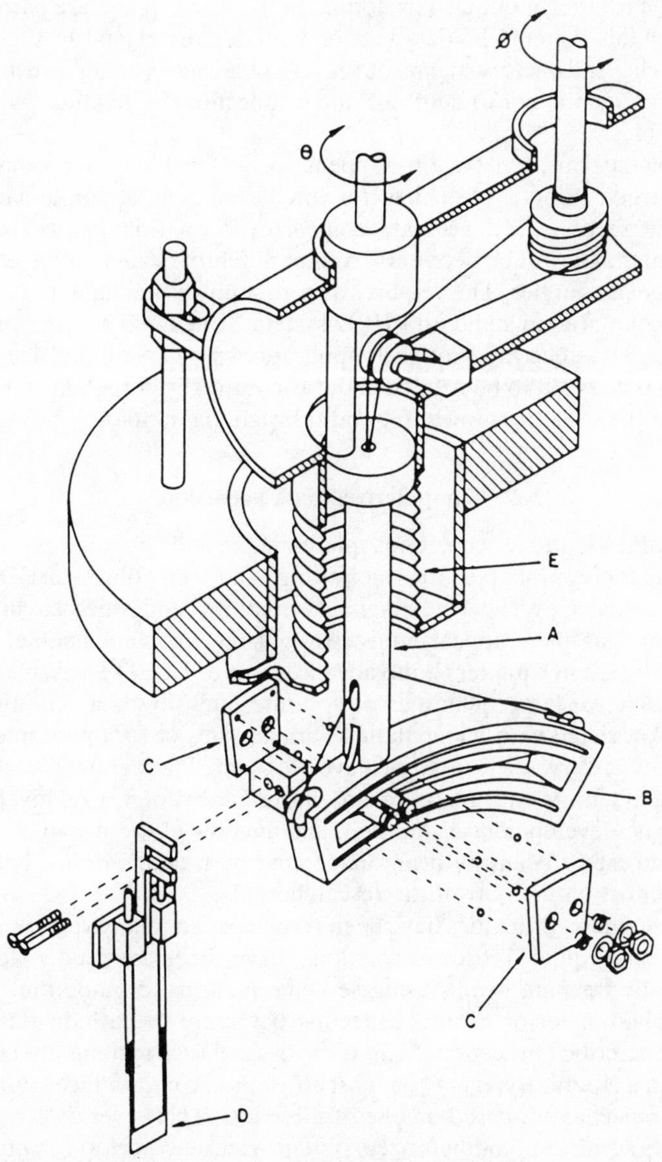

Figure 2.5. A UHV-compatible sample manipulator that allows rotation of the crystal about two mutually perpendicular axes in addition to the normal up–down, tilt and translation motions. (Reproduced by permission from the American Institute of Physics from O'Neill and Dunning[89])

to have control over both the azimuthal and polar angles of the incident electron beam with respect to the crystal. Crystal holders that enable the crystal to be rotated about an axis normal to its surface plane are particularly suitable for this. Sample holders that only enable the crystal to tilt, do not provide such a satisfactory means of setting the incident beam direction, but are generally more convenient for the connection of heating wires and thermocouples.

If the beam detection system happens to be fixed—as, for example, is necessary when using a Mott detector for the analysis of spin-polarization effects (see section 2.8), accurate control over rotation in two mutually perpendicular axes will be necessary, so that different beams can be observed over a range of energies. This requires a more sophisticated manipulator than usually encountered in standard LEED systems. The manipulator illustrated in Figure 2.5 was designed for such applications by O'Neill and Dunning,[89] and allows the crystal to be rotated about two mutually perpendicular axes in addition to the usual up–down, tilt and translational motions.

2.2 Computerized data acquisition

Although the Faraday cup, spot photometer and photograph are the traditional tools available for measuring and recording LEED beam intensities, advances in video-system technology and microcircuitry are currently making them appear unnecessarily slow and cumbersome. As will be demonstrated in Chapter 7, the advancement of LEED analysis relies on the availability of large quantities of accurate intensity data. The time and difficulty associated with manipulating Faraday cups or spot photometers by hand makes it very difficult to achieve this aim. Photographic techniques permit the intensity data to be recorded permanently in a relatively short time, but the development of the film, and analysis of the negatives using a microdensitometer beam by beam and frame by frame, still involves much time and effort on the part of the researcher.

Of course, these difficulties have been recognized for many years, and some ingenious and quite intricate solutions have been devised. Means of automatically tracking beams using servomechanisms to guide the Faraday cup spot photometer or mirrors reflecting the image of an individual beam have been described in section 2.1.3. If the basis of such systems are currently available in a laboratory, it may be most efficient to consider incorporating an automated mechanism based on one of these ideas. However, it is becoming increasingly clear that computerized systems can now perform many of the measurements possible with traditional equipment, often in a fraction of the time and not necessarily at any greater expense. The availability of complete data collection and analysis packages commercially makes this a most attractive approach to LEED analysis.

The first computerized video-camera system for LEED was developed by Heilmann *et al.* in 1976.[90] The rate at which a video system can acquire information is determined by the scan speed. In Europe the standard is 20 msec per complete frame and in America it is 16.67 msec per frame. Because TV systems scan every alternate line in one-half of the scan speed, then fill in the remaining lines across the screen in the next half it should, in principle, be possible to collect suitable intensity data every half-frame. However, the speed of early systems was limited by the ability of the interface to distribute this information and of the computer to assimilate it at such a high rate. Consequently, it was only possible to collect the intensity data at a rate several or many times slower than the theoretical maximum.

Recently, systems capable of exploiting this maximum data acquisition rate have been developed. This has been made possible by the availability of very fast analogue-to-digital converters and special interfacing capable of handling and reducing this information prior to transferring it to the computer. Details of such a system were published by Heilmann *et al.* in 1980.[91] This achievement is due not only to recent developments in computer hardware but also to careful selection of the data being gathered. A region including the diffraction beam and its immediate vicinity is initially described by means of an image projected on the TV monitor. The interface only gathers information from within this rectangular window. Each line scan of the video-camera system provides a cross-section of profile of the beam intensity, the background intensity being given roughly by the intensity values at the left- and right-hand sides of the window. The profile can therefore be background compensated by assuming that the background intensity varies linearly between the two values at each edge of the window, and by subtracting the intensity determined from this assumption at each grid point. This is repeated for each scan line within the diffraction beam window. This information may be stored directly if the spatial profile of the beam intensity is required. More often, only the integrated beam intensity is required. The full speed of the data acquisition method can be exploited in this case by summing the intensities of each line simultaneously within a fast adder contained within the interface. The only output necessary is the integrated intensity, the beam energy and current and the coordinates of the beam position. In this way, the computer is not overwhelmed with information, and it is able to construct a complete intensity–energy curve, normalized for a constant primary beam current, within a matter of seconds.

Although the process described above relates to the acquisition of a single diffraction beam, it is possible to track several beams simultaneously by defining several windows and processing data from each window sequentially.

A beam intensity at a single energy point can be measured and processed within the half-frame scan period, and once the intensities of all preselected beams have been recorded at that energy, the primary beam energy is increased. As the energy changes, so the positions of the diffraction beams (other than the specular beam) also change. The position of each window must be relocated, and this can be done automatically by the computer, provided the beam centre does not move out of the region bounded by the previous window position. This criterion is normally met if the beam energy is increased in steps of not more than 1 eV. The intensities of four beams may be recorded over a range of 250 eV in about 20 sec.

Although the time necessary to scan a complete range of energies increases linearly with the number of beams included in this way, this approach is nevertheless very rapid compared with most other methods.

The advantages of rapid data acquisition are numerous. Contamination levels during the experiment are minimized, as are the effects of electron-beam desorption. The latter is particularly important when weakly bound molecular species are being studied, as has been demonstrated when CO is adsorbed on Ni(100).[91] Accurate background subtraction is necessary for the study of surface domain structures, and video-camera techniques enable this to be carried out quickly and quite accurately. In the special case when the primary beam is incident normally on the surface, three or four beams may have identical intensity–voltage curves. This property can be used to check the correctness of the crystal alignment, since noticeable differences in the curves between one symmetry-related beam and another may sometimes be apparent even when the crystal is misaligned by a small fraction of a degree. The ability to collect and display intensity–voltage curves from a set of such beams in a matter of seconds is a considerable help in checking the effects of small adjustments. Even when the primary beam is incident off-normal but twofold symmetry remains in the pattern, this technique could be used to aid the determination of the incident angle, as described in section 2.6.

If information in addition to the beam intensities and positions is to be recovered from the diffraction pattern, the full speed of the data acquisition system may be unattainable, at least with currently available technology. For example, the spatial distribution of intensity in each beam, often called the 'beam profile', may be required if surface defects, islands or domain structures are being investigated. In this case the data-reduction process within the interface cannot be used, and the time required to store such information may be equivalent to several or many half-frame scans. Other computerized systems have been developed with such specific tasks in mind. Welkie and Lagally, for example published details in 1979[92] of a system based on the rear-view flat screen method described in section 2.3.3. A vidicon camera views the flat screen from behind, as shown in Figure 2.2, and the signal passes into an optical multichannel analyser that can digitize the image of the diffraction pattern point by point at the same rate at which the signal arrives from the camera. An intensity map comprising typically 50×50 points can be constructed in less than 2 sec, although storage on a disc memory unit may take rather longer. This is substantially slower than the selective method developed by Heilmann *et al.*,[91] but has the advantage that all beam intensities and their profiles are measured simultaneously. If high spatial resolution of particular beam profiles is required, the image area to be digitized may be reduced, enabling a resolution of $0.07°$ to be obtained—10 times better than if the full screen is viewed. For LEED intensity analysis alone, such a method would be rather inefficient, since large areas of background intensity are measured and stored. However, it does permit other techniques such as angle-resolved photoelectron spectroscopy to be used. Such applications have been noted by Weeks *et al.*, who described a similar form of apparatus, also in 1979.[93]

2.3 Different types of LEED systems

A wide variety of different forms of LEED apparatus have been developed and used in the past. The diversity of systems stems from the large number of variables available. For example, the collection of electrons may be made by a Faraday cup or by a display system (or both). If a display system is used, the screen may be flat or spherical and may be reflective or transmissive. The electrons may pass through field-free space or may be deflected by magnetic and/or electrostatic fields. To track the diffraction beams as the incident energy is varied, it may be necessary to fix the crystal position and track the diffraction beams as they track across the screen, in another situation it may be necessary to rotate the crystal to maintain a fixed direction for a given diffracted beam. If the electron beams are deflected by electromagnetic fields, it may even be possible to maintain fixed diffraction beam directions without moving the crystal. Each method has certain merits and disadvantages, and ideally one would like to use the system best suited to the particular application required. The best system for studying molecular adsorbates is not necessarily the best for studying phase transitions. The best system for studying surface geometrical structures is not necessarily the best for studying very-low-energy resonances and surface barrier effects, and so on. It would not be appropriate in a book of this nature to make an exhaustive survey of all the different types of systems developed to date, but brief descriptions of a number of significantly different types are given to illustrate the richness of the possibilities currently available. These, of course, are in addition to the many different types of measuring and recording systems (telephometric, photographic and TV-computerized) described in sections 2.1 and 2.2.

2.3.1 Hemispherical, transmissive screen systems

The main advantages of the standard LEED system are its ease of construction, and its ability to be used in conjunction with other surface-sensitive techniques. If surface crystallography is to be attempted, via the measurement of LEED beam intensities, then it is, in fact, rather inconveniently designed. Unless a Faraday cup with a computerized tracking system is employed, then the intensities must be measured indirectly from the optical brightness of the spots on the phosphorescent screen. Not only do the numerous grids cause a degree of shadowing that increases towards the edge of the screen, but the crystal holder blocks off a significant fraction of the

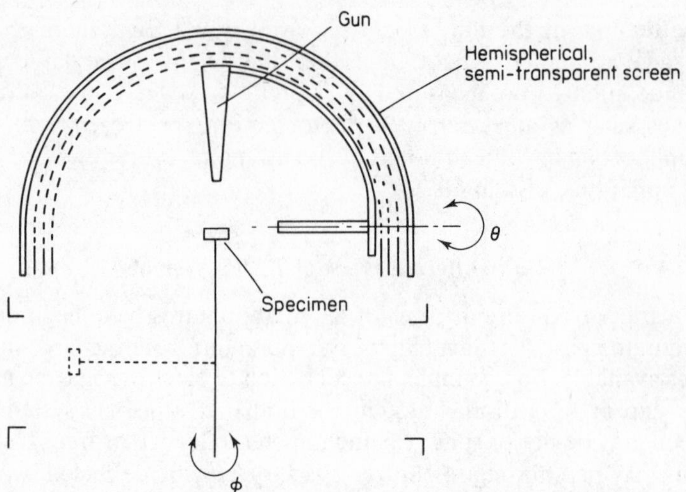

Figure 2.6. Principle of the de Bersuder LEED goniometer. (Reproduced by permission of the American Institute of Physics from de Bersuder[85])

screen. The design of the crystal holder is then a compromise between size and versatility. The observed pattern is distorted, because one observes the projection of the pattern from a spherical surface, and only a fraction of the total number of emergent beams can be observed at a time because the screen subtends an angle of not more than about 60° at the crystal surface about the axis of symmetry of the system.

Most of these disadvantages can be reduced or eliminated by using a hemispherical glass screen, making observations from the outside. Fujiwara *et al.* first used this sort of approach in 1966,[94] when they built a system with a cylindrical glass screen, and a fixed electron gun, but the first, and until very recently, the only system to incorporate successfully a spherical glass screen is that built by deBersuder in 1974.[85] The beauty of this system is that the screen is relatively very large (20 cm radius), and a miniature electron gun can be moved on an arm in the region between the crystal and the screen, as shown in Figure 2.6.

A telephotometer mounted outside of the dome can be moved to measure the intensity of any beam down to grazing emergence without problems of spot distortion of shadowing. The crystal can be mounted on a fairly substantial mount that permits accurate rotation of the crystal about its normal axis, thereby permitting the full range of polar and azimuthal incident beam directions to be explored. Apart from obvious advantages for standard intensity analysis studies, this provides the possibility of mapping intensities as a function of polar and azimuthal angles, as described in Chapter 7. The major drawback with this system is the formidable difficulty associated with constructing a large glass dome and coating it uniformly with a phosphorescent layer. The glass must first be coated with a transparent conducting layer so that impinging electrons can be removed and not cause charging of the screen surface.

2.3.2 Magnetic deflection systems

An interesting alternative approach uses the deflection of electrons by homogeneous magnetic field. The electrons can be made to impinge at normal incidence upon the surface, and the normally emergent beam does not return along the incident direction, but is deflected off towards the detection system. Such a system was built by Tucker,[95] and variants have been used by a number of researchers to enable the intensity of the specular reflection at normal incidence to be recorded. A schematic diagram of Tucker's apparatus is given in Figure 2.7. In the past, the observation of this particular beam was of special value due to the simplicity of interpretation compared to off-normal beams. Currently, the theory of scattering is sufficiently well developed that there is little advantage to be gained by using an apparatus that is specifically designed to aid the measurement of just one beam at one incident beam condition. As the energy of the incident beam is varied, so the magnitude of the magnetic field must be varied to provide suitable compensation. Consequently, this approach is now not generally used. It may, however, provide a means of observing LEED patterns in UHV systems where other diffractometer geometries are impractical, and LEED is of secondary importance to some other surface-sensitive technique.

2.3.3 Flat, transmissive screen systems

The electron-gun current necessary to provide a visible pattern on a fluorescent screen is of the order of 1 Å. With improved fluorescent materials this value may be reduced, but it may still be unacceptably high when weakly bound surface species are to be studied. If the physisorption of noble gases, or bonding of molecule species are of interest, then the dissociating or desorbing

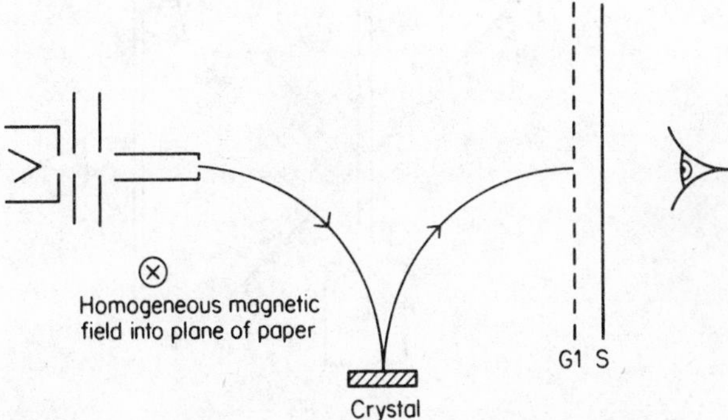

Figure 2.7. Principle of the Tucker LEED goniometer[95]

effects of the incident electron beam may be quite unacceptable. For this reason, it is desirable to reduce the electron current by several orders of magnitude, and then use an electron multiplier system to detect and enhance the resulting diffraction pattern. A Faraday cup fitted with a channeltron electron multiplier is one possibility, but the distance between sample and screen with most standard LEED systems is too small to permit the incorporation of the channeltron, and even with a computer-controlled movement of the cup, exploration of the half-space above the crystal surface is limited by the speed of mechanically scanning the region subject to the response time of the detector. Compared with display systems, this alternative is necessarily rather slow and cumbersome.

The channel-plate electron multiplier is a device that allows electron multiplication over an extended area, suitable for the visualization of diffraction patterns formed by very low electron currents. The simplest method of incorporating such a device is to use a hemispherical set of grids, as in the standard LEED system, but with a plane channel plate and optically transmitting screen behind. A plane mirror placed at 45° to the gun axis permits the resulting pattern to be observed from one side of the system. A schematic diagram of such an apparatus, based on a system used by Fain *et al*. for noble-gas adsorption studies, is given in Figure 2.8. In practice, this system uses a very similar principle to that used as early as 1959 by Scheibner *et al*.,[96] apart from the channel-plate multiplier itself and the use of hemispherical rather than flat grids.

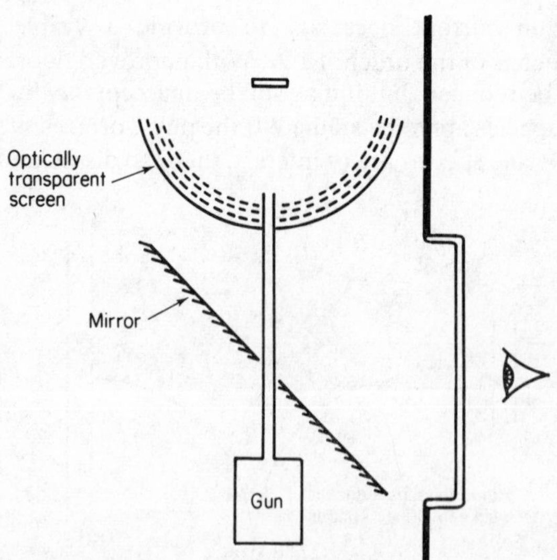

Figure 2.8. Principle of a transmissive-screen LEED goniometer

2.3.4 MEMLEED

One of the main problems with all the LEED systems that use field-free space for the propagation of the electron beams is that the pattern is only undistorted when displayed on a hemispherical screen. Moreover, if weakly bound species are to be studied, the field of view is significantly constrained by the dimensions of the channel plate (hemispherical channel plates can in principle be built, but are, at present, prohibitively expensive). If intensities are to be measured by photometric, photographic or TV-computerized system, shadowing, distortion and the movement of spots with energy must all be accurately taken into account.

The use of electrostatic deflection, however, allows the possibility of focusing all the emergent beams on to a flat focal plane of any desired dimension, and each diffraction beam is focused on to a fixed point on the plane, independent of its energy. This remarkable feat can be achieved by using the focusing principles of the mirror-electron microscope.

A high-energy electron beam is decelerated by a three-element Johanssen objective, as illustrated in Figure 2.9. The electrons are decelerated as they approach the crystal, which forms the cathode of the lens system. The incident energy is varied by varying the potential of the crystal, rather than by changing the potential of the gun. This means that a magnetic deflection field may be applied on the gun side of the objective, the magnitude of which does not need to be varied when the diffraction pattern energy is changed.

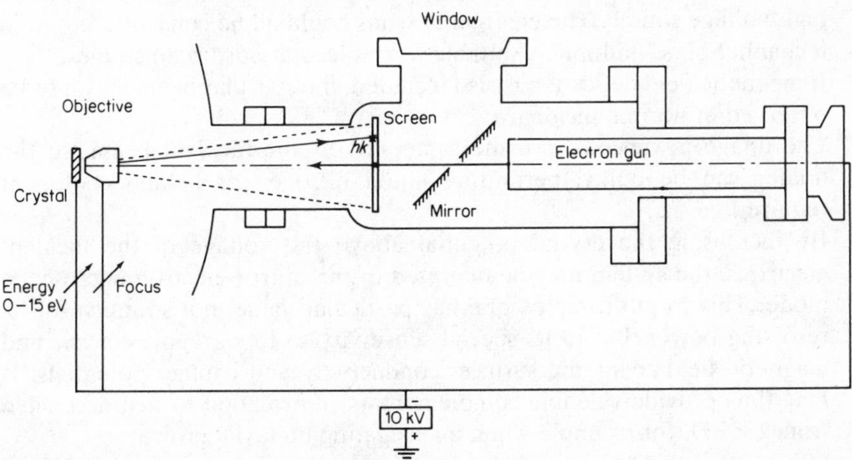

Figure 2.9. LEED apparatus based on the mirror-electron microscope principle presently used for very low energy studies but capable of extension to energy ranges suitable for LEED analysis. The flat focal plane permits efficient use of a channel plate electron multiplier, and the fact that the diffracted beams do not move across the screen creates obvious advantages for data collection. Such an apparatus may be conveniently referred to by the acronym MEMLEED. (Reproduced by permission of Taylor and Francis from Clarke[100])

Adaptation of this principle to the problem of LEED was first made by Delong and Drahos[97] and subsequently adapted by Laydevant *et al.*[98]

A dedicated LEED apparatus using this principle was successfully developed by Berger *et al.*,[99] and a simplified version subsequently marketed by Riber, SA in France. This device, which may be conveniently described by the acronym MEMLEED[100] displays an impressive number of advantages over other methods:

1. The diffraction beams do not move across the focal plane as the energy of the incident beam is varied. Intensity voltage data may be collected easily by any standard means, including by Faraday cup.
2. The cathode lens filters the emergent electrons, dispensing with the need for separate energy-filtration grids.
3. The electrons are slowed to typical low energies only within the final millimetre or so above the crystal surface—consequently, the influence of stray magnetic fields is negligible, and incident energies may be effectively reduced to zero for very low energy studies.
4. The dimensions of the beam spot on the sample depends on the lens aperture, but, by using field-emission tip sources or similar, may be reduced to a few micrometres (μm); to be compared with the 0.5 mm minimum usually possible. This provides an opportunity for studying very small crystallites, rather than the large single crystals normally required.[100]
5. All the diffraction beams are focused on to a single plane. The dimensions of the pattern are determined by the crystal periodicity and the initial high-voltage source. The emergent beams could all be concentrated on to a channel-plate multiplier, suitable for molecular adsorbate studies.
6. If magnetic field deflection is also included, the specular beam may also be observed at normal incidence.
7. The image is viewed from the same side as the incident beam, so the holder can be bulky, permitting liquid nitrogen or helium cooling if required.
8. By increasing the crystal potential above the voltage of the incident electrons, the system may be operated in the mirror-electron microscope mode. This form of microscope has particular value, not so much for its resolving power, but in its special sensitivity to topography, electric and magnetic fields near the surface, conductivity and contact potentials. It may thus provide valuable complementary information to that accessible from LEED, for example when investigating epitaxial growth.

Needless to say, such advantages are only attainable in the face of significant experimental difficulties. Aberrations will remain small, and the positions of the diffraction beams upon the image plane only if the high-tension source has a voltage nearly two orders of magnitude greater than the required maximum incident beam voltage. Thus, if intensities are to be measured up to say, 250 eV, a high-tension source of at least 20 kV would be required. This clearly introduces significant demands on the quality of the electrical insulation. Secondary electrons are not filtered out but instead are concentrated about the specular beam direction. This may hinder accurate measurements of the specular beam intensity. Because the crystal forms part of the objective system, it cannot be tilted far from normal without introducing unacceptable aberrations. Nevertheless, in practice, tilts of up to 20° appear to be possible without undue deterioration in performance.

(See also note in proof on page 115.)

2.3.5 High-resolution LEED

Usually, the intensity of a diffraction beam varies relatively smoothly with changing energy: loss processes limit peak widths in intensity–energy curves to about 4 eV (see Chapter 3) and the overlapping of adjacent peaks may cause many peaks to be even wider than this. The energy spread of the gun, about 0.5 eV (see section 2.1.4), normally sets the limit to the energy resolution of typical LEED systems, and this is adequate for distinguishing most features of intensity–energy curves.

At very low energies, however, surface resonances may be observed. These are associated with electron propagation subject to minimal energy loss.

Consequently, intensity peaks caused by surface resonances may have narrow half-widths, less than 1 eV. Surface resonances can be predicted from the surface band structure, and may occur wherever the electron wavelength is in resonance with the two-dimensional periodicity of the surface. The comparison of predicted surface resonance band structures and observations may convey useful information concerning lateral surface structures (i.e. parallel to the surface). Surface resonances are also very sensitive to the size, location and shape of the surface barrier, and have been used extensively in the investigation of the surface barrier of metals.[101] However, such narrow peaks cannot be easily distinguished if the resolution of the LEED system is limited by the gun to about 0.5 eV.

It is, however, not necessary for the resolution of a conventional LEED system to be limited by the energy spread of the primary electron beam. If the potential on the retarding grids is modulated with a sine-wave, a lock-in amplifier is tuned into this frequency and the electron gun potential is ganged together with the grids, the resolution may be made independent of the primary beam energy spread. Provided the modulating voltage is kept small, the resolution will be determined instead by the grid optics. Taylor[102] has shown that the resolution of a four-grid analyser is about 0.5 per cent, giving a theoretical limit of 50 meV at a primary beam energy of 10 eV. Improved noise filtration, at the expense of a lower signal intensity may be achieved by a double modulation method, as described by Price.[103] In this latter method, the retarding grids are modulated with square waves, but the gun energy is also modulated and the lock-in amplifier provides an output proportional to the derivative of the electron signal in the limit as these modulations tend to zero. This approach is similar to that commonly used in AES.

If higher energy resolution is required, a radically different apparatus is necessary. Higher resolution is not normally necessary for carrying out conventional intensity–voltage analyses, but may be desirable if energy-loss mechanisms are to be studied.

Many energy-loss mechanisms are relatively weak, and usually the associated momentum transfer will be insufficient to deflect the impinging electron through more than a few degrees. If an electron that undergoes an inelastic event (i.e. energy-loss process) is to be scattered back from the surface, it must also undergo an essentially elastic reflection, in other words, the sort of scattering event that contributes to the LEED process. Consequently, inelastic scattering is found to be closely linked to elastic diffraction, and may be described within the compass of inelastic low-energy electron diffraction or ILEED.

Wendelken and Propst[104] published details of a high-resolution instrument that uses double-pass 127° energy selectors both to produce a high-resolution beam and also to collect and analyse the resulting beams. A diagram of their apparatus is given in Figure 2.10. Double-pass analysers are very effective in preventing the eventual emission of rejected electrons (a problem with single-pass 127° analysers), and they permit energy losses to be measured with an accuracy 0.005 eV (~5 meV). Angles may be set with an accuracy better than 0.1°.

Although this form of monochromator-spectrometer is commonly used for high-resolution electron energy loss spectroscopy, it may of course be used in the study of surface resonance features if desired.

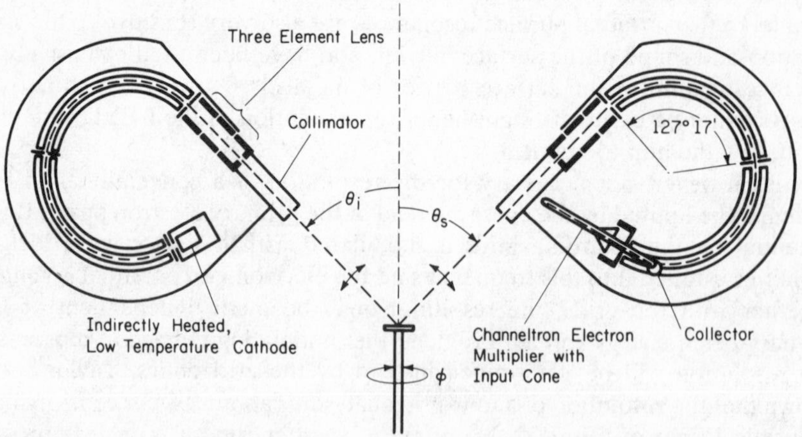

Figure 2.10. Schematic diagram of a high-resolution, low-energy electron diffractometer. Double-pass 127° energy selectors are used to reduce problems from internal scattering. The collimators maintain angular resolution under those conditions in which the lens elements do not maintain focus. (Reproduced by permission of the American Institute of Physics from Wendelken and Propst[104])

2.4 Detection limits of a diffractometer

The resolving power of a diffractometer is not a simple quantity to define. The quality of the detected signal is affected by fundamental properties of the electrons themselves, details of the source and the size and perfection of the crystal. Consequently, a variety of different approaches and concepts have been derived and used in the past, none of which necessarily give a complete and accurate picture of the phenomenon.

In practice, it is desirable to have a measure of the largest diffraction grating periodicity that can be resolved directly with a given instrument. This is necessary if one is to be able to estimate the average size of islands or domains from LEED beam characteristics, or if one wishes to know, for example, the lowest density of steps that can be detected.

Two recent and commonly adopted approaches to the determination of diffractor resolution are those of Pendry[105] and Park *et al.*[106] Pendry defined a 'coherence length' as the distance across the surface within which atoms could be considered to be illuminated by a single plane wave, and beyond which the phase relationship of the illuminating waves is arbitrary. The mean square deviation of the phase difference between two points in the surface is expressed as a function of the energy spread of the beam and its angular dispersion, and the coherence length defined as the separation at which the mean square deviation in phase becomes π^2, since at this condition the phase relationship between the points vanishes. Park *et al.* [106] on the other hand, defined a 'transfer width' which is analogous to the 'coherence length', but derived from an instrument response function, and is defined as the range over which the instrument can behave as an interference detector.

Comsa[107] has critically examined these and other approaches and derived additional formulae for the transfer width and angular spread which he shows compare well with results from molecular beam diffraction experiments.

The important point made by Comsa is that the diffraction process occurs as a result of the interference of an electron with itself, and the diffraction pattern is then obtained by the superposition of probability patterns from individual particles. This leads to formulae which differ from those of Pendry. If λ is the wavelength, E the energy and ΔE the energy width of the incident beam, and θ_i and θ_f are the incident and emergent beam polar angles, with an angular spread Δ_θ, then the transfer width w is given by

$$w = \lambda/[(\Delta_\theta \theta_f)^2 \cos^2 \theta_f + (\sin \theta_i - \sin \theta_f)^2 (\Delta E)^2/E^2]^{1/2} \qquad (2.1)$$

The dominant contribution to the transfer width is the finite angular size of the electron source Δ_θ, which is typically 1°. As a representative value consider normal incidence at an energy $E = 100$ eV and diffraction beam angle $\theta_f = 45°$. If the energy spread is zero, equation (2.1) gives $w = 99°$. If a typical energy spread $\Delta E/E = 4 \cdot 10^{-3}$ at $E = 100$ eV is included, $w = 26°$. The influence of the energy spread is clearly much less than that resulting from the finite source dimensions.

The transfer width originally defined by Park *et al.*[106] was defined as the width of a transfer function, which in turn is the Fourier transform of an instrument response function. Typically, these functions are Gaussian, so the electron wave will be correlated in phase well beyond the limits defined by the 'transfer width'. If the crystal itself contains domains or islands of a comparable width, these will also affect the beam width, and the measured signal will be a convolution of beam and surface effects. Wang and Lagally have made an extensive study of such convolution effects[108] and have clearly demonstrated the importance of correlations beyond the transfer width. As an example, they note that if a lattice with dimensions similar to the transfer width (~100 Å) is used to model the diffraction from a real lattice, and the resulting width is matched to the experimental width without proper convolution of the instrument response function, the calculated domain size will be too small by at least 50 per cent.

Resolvability will depend on the type of domain, and therefore the transfer width, a single number, is not sufficient to describe the resolving power of a LEED instrument. Lu and Lagally[109] propose a new parameter—the minimum angle of resolution. This can be defined in terms of the full width at half maximum (FWHM) of the instrument beam profile, b_T, and an estimate of the percentage error in the measurement of this quantity (X per cent). If the actual signal function $I(\theta)$ has a FWHM b_I and the instrumented response function $T(\theta)$ has a FWHM b_T, then the measured signal function $J(\theta)$ will be given by the convolution

$$J(\theta) = T(\theta) * I(\theta) \tag{2.2}$$

and the measured FWHM, b_J, is given by

$$b_J^2 = b_T^2 + b_I^2 \tag{2.3}$$

Neither b_T nor b_J can be measured precisely and therefore, neither can b_I. Assuming that the uncertainties in $T(\theta)$ and $J(\theta)$ are the same, then it can be shown that the uncertainty in b_I will lie between zero and

$$\theta_{\min} = 2b_T(X \text{ per cent})^{1/2} \tag{2.4}$$

This minimum angle may then be used as a measure of the resolving power of a LEED instrument. Clearly the resolving power may be improved either by improving the instrument to reduce b_T or else by repeating the measurement numerous times and taking the average, thereby reducing the uncertainty, X. Application of this concept depends on the specific surface features being examined, as disussed in some detail in their paper by Lu and Lagally.[109]

Peak widths are important diffraction features in the study of surface defects, but it is the peak intensities which are of prime interest in most crystallographic studies. LEED intensity calculations are currently based on the assumption of perfect coherence or infinite transfer width, which we have now indicated is not strictly true. Electron energy loss processes, however, also serve to restrict the mean free path and these are included in the theory. Given that the magnitude of these loss processes is usually set to match experimental data, some compensation for imperfect coherence effects is automatically included. It is not possible to derive a simple analytical expression which can predict the effects of finite transfer width on multiple scattering intensities, because the net result is due to interference effects, and the removal of certain contributions is as likely to result in an increase in final intensity as a decrease. Heinz[110] has made a detailed study of the effect of finite transfer width on LEED intensity calculations by looking at the consequences of restricting path lengths within planes of atoms. If path lengths are restricted precisely to typical transfer widths then the effects may be significant but, as we have already noted, the transfer width is only the width of an approximately Gaussian distribution, so that in effect contributions should be included from a wider area. Consequently, modifications to beam intensities would probably be smaller than those calculated by Heinz.

2.5 Magnetic field cancellation

An electron beam is easily deflected by magnetic fields, and so it is important that stray magnetic fields be reduced to a minimum or cancelled out by the introduction of a compensating field. Stray magnetic fields have a number of undesirable effects—within the electron gun, alterations in electron paths will reduce the efficiency and introduce aberrations. The electron beam becomes curved so that its incident angle is uncertain, and its position shifts across the face of the crystal as the energy changes. Curvature of the diffracted beams is less important since the diffraction process is unaffected and the only effect is a distortion of the diffracted pattern.

If the electrons have a potential V volts and travel in a direction perpendicular to a magnetic field of B gauss, then the radius of curvature R of the beam will be approximately

$$R = 3.37V/B \text{ cm} \tag{2.5}$$

If the beam path length is l cm and the curvature is not large, then the deflection d will be approximately

$$d = l^2/2R \text{ cm} \tag{2.6}$$

A typical distance from LEED screen to crystal is 7 cm. In this case

$$d = 7.26B/\sqrt{V} \text{ cm} \tag{2.7}$$

Normally, it will be the earth that provides the major unavoidable magnetic field. Its strength and direction can be measured using a Hall probe, or alternatively reference may be made to tabulated data for the relevant geographical locality.

Very roughly, B will be about 0.5 gauss. If V is 100 V then the resulting deflection will be about 0.36 cm. At lower energies, the deflection would be even greater.

The magnetic field may either be eliminated by constructing the chamber from a metal with very high permeability, or else some compensating magnetic field be applied to cancel the effects of the earth's field. The former way sound an attractive proposition but in practice is expensive and not always successful, since the magnetic shield may be pierced by numerous windows and ports which may actually concentrate the remaining magnetic fields in certain directions. It is important for reasons given above, that the gun also be well shielded.

The shielding factor for a cylindrical shield in a static field is given by the formula

$$\text{shielding factor (cylindrical shield)} = \frac{H_{\text{out}}}{H_{\text{in}}} = \frac{\mu t}{2d} \qquad (2.8)$$

where H_{out} = the field to be shielded, H_{in} = the field inside the shield, μ = material permeability; t = material thickness and d = cylinder diameter.

For effective shielding, a material of very high permeability must be used. Mu metal is a material that possesses the necessary properties, and the vacuum chamber may be constructed entirely from this material. If a chamber of stainless steel already exists, an inner sleeve of Mu metal may be inserted to provide the necessary shielding. Advantages of Mu metal chambers are that they provide shielding from variable magnetic fields, and that the whole region enclosed should be field-free. By contrast, magnetic field cancellation cannot reduce the effects of ac fields, must be regularly rechecked in case the external field is slow-varying and in general cannot cancel precisely any field that varies spatially throughout the chamber region. Despite these apparent disadvantages to the field cancellation technique, it must be noted that it is often adequate for the purposes of LEED analysis, provided care is taken to optimize the arrangement.

For successful field cancellation, the applied field must be as uniform as possible over the whole volume in which the electrons move. Electrons are only deflected by the component of the magnetic field perpendicular to their direction of motion and so, if we only concern ourselves with the incident electron beam, then it is only necessary to cancel stray magnetic fields normal to this axis, i.e. in the y- and x-directions as given in Figure 2.11. To reduce deflections of the diffracted beams, cancellation in the third (z)-direction may be considered. Compensatory magnetic fields are provided by passing dc currents through pairs of coaxial coils, or Helmholtz coils. The magnetic field between two coils will be most uniform when the coils are separated by a certain distance known as the Helmholtz spacing.

For circular coils, this occurs when the separation distance, S, is half the diameter of the coils. Large circular coils are not particularly easy to construct and are rather inconvenient in that they restrict access to the apparatus. It is quite common, therefore, to use square coils. If a square coil has sides of unit

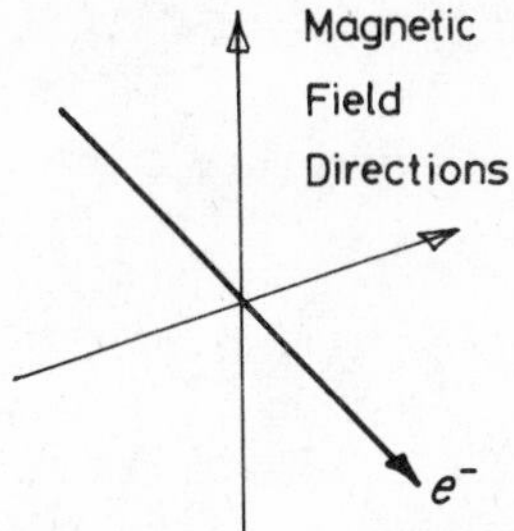

Figure 2.11. To influence the path of an electron, it is only necessary to apply magnetic fields in directions perpendicular to the original electron trajectory

length, then the Helmholtz spacing S is found to be 0.5445. Firester[111] has made computer calculations of the magnetic field duration as a function of x, y and z for square Helmholtz coils and these are given in Figure 2.12. The axial field deviation E in this figure is defined as follows:

$$E = \frac{B_z(x, y, z) - B_z(0, 0, 0)}{B_z(0, 0, 0)} \tag{2.9}$$

The field at the centre of the pair of square Helmholtz coils is

$$B_z(0, 0, 0) = 1.629) \frac{NI}{S} \text{ gauss} \tag{2.10}$$

where N = the number of turns on each coil, I = the current in amperes and S = length of one side of a coil in centimetres.

Various options are available—for example, the coils may be designed to meet the Helmholtz spacing requirement, and built of such a size that they fit around the apparatus without excessively restricting access. Alternatively, the coils may be made so large that the Helmholtz criterion need not be fulfilled, and the coils are fitted together to form the edges of a cube. If S is greater than about 2 m, then the resulting field may be very uniform and access to the diffractometer is simply gained by walking through!

If there are ac fields present, then some form of shielding may be necessary. The shielding factor for ac fields will be rather greater than that given in equation (2.8) due to the reflection and absorption of electromagnetic waves at the shield. Nevertheless, identifying and shielding the source of the fields is probably the best solution.

Adjustment of the compensating magnetic field is most easily achieved by noting shifts in position of the specularly diffracted beam with energy. At high energies, the effect of the stray magnetic field will be much smaller than at very low energies. If, for example, a spot photometer or Faraday cup is aligned with the specular beam at 400 or 500 eV, when the energy is reduced to 20 or 10 eV, any residual field will cause the beam to shift away from that position. Manipulating the magnetic fields to restore the maximum beam intensity reading should improve the quality of the compensation. Alignment should be rechecked at high energies, and the magnetic fields at low energies,

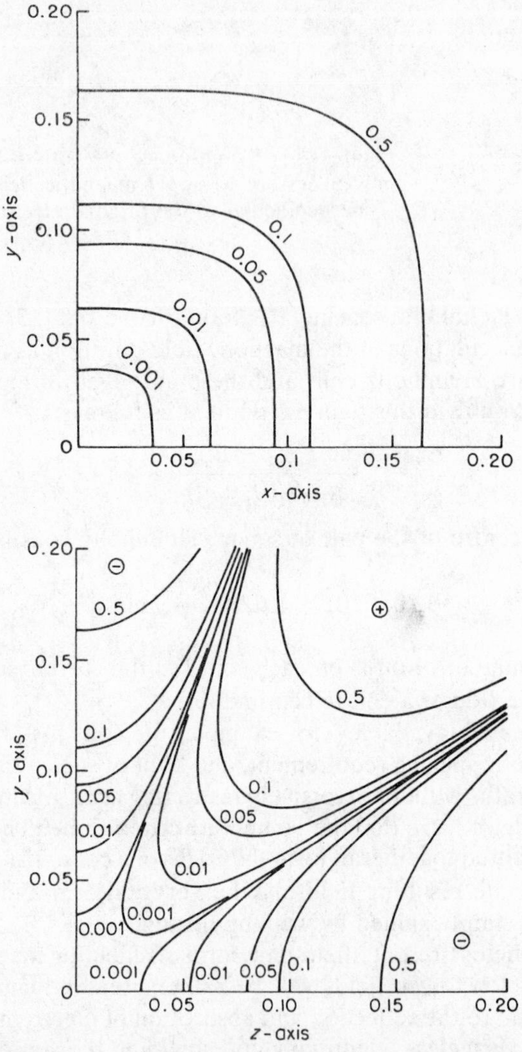

Figure 2.12. (a) Contours of constant axial field deviation in the y–x plane (normalized units—1×1 coil system). (b) Contours of constant axial field deviation in the y–z plane: ⊖ indicates that the field at these lobes is less than the field at the origin; ⊕ indicates that the field at those lobes is greater than the field at the origin (normalized units—1×1 coil system). (Reproduced by permission of the American Institute of Physics from Firester[111])

until no deviations are obtained down to the lowest usable energy. The necessity to reduce stray magnetic fields to a minimum cannot be overemphasized—uncertainties or variations incincident beam angle are often cited as a major source of discrepancies between experiment and theory or, indeed, between different experimental observations.

2.6 Determination of the incident beam angle

Even when residual magnetic fields are negligible, the determination of the incident beam angle is not always an easy task. The direct approach is to rely on mechanical settings on the goniometer to give a direct reading of, say, the polar angle. The indirect approach uses information obtainable from the diffraction beams themselves.

For a direct measurement it is first necessary to calibrate the zero of the goniometer scale by determining the position at which the crystal surface is precisely normal to the incident beam direction. Even when this has been achieved by the precise matching of equivalent beam intensities, it is still necessary to be certain that there is no slack in the mechanical movement which will affect the readings. Direct readings from a calibrated knob are usually insufficiently accurate (a minimum accuracy of $\pm 0.5^\circ$ is required, and if possible measurements should be within $\pm 0.1^\circ$), but the reflection of a sharp light source or laser from a small reflector mounted on the goniometer may be used to magnify and thereby improve the accuracy of the measurements.

Heating the crystal for cleaning purposes often causes parts of the crystal support to expand and thus change the alignment. For this reason, and the fact that direct settings are not always satisfactory, especially for the measurement of azimuthal angles, it is more usual to measure the angles indirectly via the displayed diffraction pattern.

A variety of methods are available which make use of photographs of the diffraction pattern. The introduction of direct computer imaging systems should permit analogous methods to be incorporated in automatic angle-determining procedures.

A method proposed by Taylor[112] involves the geometrical relationships between the image of the diffraction beam and the diffraction angle necessary to produce such a pattern. Referring to Figure 2.13, it can be shown that the distance of the image from the central axis, y_2 is given by

$$y_2 = \frac{br \sin\theta + c}{a - 2r \sin^2(\theta/2)} \tag{2.11}$$

where c is a correction factor which takes into account refraction through the window. In practice, c is only about 1 per cent of $r \sin\theta$ and so may be ignored. The ratio b/a is the magnification, which can be readily determined by taking a photograph of a scale placer perpendicular to the axis of the lens and separated from the film by precisely the same distance as the centre of the

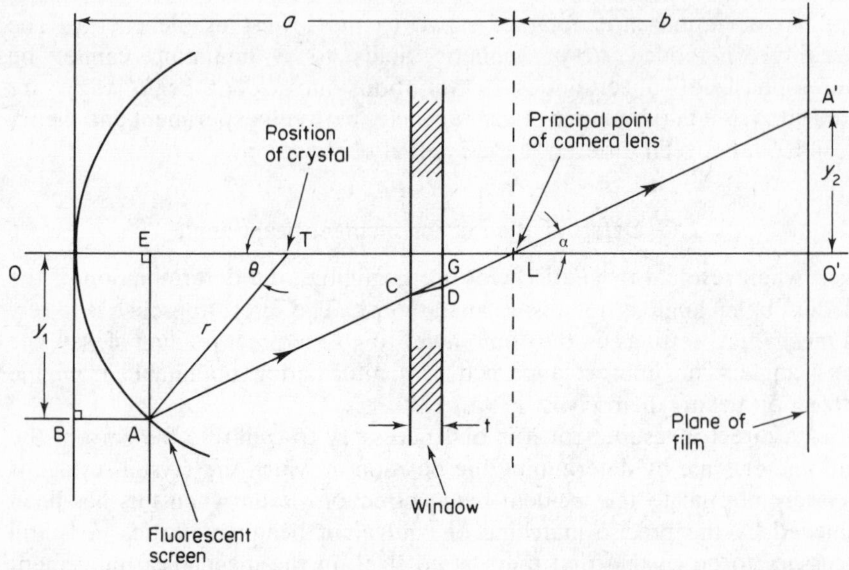

Figure 2.13. Geometrical construction useful for the determination of the incident beam angle. (Reproduced with permission from Taylor[112])

phosphorescent screen, O. When the camera is focused on O, this distance is $(a + b)$, so from these two quantities the values of a and b may be determined. Here, r is known from the physical characteristics of the screen, and so measurement of y_2 permits θ to be determined from equation (2.11). This provides a simple, rapid and relatively accurate method, although it does depend on the crystal being positioned at the centre of curvature of screen, a condition that is not always easy to check or maintain.

An alternative method, proposed by Cunningham and Weinberg,[113] is also applicable if the crystal is not at the centre of the screen, and claims absolute accuracies better than 0.1° for both polar and azimuthal angle determinations.[114] A clear sheet of plastic marked with angles is used to measure, from a photograph, the angle subtended by the (0, 0) beam and any other (h, k) beam at the centre of the screen. A computer program solves iteratively a general relationship between the diffracted beam wave vector components and the determined parameter in order to determine $[\theta, \phi]$. An advantage of using data from a general beam (h, k) is that numerous different measurements can be made, corresponding to each visible beam, and this permits mean values of θ and ϕ to be determined with a high degree of accuracy. The mathematics are too lengthy to discuss here, but the program is available from the authors referred to above.

Finally, there is a method devised by Clarke,[15] which has the advantage that it requires no photographs, so can be used to provide an accurate estimate of the angle whilst an experiment is in progress. Despite its rather limited range of applicability, the advantage of being able to measure angles rapidly and directly from visual observations is evident. The basis of the method is the strong dependence of the diffracted beam angle on energy. The method is best described by a specific example. Consider a clean, fourfold symmetric surface. At normal incidence the specular or (0, 0) beam emerges along the incident beam direction and the diffraction beam likewise exhibits fourfold symmetry, in both the two-dimensional arrangement of spots on the screen and their intensities. In the special case when the crystal is rotated such that the incident beam is off-normal but the azimuthal angle defines a high symmetry direction (in this example, any multiple of $\pi/4$) then there will be a mirror plane in the diffraction pattern and there is only twofold symmetry in both position and intensities. As the energy of the incident beam is increased, so the non-specular spots will move across the screen towards the specular spot. It is thus possible for any diffraction beam which lies on the mirror plane to pass, at one particular energy, back along the central axis of the grids. In this case, a fourfold symmetric arrangement of spots will be restored, although not, in general, in their intensities. Even if the energy at which the symmetry condition occurs is identified purely by eye, it is found that this can be identified within about $\pm$ 4eV. A more sophisticated means of checking symmetry about the central axis using a spot photometer or Faraday cup could be used to improve this accuracy much more. The precise magnitude of the angle, and the accuracy with which it is thus determined can be derived as follows.

Consider the energy at which the first-order diffraction beam, indexed ($\bar{1}$, 0), is diffracted back along the incident beam direction. This is illustrated in Figure 2.14. In this case, if $k_{\parallel}^{\circ}$ is the component of the incident wave vector $\mathbf{k}^{\circ}$

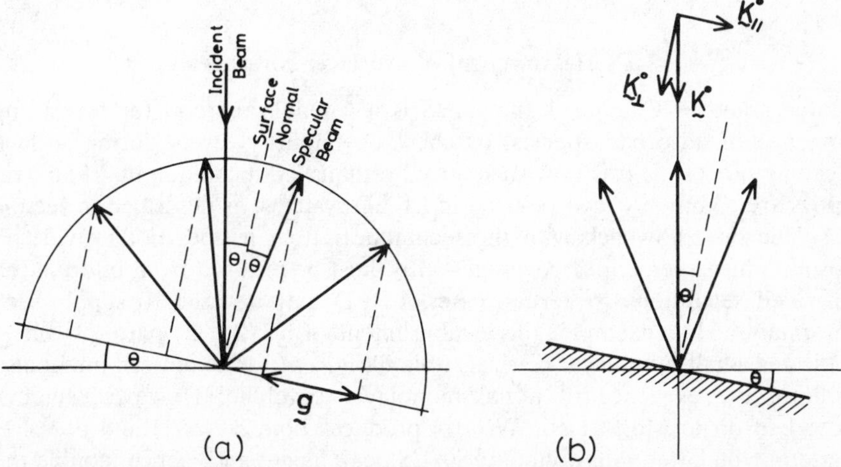

Figure 2.14. (a) Ewald sphere model of the diffraction beams produced by a crystal face tilted from the incident beam direction; **g** is a reciprocal lattice vector. (b) Notation used in the text, illustrating the energy at which the ($\bar{1}$, 0) beam returns along the incident beam direction. (Reproduced by permission of North Holland Physics Publishing from Clarke.[15])

parallel to the surface and $\mathbf{g}_i$ is a reciprocal vector of the surface lattice, λ is the wavelength and θ the incident polar angle, then

$$2k_{\parallel}^{\circ} = \mathbf{g}_i \tag{2.12}$$

where

$$k_{\parallel}^{\circ} = |\mathbf{k}^{\circ}| \sin\theta = 2\lambda \sin\theta \quad \text{and} \quad |\mathbf{g}_n| = 2\pi n/a_0$$

where a_0 is the real lattice dimension in nanometres (nm) and n is the order of diffraction. From this and equation (1.1) (note: 1 nm = 10 Å)

$$\sin\theta = \frac{n\ 150.4^{1/2}}{2a_0 E(\text{eV})} \tag{2.13}$$

From this, θ may be derived quite accurately. As an example, taken from a study Mo(001), for which $a_0 = 0.3147$ nm, symmetry about the $(\bar{1}, 0)$ beam was obtained for an energy 200 ± 4 eV. Inserting these values, we derive the value $7.93 \pm 0.08°$.

For clean surfaces, this method is limited to polar angles greater than about 5° since the energies necessary to make a beam satisfy the symmetry condition at very small incident angles becomes excessively high and the accuracy with which the energy can be determined deteriorates. However, if there are fractional-order beams present on the mirror plane than these may be used to provide an additional symmetry condition at lower energies. If several different beams pass through the central axis within a reasonable energy range, then they may each be used to estimate the angle, thereby obtaining a more accurate mean result. In general an accuracy of about ±0.1° should be obtainable.

2.7 Measurement of overlayer coverages

It was noted in Chapter 1 that AES is a valuable method for measuring coverage of adsorbed species, or checking impurity levels during surface cleaning processes. It is now such a well-established technique that four-grid optics are commonly incorporated in LEED systems, as described in section 2.1. The one drawback with the technique is that it does not provide an absolute measurement of coverage—this must be inferred from information provided separately. In certain cases, LEED may be used to supply such information. For example, the establishment of a $c(2 \times 2)$ pattern usually indicates a half-monolayer coverage. Similarly, a $p(2 \times 1)$ pattern might also indicate the existence of a half-monolayer coverage. However, sulphur adsorbed on to Mo(001) or W(001) produces both $c(2 \times 2)$ and $p(2 \times 1)$ patterns, the latter with double the AES peak height and, hence, double the coverage of the former.[11, 115] This paradoxical behaviour could be resolved by assuming that the $p(2 \times 1)$ pattern actually involved a second atom within the unit cell, and was thus present as a complete monolayer (see Figure 1.31). The alternative interpretation, that the $c(2 \times 2)$ structure comprised islands which covered half the surface, leading to a quarter-monolayer coverage, could be dismissed on the basis of two separate tests. Firstly, it is found that CO normally bonds strongly to Mo(001), but is completely prevented from bonding when the $c(2 \times 2)$S pattern is formed. If there were patches of uncovered surface, then some CO could be expected to adsorb, and be detected by AES. (This was checked by deliberately producing a $c(2 \times 2)$S structure without adsorbing sufficient S to form a continuous adlayer; in that case, a small quantity of CO could, indeed, be adsorbed.) Secondly, dynamical LEED analysis of the $c(2 \times 2)$S pattern closely matched the experimental intensity profiles. For certain beams at certain energies, near-zero intensities occurred for the $c(2 \times 2)$S structure where a maximum existed for the clean surface (see Figure 2.15).

Clearly, if the $c(2 \times 2)$ pattern had only been produced from islands, then a superposition of intensities from the covered and clean portions of the surface would lead to finite intensities where in fact clear-cut minima were found.

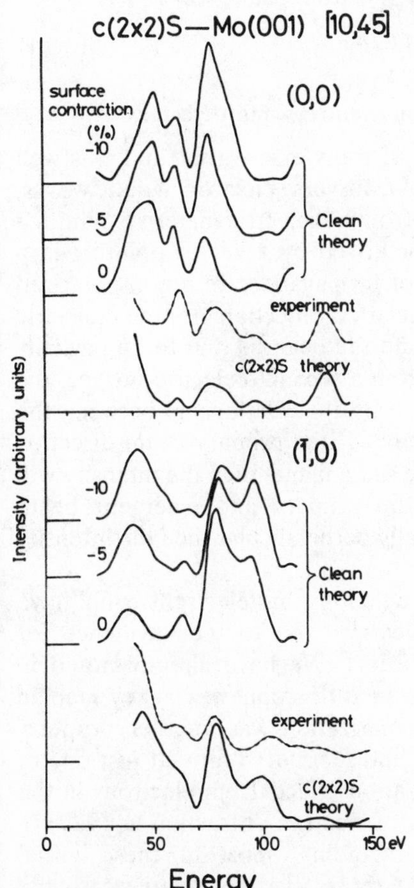

Figure 2.15. Comparison of theoretical $I(E)$ curves for clean surface and $c(2 \times 2)S$ overlayer showing that differences between experiment and theory are not simply the results of areas of clean surface between $c(2 \times 2)S$ islands. (Reproduced by permission of North Holland Physics Publishing from Clarke.[11])

From this specific example, it may be seen that a variety of methods may be used to assist in the identification of precise coverages.

The use of radioactive tracers is one method which has been used occasionally to provide quantitative coverage measurements. Bénard and Laurent[116] used radioactive sulphur in 1956 to measure adsorbate coverages on a copper surface, and this has been used more recently on several occasions.[117,118] CO on Mo(001) has also been studied using radio tracer techniques.[119]

Limitations on the availability of radioactive components, and the fact that AES is now so commonly available mean that radiotracer techniques are rarely use in present-day surface studies. Nevertheless, they may be of value in cases where substrate AES peaks overlay or obliterate important overlayer peaks, and they can be used to verify the linearity of coverage with AES output.

It should be noted that the coverage of an overlayer species is proportional, not directly to the height of the relevant AES peak (I_0), but to the ratio of the peak heights of overlayer species and any substrate peak (I_s):

$$\text{coverage} \propto (I_0/I_s) \tag{2.14}$$

2.8 Spin-polarization of electrons: sources and detectors

Polarizability is a characteristic property of transverse waves, and it is well known that a beam of light, being a flow of transverse electromagnetic waves, can be partially polarized when reflected from a dielectric material. Within a beam of light each photon can be characterized by a vector pointing in a direction perpendicular to the direction of propagation. In a non-polarized beam these vectors are uncorrelated, but after reflection from a dielectric surface one direction may predominate and the beam is said to be partially polarized. If this beam is then reflected from a second dielectric surface, the intensity of the resulting beam will depend on the relationship between the axis of polarization and the scattering plane. If the primary beam direction and the emergent beam both lie within the same plane, then the intensity will be a maximum (see Figure 2.16). If the primary and emergent beam directions life within planes that are mutually perpendicular then the intensity will be minimized.

Following in the wake of theoretical predictions that electrons could have wave-like properties, experiments were carried out to seek evidence for properties analogous to those of light beams. We have already noted in Chapter 1 that the observation of electron diffraction was a key step in confirming those predictions. Electron polarization was another property sought at about the same time. Davisson and Germer[120] formed just one of many research groups who sought polarization effects from electrons in the years immediately following the discovery of electron diffraction in 1927. In the absence of a more detailed understanding, most of these initial experiments were similar in approach to those used to demonstrate the

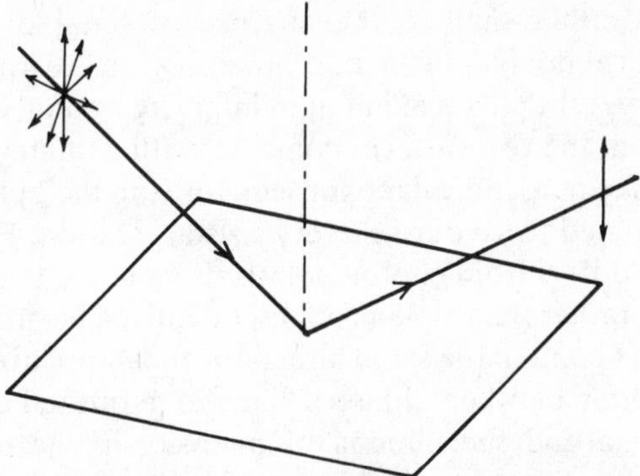

Figure 2.16. Polarization of light by reflection. Electron spin polarization differs fundamentally from this picture because electrons can only be polarized in one of two mutually opposite directions

polarization of light, using double-diffraction methods.[121] In addition, attempts were made to polarize electron beams by transmission through magnetized iron foils[122] or by means of transverse magnetic fields.[123]

Despite the attention of many researchers, no conclusive evidence was found. This can be ascribed to a lack of recognition of fundamental differences between photon and electron polarization, as well as incomplete understanding of the interaction of polarized electrons with matter. Continuing theoretical work provided new insights that could be used in the development of further experiments. In particular, Mott[124,125] studied the scattering of electrons by isolated atomic nuclei, and derived a number of conditions necessary for the production of observable polarization. He concluded that the electron beams should be scattered through an angle of about 90°, and this should be effected by single scattering from a heavy nucleus (atomic number similar to 137). Multiple scattering though smaller angles from a series of nuclei would have a depolarizing effect. These conditions may be met by scattering from a thin gold foil and a very high energy (100 kV or more) electron beam. Nevertheless, a considerable number of separate investigations were carried out before the first successful observation of spin-polarization by Shull *et al*. in 1943 who used a beam accelerated to 400 keV energy scattered by transmission through two gold foils.[126] This approach has been subsequently developed into a standard experimental apparatus for detecting the polarization of electrons, known as a Mott detector.[127]

If polarization effects could only be produced by very high energy electron interactions, then polarized electrons would have little or no part to play in the study of surfaces. Fortunately, the inconclusiveness of early experiments was due to a fundamental flaw in the understanding of the nature of spin-polarization rather than to the absence of suitable interactions. The analogy with optical double-diffraction processes, as described earlier in this section, would suggest that a maximum in intensity would occur wherever the scattering planes at the two surfaces coincide, with a minimum when the two planes are mutually perpendicular. Consequently, in the optical situation, the intensity is modulated twice during every crystal rotation. However, electron spin-polarization differs from photon polarization in that it is characterized by a single direction rather than a plane. Consequently, a beam is polarized when all the spin vectors point in the same direction, and unpolarized when they are distributed randomly between either of the two permitted orientations. As a result, the maxima and the minima of intensity in the double-diffraction experiment are obtained whenever the crystal is rotated through 180° rather than 90°. In a recent re-analysis of Davisson and Germer's data, Kuyatt[128] has shown that, contrary to their own conclusion, spin-polarization might indeed have been observed in 1929. However, theoretical calculations made subsequently by Feder[129] show that conditions of the experiment did not warrant such a large asymmetry as observed, and that alignment errors as small as 0.5° could have generated the observed results. Clearly, this approach requires extreme care in execution and interpretation. The first genuine demonstration that double diffraction could indeed produce spin-polarized effects was made as recently as 1979, when Kirschner and Feder[130] used a very elaborate experimental apparatus to detect spin-polarization by double diffraction at low energies from W(001) crystal surfaces.

Spin-polarization effects have been observed in surface studies since about 1968. In that year, Palmberg *et al.*[131] observed exchange reflections with half-order indices in LEED patterns from antiferromagnetic NiO below the Néel temperature. This is a special case in which non-polarized electron beams could be used to detect spin-orbit periodicity that differs from the structural periodicity of the crystal, even without detecting the polarization of the diffracted beams. EuO is another material found to exhibit extra reflections attributable to spin exchange terms.[132]

Spin-polarization in photoelectrons was first successfully observed in 1969, when Busch *et al.*[133] detected polarization effects in electrons emitted from thin films of ferromagnetic Gd. An important discovery was made in 1975 when Pierce *et al.*[134] used polarized light to stimulate the emission of spin-polarized electrons from a GaAs surface, covered with CsO_2 in order to obtain negative electron affinity (NEA). This not only demonstrated that spin-polarized photoemission could be obtained from non-magnetic materials but also provides a relatively versatile means of providing an intense source of polarized electrons.

Spin-polarized low-energy electron diffraction (SPLEED) experiments may be performed by directing a beam of unpolarized electrons at a surface and detecting the polarization of the diffracted beam by means of a Mott detector or by using the method of double diffraction. Alternatively, a polarized primary beam may be used, and the resulting diffraction beam intensities measured using conventional means of detection.

The earliest conclusive demonstration of spin-polarization effects in LEED was published by O'Neill *et al.*[135] using a standard thermionic electron gun and a Mott detector. Mott detectors are rather massive objects that must necessarily have a fixed orientation with respect to the high-vacuum chamber in which the LEED experiment is to be performed. The observation of energy–intensity profiles must therefore be effected by moving both the gun and crystal in a carefully controlled manner so that the relevant diffraction beam from the crystal is directed towards the detector at all energies. An experimental apparatus suitable for this is illustrated in Figure 2.17.[136]

In the double-diffraction experiment as illustrated in Figure 2.18[130] a similar constraint is imposed. However, as an alternative to intensity–voltage curves, structural analysis may be based on information gained from rotation diagrams in which the primary energy and incident angle are fixed, and the polarizer crystal is rotated azimuthally, about its surface normal. Rotation diagrams using a primary beam near normal incidence tend to be rather insensitive to azimuthal angle, but when the incident beam angle is large, as in

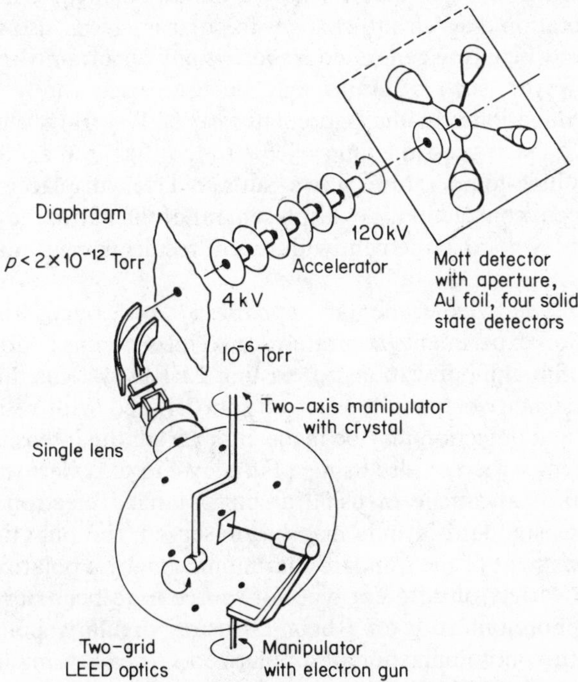

Figure 2.17. Polarized LEED experiment, using an unpolarized electron source, and a Mott detector to measure the polarization of the diffracted beams. (Reproduced by permission of The American Physical Society from Muller *et al.*[136])

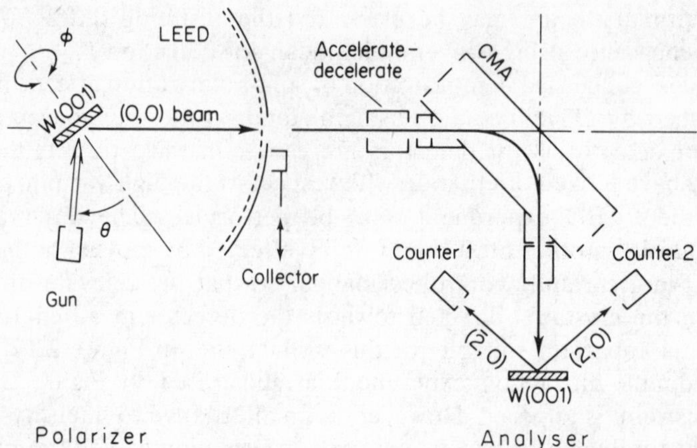

Figure 2.18. Schematic of the experimental set-up of a double-scattering LEED experiment. (Reproduced by permission of the American Physical Society from Kirschner and Feder[130])

the double-diffraction experiment, sensitivity may be high. Unfortunately, sensitivity of rotation diagrams to changes in polar angles is also very high[137] and so errors may be easily generated if there is any uncertainty in this angle. With care, however, such diagrams may be used successfully for surface structure determinations. In the particular case of W(001) surfaces, Feder and Kirschner[138] have used this approach to identify the magnitude of the clean W(001) high-temperature phase surface layer displacement. Their value of 7 per cent contraction (i.e. displacement of the surface layer towards the bulk) is in very good agreement with recent results using standard means of analysis.[32]

In general, the experimental apparatus necessary to perform double-diffraction experiments accurately, or to incorporate a Mott detector, will require complex modifications to existing LEED systems. Most LEED systems are designed so that the electron gun is fixed with respect to the display screen, and detection may be made either from the intensities of spots on the fluorescent screen or else using a Faraday cup to collect the electrons directly. A major advantage of using a spin-polarized electron source for SPLEED is that standard layouts may be preserved, the only modification being the replacement of the standard thermionic gun by a polarized electron source. A wide variety of different types of source have been developed, for example: the photoionization of Rb or Cs using circularly polarized light (Fano effect); the photoionization of a polarized Li beam using unpolarized light; optically pumped He discharge; field emission from W tips coated with magnetized EuS; and photoemission from ferromagnetic EuO. We noted above that photoemission from GaAs also provides a good source, and is now used in a number of laboratories. A NEA source has been described in detail by Pierce *et al.*[139] and in the same paper they provide a brief comparison with

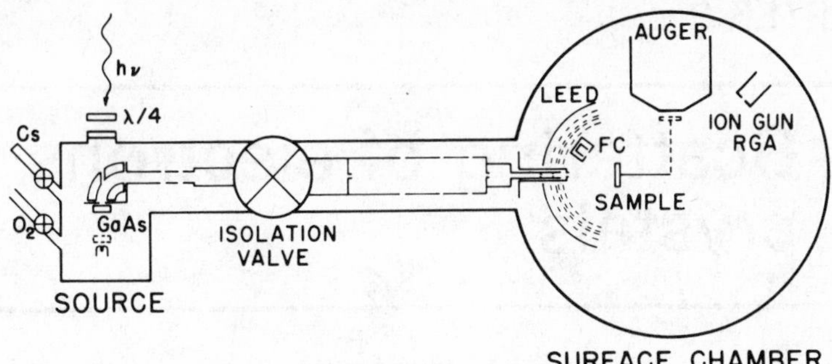

Figure 2.19. Schematic diagram of a polarized LEED experiment, utilizing a GaAs crystal as a source of polarized electrons. (Reproduced by permission of the American Institute of Physics from Pierce *et al.*[139])

most other polarized electron sources. Positive electron affinity (PEA) GaAs surfaces may also be used with comparable success, but recent improvements appear to favour NEA GaAs as the optimal source currently available. A schematic diagram of a UHV chamber equipped with standard LEED optics and a GaAs polarized electron source is given in Figure 2.19.

Note in Proof

An apparatus based on the MEMLEED principles described in section 2.3.4 has now been completed (Dupuy, J. C., Foster, M. S., Campuzano, J. C., and Willis, R. F. submitted to *J. Royal Microscopy Society*, 1985). Using a tungsten/zirconium thermal field emitter gun operated at 2000 °C in the range 7–12 kV, a spot diameter of only a few μm can be produced (see page 84), and LEED can be carried out without any shifting of the diffraction beam positions over the range 0–160 eV. The coherence length of the beam is greater than 1000 Å, in contrast with the typical value of about 100 Å for conventional LEED systems. The study of surface steps, islands and phase transitions may thus be considerably enhanced; it may also be possible to use Fourier transform techniques successfully in the interpretation of LEED data.

CHAPTER 3

Scattering of electrons by crystals

3.1 Introduction

An atom can be thought of as a tightly packed nucleus of neutrons and positively charged protons surrounded by a diffuse cloud of orbiting electons, each of which carries a negative charge. In a stable, isolated atom the number of protons and electrons must be equal, in order to preserve total charge neutrality. The nucleus has a radius of roughly 10^{-15} m, whereas the electron cloud extends perceptibly to a radius of 10^{-10} m or more. Consequently, if an electron is directed towards the atom, the force exerted on the electron will vary with distance from the nucleus. The predominant interaction is the direct electrostatic or Coulomb interaction. Like charges repel one another, so the incident electron will experience repulsive forces from the constituent electrons of the electron cloud, but will experience an attractive force towards the nucleus.

When far from the atom, these Coulomb forces will essentially cancel each other, and the major remaining interactions are those that result from the fact that the electron cloud has finite dimensions, and may oscillate with time; these are second-order effects, and cannot be discussed in detail here. As the electron passes through the electron cloud, the screening effect of the orbiting electrons is reduced, and the incident electron experiences more strongly the attractive force of the nucleus. This force is most conveniently described in terms of a potential. Because the forces are quantum-mechanical in nature, simple analogies with Newtonian physics are inadequate to describe the process, and a particularly important aspect is the fact that an electron may be partially reflected from a steep change in potential, even when the potential is attractive. More precisely, we may say that there is a finite probability that an electron may be reflected or scattered in any particular direction. This scattering arises from the interaction of the electron with the potential of the atom rather than from any simple type of hard-sphere scattering due to collisions with massive particles.

The scattering of high-energy electrons can be described by a fairly simple approximation (Born approximation) because the incident electron is relatively weakly perturbed by the orbital electrons and the predominant interaction is with the symmetric nuclear Coulomb potential. By comparison, the interaction of low-energy electrons with matter is very complex, because significant interactions occur with the many electrons that comprise the orbital cloud. In this chapter, we outline the methods commonly used in LEED analysis to describe the scattering of electrons from atoms within a crystal, and discuss some of the main approximations adopted.

3.2 Ion-core Potentials

The potential of an isolated atom cannot be solved exactly because each of the bound electrons interacts with all the others, and consequently one is faced with a many-body problem. It is usual to use a 'single-electron' approximation such as that devised by Hartree and subsequently improved by Fock. The basis of the method is to guess the wavefunctions of all the electrons in the system, and then consider the potential that an individual electron would experience as a result of all the others plus the nucleus. Applying the Schrödinger equation with this potential, a new wave function for the electron can be calculated. This is repeated for each successive electron until a complete new set of wave functions is obtained. The whole cycle is then repeated as often as is necessary to obtain a self-consistent result, i.e. one in which a further iteration would produce no change in the direct Coulomb potential between electrons, ignoring the fact that there are additional correlation effects between electrons that will tend to keep them apart. In fairness to Hartree, this omission was certainly not an oversight, but rather a practical limitation set by the fact that computers were not readily available at the time, and it was his father who was performing all the calculations by hand. Another important contribution to the electron–electron interaction is the indistinguishability of the electrons. The Pauli principle states that no more than one electron may occupy a particular state, and it is a corollary that for electrons, which have half-integral spin, the wave function must change sign whenever the positions of two electrons are interchanged. In practice, this means that electrons of like spin will tend to repel each other (in addition to the Coulomb interaction). This repulsive energy is included as an 'exchange energy' term in the Hartree–Fock method.

Atomic potentials for all the elements have now been calculated using variants of the Hartree–Fock method. Compilations of results by Herman and Skillman,[140] Clementi[141] and Clementi and Roetti[142] provide sufficiently detailed information for most LEED calculations. We shall now consider how the scattering potential of atoms within a crystal may be derived from individual atomic data. For heavy atoms, the potential may be sufficiently strong that additional relativistic terms must also be included. This effect will be dealt with separately in section 3.2.4.

3.2.1 The muffin-tin approximation

In the atomic state, the electrons are 'bound' to the nucleus, the binding energy being given by the potential as described above. When atoms are brought together, electrons from neighbouring atoms may interact, and the most weakly bound electrons may be able to detach themselves from their parent atom. In the case of a solid crystal, these 'outermost' electrons may become totally free to travel within the confines of the crystal. The more tightly bound electrons will remain localized in the vicinity of their associated nuclei, and the crystal may be visualized as an array of *ion cores* embedded in a 'sea' or 'gas' of non-localised electrons. This situation is basic to solid-state physics and chemistry and well-known methods exist to calculate the resulting energy levels of the electrons in the crystal. However, the electron energies we are concerned with for LEED are well above those dealt with normally in solid-state theory, and so we cannot draw our analysis exclusively from such theories.

We have already noted that the crystal comprises ion cores embedded in a 'sea' of electrons. To produce a mathematical theory we generally assume that the ion-core regions are perfectly spherical, and that the potential between these has a constant value. This is known, rather quaintly, as the 'muffin-tin' approximation. The boundary of each ion-core region is defined by its muffin-tin sphere and if the constant potential between the ion cores is used as the basic level from which to describe the rest of the features of the crystal potential, then it is called the 'muffin-tin zero'. This approach generally gives the most satisfactory results when the muffin-tin spheres are defined to be as large as possible without actually overlapping. In the limiting case of just touching at their boundaries, they may be alternatively described as the inscribed spheres of the Wigner–Seitz cells.

3.2.2 Calculation of ion-core potentials

The energy levels of the tightly bound electrons that comprise the ion core will be not unlike those of an equivalent isolated atom. Consequently the atomic wave functions provide a good starting-point for a calculation of the ion-core potential within a crystal. To begin with, consider the Coulomb interaction between electrons and nuclei.

The atomic charge density is given by the sum of the squares of the wave functions from all the occupied levels

$$\rho_0(r) = \sum_{\text{occupied}} |\psi_l|^2 \tag{3.1}$$

The equivalent potential $U_C(r)$ in three dimensions is found by solving Poisson's equation

$$\nabla^2 U_C(r) = -8\pi\rho_0(r) \tag{3.2}$$

The Coulomb potential at a distance r from an ion core is then given by the sum of the nuclear potential (total charge Z) and the negative electron contribution

$$V_C(r) = \frac{2Z}{r} - U_C(r) \tag{3.3}$$

The total Coulomb potential at an ion core is given by the sum of contributions from all its neighbours. This summation involves a mathematical process that crops up again and again in LEED theory; that is, the expansion of a spherically symmetric function about one origin in terms of spherically symmetric functions about some other origin. In this case it is necessary to express the Coulomb potential $V_C(r_j)$ centred on ion core j in terms of spherical harmonics $Y_{lm}(r_i)$ centred on core i which we shall consider to be the origin of our coordinate system.

A significant simplifying feature of the muffin-tin approximation is that the result of adding contributions from neighbours about any particular ion core must be a spherically symmetric function. Non-spherical contributions are assumed to cancel out and can therefore be ignored. The mathematical treatment of this process has been described in detail by Loucks.[143]

The spherically symmetric Coulomb potential $V_C(r)$ about a particular ion core within the muffin-tin radius is given as the sum of the atomic potential $V_0\,(r)$ plus the contributions from each of the neighbours $V_i(r_i)$. Loucks introduces a convenient notation $V_0(a_i\,|\,r)$ to describe the spherical contribution in terms of radius r about the ion core of interest due to another ion core separated by a distance a_i from it. Thus

$$V_C(r) = V_0(r) + \sum_{i}^{\text{neighbours}} V_0(a_i|r) \tag{3.4}$$

The imposition of spherical symmetry means that the separation distance a_i is sufficient to describe the position of a neighbour. Because several neighbours may be separated from the core of interest by the same distance, the disposition of these neighbours is often described in terms of 'shells' containing the relevant number of contributions.

Having determined the radial distribution of the ion-core potential within the muffin-tin sphere, it is necessary to consider the region of constant potential between the spheres. It is not sufficient to give this reason the same potential as that reached at the boundary of the muffin-tin spheres, because this would imply a net loss of charge from the crystal. A commonly used scheme is that of Mattheiss,[144] in which total charge is conserved by raising the interstitial potential to an appropriate level—this introduces a small step in the potential at the muffin-tin boundary, but this is less serious than losing charge.

To illustrate the procedure, consider the two-dimensional section through a plane of ion cores given in Figure 3.1(a). Ideally, the potential along section OO′ would continue to rise in the interstitial region, if we continued to treat this region in a similar manner to that within the sphere (Figure 3.1(b)). However, it is necessary to select a single potential representative of the mean potential within this region. Figure 3.1(c) shows an exaggerated view of the potential near the sphere boundary. One way of determining the mean potential is to sample the interstitial region and take an average. However, it is simpler, and generally quite adequate, to consider the spherically symmetric volume centred on the ion core that has the same total volume as the Wigner–Seitz cell (the total volume of the crystal divided by the number of ion cores, i.e. the mean ion-core volume assuming no free space within the crystal). If v is the mean ion-core volume, then the Wigner–Seitz radius r_0 is given by

$$v = \tfrac{4}{3}\pi r_0^3 \tag{3.5}$$

So the average potential in the interstitial region is given by the integral of the total ion-core potential $V_T(r)$ between the muffin-tin radius r_{MT} and r_0, which is easily shown to be

$$V_{avg} = 3\int_{r_{MT}}^{r_0} \frac{V_T(r)r^2}{r_0^3 - r_{MT}^3}\,dr \tag{3.6}$$

The magnitude of this step will be minimized by choosing r_{MT} to be as large as possible, and is generally not more than a few hundredths of an electron-volt. A potential calculated for Mo, which may be expected to have a relatively large overlap, is illustrated in Figure 3.2. At the centre of the ion core, the potential goes to $-Z$ volts (where Z is the atomic number of element), so even on the scale of this figure, the step is seen to be very small.

The spherical symmetry of the core is imposed even if the isolated atom did not have a closed shell of electrons to give a symmetric starting condition. For example, Ni has effectively 0.94 electron in the d-band; this can be introduced into the calculation by assuming that all 10 orbitals are equally occupied by 0.94 electron.

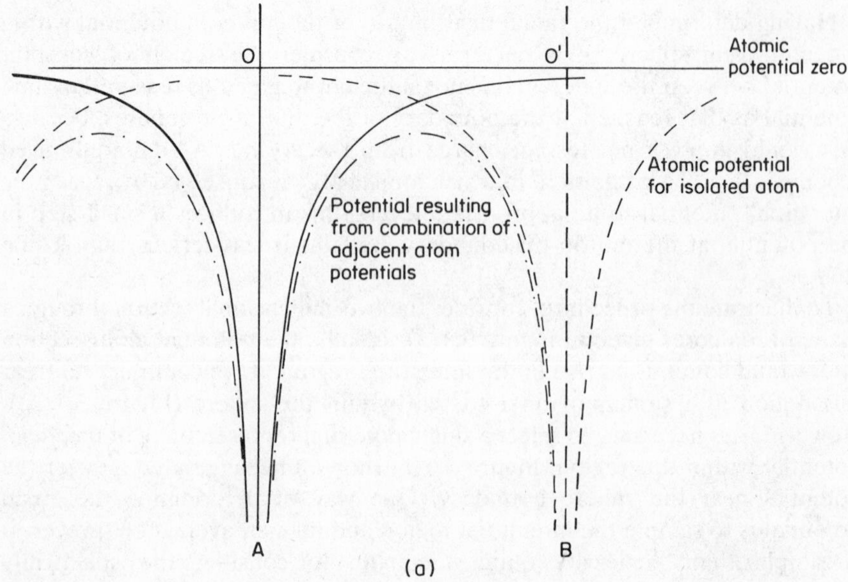

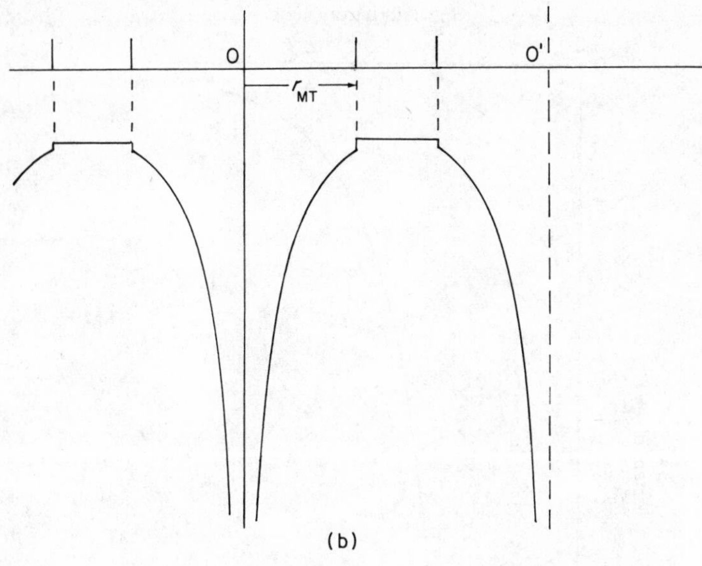

(b)

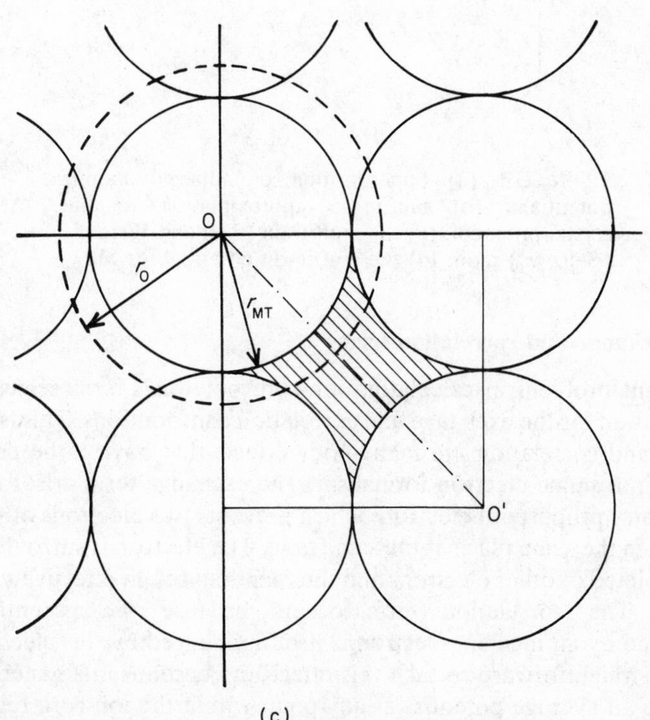

(c)

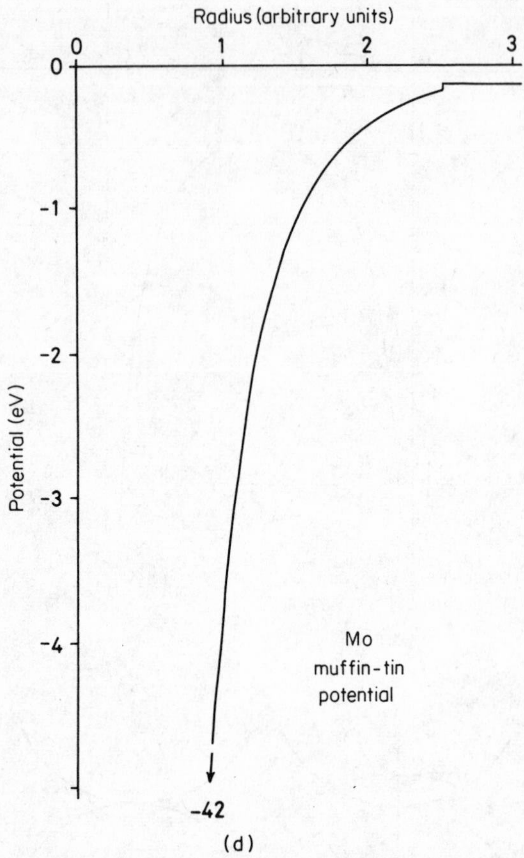

Figure 3.1. (a) Superposition of adjacent atomic potentials; (b) muffin-tin approximation to the crystal potential; (c) the muffin-tin (r_{MT}) and Wigner–Seitz (r_0) radii; (d) the muffin-tin potential for Mo.

3.2.3 Exchange and correlation terms

A persistent problem in calculating ion-core potentials is to select a good approximation for the exchange and correlation contributions. This is because exchange and correlation are many-body effects that have to be dealt with from within a single-electron formalism. The exchange term arises from the antisymmetry property of electrons which prevents two electrons of the same spin being in the same place at the same time. The electron is surrounded by a region depleted of other electrons and thus experiences an effectively lowered potential. The correlation term occurs because the potential field encountered by an incident electron is itself influenced by this electron. It is relatively straightforward to take this effect into account in a general sense, calculating an average potential at any point within the ion-core region, but such a procedure does not take into account the local perturbations in

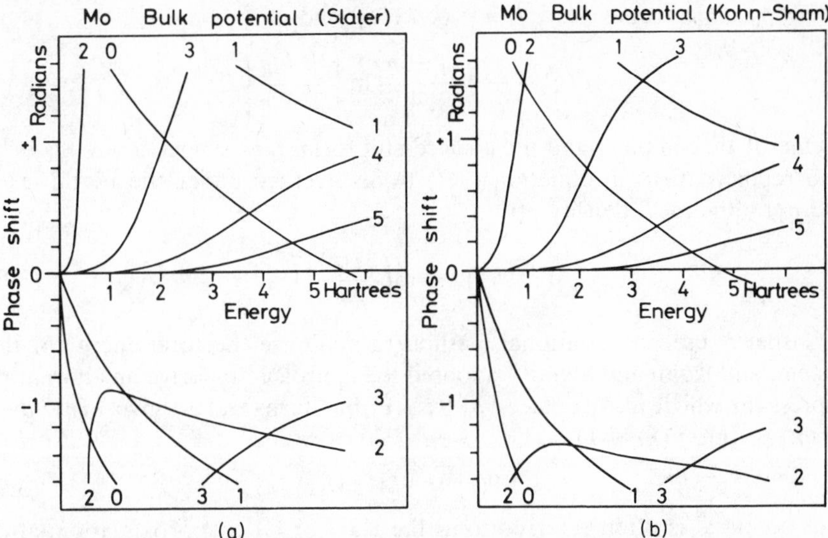

Figure 3.2. Phase shifts for Mo, calculated using (a) the Slater ($\alpha = 1$) approximation; (b) the Kohn–Sham ($\alpha = 2/3$) approximation. (Reproduced by permission of North Holland Physics Publishing from Clarke.[15])

electron density that surround the individual electron. In brief, the Coulomb repulsion between the electrons will push ion-core electrons away from the immediate vicinity of the incoming electron. This creates a small region around the electron that has an effectively lower electron density than it would otherwise have if the electron density were completely homogeneous. This region will therefore have a slightly less negative net charge, and so is often called the 'correlation hole'. As with the exchange effect, the potential experienced by the electron is slightly less repulsive than would be calculated if this behaviour were not included.

A formulation of the exchange contribution using the Hartree–Fock approximation has been given by Andersson and Pendry.[145] In this approximation, the exchange is described in a non-local form that must be solved self-consistently: a relatively long and complex procedure. 'Non-local' means that individual interactions between electrons are taken into account, rather than relying on smeared-out average values assumed to occur at any point or locality within the ion-core region ('local approximation').

Clearly, a local approximation, in which the potential can be described by a simple function of position $V(r)$ would be much simpler to handle. A group of relatively successful local approximations have been derived from Dirac's solution[146] of the Hartree–Fock equations for a uniform electron gas. The exchange potential is then

$$V_X(k) = -4F(\eta)\left(\frac{3\rho_0}{8\pi}\right)^{1/3} \tag{3.7}$$

where ρ_0 = electronic density, $\eta = (k/k_{\text{Fermi}})$ and

$$F(\eta) = \frac{1}{2} + \frac{1-\eta^2}{4} \ln \left| \frac{1+\eta}{1-\eta} \right|$$

One of the simplest and most successful forms was suggested by Slater[147] who replaced $F(\eta)$ in equation (3.7) by its average under the Fermi level and ρ_0 by the local density $\rho(\mathbf{r})$:

$$V_X^{\text{Slater}}(\mathbf{r}) = -3\left(\frac{3\rho(\mathbf{r})}{8\pi}\right)^{1/3} \tag{3.8}$$

Gaspar[148] used a variational method to minimize the total energy of the system, and Kohn and Sham[149] adopted the approach to derive an alternative expression which also replaces ρ_0 by $\rho(\mathbf{r})$ but turns out to give $F(\eta)$ when $k = k_{\text{Fermi}}$ (i.e. $F(\eta) = 1$):

$$V_X^{\text{KSG}}(\mathbf{r}) = -\tfrac{2}{3} V_X^{\text{Slater}}(\mathbf{r}) \tag{3.9}$$

This potential is often referred to as the KSG or GKS approximation, after the names of the major contributors to its development.

Given the spread presented by reasonable theoretical models, an obvious alternative is the 'α exchange' or $X\alpha$ method, in which the coefficient is replaced by a variable

$$V_X^{\alpha}(\mathbf{r}) = \alpha V_X^{\text{Slater}}(\mathbf{r}) \tag{3.10}$$

The Slater and KSG approximations can then be considered to be special cases of the $X\alpha$ potential, derived by imposing certain further assumptions.

There are many examples of the use of $X\alpha$ potentials in LEED calculations, and to a good first approximation they can be assumed to be adequate for most purposes. They are not, however, designed adequately to model regions in which the electron density varies rapidly, and such regions may be probed by the incident LEED beam, particularly as the incident energy is increased to several hundred electron-volts. Consequently, a number of further modifications and refinements have been made and tested with the intention of improving the accuracy of structural analysis using LEED. One such modification is that developed by Liberman.[150] He assumed that the exchange contribution should also depend on the momentum of the electron, k. Retaining the assumption of a free-electron gas, one may write

$$k = (E - V_C - V_X)^{1/2} \tag{3.11}$$

and this may be used to derive the exchange term

$$V_X^{\text{Lib}} = -8F(\eta)\left[\frac{3\rho}{8\pi}\right]^{1/3} \tag{3.12}$$

where $F(\eta)$ is obtainable from the self-consistent dolution of the expression

$$\eta = \frac{k}{k_{\text{Fermi}}} = E - V_C + 8F(\eta)\left[\frac{3\rho}{8\pi}\right]^{1/3} \tag{3.13}$$

Slater *et al.*[151] pointed out that this expression does not give the value $\eta = 1$ at the Fermi energy and suggested that a better expression would be

$$\eta = \frac{\left[E - V_C + 8F(\eta)\left(\frac{3\rho}{8\pi}\right)^{1/3}\right]^{1/2}}{\left[E_{Fermi} - V_C + 4\left(\frac{3\rho}{8\pi}\right)^{1/3}\right]^{1/2}} \tag{3.14}$$

Both the Liberman[150] and Slater *et al.*[151] expressions require the calculation of fresh exchange terms for each orbital, which adds slightly to their complexity when compared with $X\alpha$-type potentials.

A comparison of all these different local potential schemes has been made by Slater *et al.*[151] in the calculation of band structures for the Cu^- ion. The value of α used in the $X\alpha$ scheme was determined by finding that which minimized the total energy. In fact, Schwarz[152] has carried out this procedure for the whole range of elements up to $Z = 90$, and found that the optimum value for α lay in each case between 0.7 and 0.77 (not far from the KSG value of 0.66).

A comparison between the Hartree–Fock non-local and Slater free-electron approximations was made by Mattheiss in his study of the solid Ar band structure. He found that the free-electron approximation underestimates the exchange at large values of radial distance r for all occupied states. However, when the results of LEED calculations are compared with experimental results, it is clear that the free-electron model actually provides a better approximation than that of Hartree–Fock. With hindsight, this can be attributed to the neglect of correlation effects, since the Hartree–Fock approximation completely neglects electron correlation, and correlation is much more important for electrons with energies typical of LEED experiments, than for valence electrons. Curiously, all the approximation methods described above are designed to treat the exchange terms only, and the only reason that they provide satisfactory effects are partly included as a result of the approximations inherent in their starting equations. Neglect of any specific treatment of correlation effects reflects the fact that all these potentials were originally developed for band-structure calculations, which only apply at low energies (near the Fermi level).

Although it does not appear to have been used for the LEED problem, it is possible to include correlation effects explicitly to 'correct' the Hartree–Fock potential. A disincentive to the exploitation of increasingly complex potentials of this type has been the high degree of success associated with the use of the much simpler local potential schemes. Provided the electron density is a relatively slow-varying function of position these non-local schemes are reasonable; this applies in the case of metals, alloys and small-gap insulators in which the free-electron assumption is relatively good. However, near the surface of atoms, or in the overlap region of molecules, this basic assumption will not necessarily remain valid. As LEED methods are applied increasingly to more complex molecular adsorbates, a new look at local potential methods may become necessary.

Despite the considerable amount of work devoted to the quest for better band-structure potentials, it must be concluded that no single scheme has been devised that stands out as distinctly better than the others. The Hartree–Fock scheme has been frequently used as the standard with which to compare the accuracy of simpler, local potential schemes, but even then, the comparison of potentials seems to depend at least partially on the particular behaviour one is interested in; one potential may give better agreement overall for charge densities, another may agree more closely with the value of the total energy. In any case, the Hartree–Fock scheme, with its complete neglect of correlation effects, has been found to be quite unsuitable as a potential for use in LEED calculations. Therefore, closeness of agreement with Hartree–Fock does not necessarily constitute a good criterion for the solution of the ion-core potential in LEED. We do not have a good *a priori* criterion for such a potential, but are obliged to use various plausible potentials in full LEED calculations and see which give the best agreement with the experimental results.

Ultimately, the suitability of a potential can only be judged by its success in allowing close agreement between theoretical and experimental LEED beam intensities. In the case of the local density approximation, as used in the Slater and KSG schemes, quite satisfactory agreement between LEED theory and experiment can be obtained despite the fact that such schemes have little theoretical validity when applied to regions where the electron density is highly non-uniform, as found near the ion core. The self-consistent Hartree–Fock potential, although less satisfactory overall as a potential for LEED calculations, is nevertheless a superior scheme for describing these core electrons. Echenique[153] constructed a hybrid potential and amalgamated the Hartree–Fock scheme to describe the core electrons with a local density approximation to describe the valence and conduction electrons. Instead of the KSG potential, he used a more sophisticated scheme described by Hedin and Lundqvist[154] in which a local potential is defined in terms of the electron gas self-energy by means of a Wenzel–Kramers–Brillouin (WKB) approximation. In practice, the potential is treated almost as simply as in the KSG scheme (using $\alpha = \frac{2}{3}$), but with the inclusion of correction terms tabulated by Hedin and Lunqvist, giving a more precise description of the exchange and correlation term.

Echenique compared 10 differently constructed potentials for Na, based on the methods described above, and found that similarities between the individual ion-core scattering properties were reflected in similarities between LEED $I(E)$ curves obtained from those potentials. For example, the potential formed by an amalgam of Hartree–Fock and Hedin–Lundqvist gave similar ion-core scattering to that produced by Slater, whilst a potential using the Hedin–Lundqvist scheme throughout was quite similar to that obtained from the KSG scheme. Likewise, $I(E)$ curves from the former pair of potentials were similar to each other, as were $I(E)$ curves obtained from the latter pair, but curves from one pair differed in particular respects from those from the other pair.

In the particular case of Na, the combined Hartree–Fock/Hedin–Lundqvist (HF/HL) scheme was found to give the best agreement. However, the theory is only really justifiable for nearly free electron (NFE)-type materials and does not include any specific prescription for d-electrons, as occur in transition metals. Subsequent studies of Al, Ag, Cu and Ni by Echenique and Titterington[155] revealed that the quality of agreement of HL/HF is not particularly good for Ni or Cu, better agreement being obtained with the simple Slater theory. For Ag, there was little to choose between the schemes, some beams appearing to be better reproduced using one scheme, other beams better using another scheme. For Al, the weakest scatterer of the set, all the theories gave very similar results.

We may conclude that this last potential is probably the most suitable yet devised for LEED calculation, but even this, despite all its complexity, does not manage to produce agreements that are consistently better than those obtainable using simple Slater or KSG methods. For most purposes, these latter potentials may well be quite adequate.

3.2.4 Relativistic contributions

The Schrödinger equation is an adequate basis for the description of electron scattering only if relativistic effects can be ignored. At typical LEED energies the velocities of the electrons are far from the speed of light. Nevertheless, if the scattering nuclei are heavy (e.g. W or Au) then relativistic effects may be encountered as perturbative corrections to the non-relativistic solutions. In order to meet the mathematical requirements of relativistic theory, the governing equation needs to be symmetrical in time and space, but the Schrödinger equation is unsymmetrical in that it contains the second derivative with respect to space but only the first derivative with respect to time. Dirac[146] succeeded in reconciling these factors, but at the expense of hypothesizing the possible existence of negative-energy particles, or holes. Moreover, both electrons and these positively charged holes (known as positrons) have two states that can be distinguished by applying a magnetic field, and the property that determines their interaction is called the spin (**S**). Methods of producing and detecting electrons with spins preferentially aligned in one direction (i.e. polarized) have been discussed in section 2.8. Dirac's equation can be written in a form that shows explicitly its relationship to the Schrödinger equation:[156]

$$\left[\frac{p^2}{2m} - \frac{p^4}{8m^3c^2} + V - \frac{h^2}{4m^2c^2}\frac{\mathrm{d}V}{\mathrm{d}r}\frac{\partial}{\partial r} + \frac{1}{2m^2c^2}\frac{1}{r}\frac{\mathrm{d}V}{\mathrm{d}r}\mathbf{S}\cdot\mathbf{L}\right]\psi = E'\psi \quad (3.15)$$

The first and third terms are familiar from the non-relativistic Schrodinger equation. The second term may be regarded as a relativistic mass correction to the first, kinetic energy, term. This follows from the following relativistic equations:

$$E' = E - mc^2 = (c^2p^2 + m^2c^4)^{1/2} - mc^2 \simeq \frac{p^2}{2m} - \frac{p^4}{8m^2c^2} \qquad (3.16)$$

The last term on the left-hand side of the equation describes the interaction between the intrinsic spin angular momentum of the electron **S** and the magnetic field **L** set up by the orbiting electrons about the nucleus. The fourth term does not have a simple derivation but is sometimes referred to as the 'Darwin' term, and may be considered to be a relativistic correction to the potential energy term, V.

If the incident beam is unpolarized, so that there are as many electrons with spin up as down, then to a first approximation the net effect of the last term will be averaged out. The remaining correction terms, the second and fourth in the equation, are both negative and thus lead to a net shift in the energy of the solution E' with respect to the non-relativistic case. This can be demonstrated from LEED by comparing intensity–voltage curves calculated using relativistic potentials with those calculated using non-relativistic potentials. To a first approximation, the shapes of the curves are quite similar, but in the case of the relativistic curves, the principal features occur at somewhat lower energies (by as much as 4 eV).[32,157]

This is also manifest in the corresponding phase shifts: for a given incident beam energy, the relativistic phase shifts will appear to be displaced positively in energy compared with non-relativistic phase shifts.

Spin-polarization effects may be included, either to interpret the results of specific spin-polarization LEED experiments, or else in an attempt to improve the quality of non-polarized beam studies. From equation (3.12), it can be seen that the electron spin **S** appears in the scattering potential as a spin-orbit coupling term, **S** · **L**. An electron may approach the crystal with its spin up or down, and may either maintain the same spin orientation following the scattering process, or else be 'flipped-over' to the opposite orientation. This introduces four possible new values for the potential, corresponding to

$$\mathbf{S} \cdot \mathbf{L} = \pm\tfrac{1}{2}(l + \tfrac{1}{2} \pm \tfrac{1}{2}) \qquad (3.17)$$

A single phase shift δ_l, corresponding to each value of the orbital angular momentum, l, is insufficient to incorporate electron-spin effects. Instead, the total angular momentum $j = l \pm \frac{1}{2}$ must be used, resulting in a pair of phase shifts $\delta_l^{\pm}$.

In a non-polarized diffraction experiment, incident electrons with spin up will experience a different scattering potential from those with spin down, and in principle one may calculate the intensity–voltage curves for each spin orientation in turn, and then average the results. In practice, however, it is found that the resulting curves are almost identical with those obtained by making a single calculation assuming spin-averaged phase shifts[158]

$$\delta_l(s) = (\delta_l^{+} + \delta_l^{-})/2 \qquad (3.18)$$

For third-row transition metals and heavier, the inclusion of an averaged relativistic potential can provide a small improvement over a non-relativistic potential.

If specific polarization effects are being investigated, by using a polarized beam of electrons, or by detecting polarization in the scattered beams, the inclusion of the spin-orbit term in the potential is essential. This can be accomplished by modifying the appropriate t-matrices, as described in section 3.3.2.

3.3 Phase shifts

The dynamics of electrons are most conveniently described by their wave functions. A wave function ψ is a function throughout space that can be specified for every particle within a system. It is a complex number (in the mathematical sense), and is often written in the form

$$\psi = a\exp(-ib) \tag{3.19}$$

where a is the amplitude and b the phase of the wave. The intensity of a beam can be related to the probability of an electron being at a certain place, and this is given by the product $\psi^*\psi$ where ψ^* is the complex conjugate of ψ. This means that, if $\psi = a\exp(-ib)$ then $\psi^* = a\exp(ib)$ and $\psi^*\psi = a^2$. This corresponds directly with the classical notion that the intensity of a beam is equal to the square of the amplitude. If no energy is lost during the scattering process (elastic scattering) then the total intensity may not change, and the phenomenon may be described entirely in terms of changing phases of the component wave functions.

3.3.1 Relationship between phase shifts and scattering potentials

Wave functions are related to the scattering potentials by the Schrödinger equation. The principal benefit of the muffin-tin model is that the resulting value for the ion-core potential is spherically symmetric and can thus be expressed purely in terms of a radial function $V(r)$, where r is a measure of distance from the ion-core centre. This permits the wave function solution to be expressed as the product of three functions, each the solution to a separate equation. One solution is a function of azimuthal angle $\Phi_m(\phi)$ about a prespecified coordinate system, another is a function of polar angle with respect to that same coordinate system, $\Theta_{lm}(\theta)$ and the third is a radial function $R_l(r)$, which is dependent only on the distance from the origin of the coordinate system and is centred on the ion core. The three equations are as follows:

$$\frac{\mathrm{d}^2\Phi_m(\phi)}{\mathrm{d}\phi^2} + m^2\Phi_m(\phi) = 0 \tag{3.20}$$

$$-\frac{1}{\sin\theta}\frac{\mathrm{d}}{\mathrm{d}\theta}\left(\sin\theta\frac{\mathrm{d}\Theta_{lm}(\theta)}{\mathrm{d}\theta}\right) + \left[\frac{m^2}{\sin^2\theta} - l(l+1)\right]\Theta_{lm}(\theta) = 0 \tag{3.21}$$

$$\frac{1}{r^2}\frac{\mathrm{d}}{\mathrm{d}r}\left(r^2\frac{\mathrm{d}R_l(r)}{\mathrm{d}r}\right)+\left\{\frac{2\mu}{\hbar^2}[E-V(r)]-\frac{l(l+1)}{r}\right\}R_l(r)=0 \quad (3.22)$$

These three equations have well-documented mathematical forms, and their solutions can be found in many appropriate texts. The first two equations do not contain the variable $V(r)$ and so have unique solutions for a given combination of l and m. In these equations, l is the angular momentum of a particular electron trajectory or orbital and m is the component of that momentum with respect to the $\theta = 0$, $\phi = 0$ direction. The third equation contains $V(r)$ and so its solution will contain a function of the individual ion-core scattering potential.

The solution to equation (3.20) is given simply by the exponential

$$\Phi_m(\phi) = \exp(im\phi)$$

The solutions to equation (3.21) are more complex, and are known as associated Legendre functions, often written in the form $P_l(\cos\theta)$. The product of these two solutions is independent of the scattering potential, and provides an expression of the angular distribution of the scattered waveform. This product is also a function of l and m, and is known as a spherical harmonic $Y_{lm}(\Omega)$:

$$Y_{lm}(\Omega) = \Phi_m(\phi)\Theta_{lm}(\theta) = P_l(\cos\theta)\exp(im\phi) \quad (3.23)$$

where Ω may be used to describe any angle in three-dimensional space, in terms of θ and ϕ.

The radial solution $R_l(r)$ contains information about the ion-core potential $V(r)$, but we can see from equation (3.22) that a separate solution will exist for each possible value of l. We shall briefly consider the manner in which the solution is obtained, because a variety of equivalent solutions may be derived, each having certain advantages and disadvantages when applied to the LEED problem. To transform equation (3.22) into a mathematically soluble form, it is usual to change variables: choosing $\rho = kr$, where $k = \sqrt{2[E - V(r)]}$}, and defining the function $f_l(\rho) = R_l(r)/\rho$ one obtains the 'spherical Bessel differential equation'

$$\left\{\frac{\mathrm{d}^2}{\mathrm{d}\rho^2}+\frac{2}{\rho}\frac{\mathrm{d}}{\mathrm{d}\rho}+\left[1-\frac{l(l+1)}{\rho^2}\right]\right\}f_l(\rho)=0 \quad (3.24)$$

This equation has a number of possible particular solutions. It is a second-order differential equation, so there will be two separate terms with their associated coefficients. For the region outside the muffin-tin sphere, the solutions frequently used in LEED are the spherical Hankel functions of the first and second kind $h_l^{(1)}(\rho)$ and $h_l^{(2)}(\rho)$ respectively, from which we obtain

$$f_l(\rho = \rho\,[c_l^{(1)}\,h_l^{(1)}\,(\rho) + c_l^{(2)}\,h_l^{(2)}\,(\rho)] \quad (3.25)$$

where $c_l^{(1)}$ and $c_l^{(2)}$ are constants. The potential in this interstitial region is constant, and can be defined as zero, so that $\rho = r\sqrt{(2E)}$. In their asymptotic limits (as r becomes very large) the Hankel functions converge towards simple exponential forms

$$h_l^{(2)}(r) \xrightarrow[r\to\infty]{} \frac{i^{l+1}\exp(-ikr)}{kr} \tag{3.26}$$

and

$$h_l^{(1)}(r) \xrightarrow[r\to\infty]{} \frac{i^{l+1}\exp(+ikr)}{kr} \tag{3.27}$$

Far from the ion core, they may therefore be considered as outgoing and incoming spherical waves.

In an elastic scattering process, no energy is lost and thus only the phase but not the amplitudes of these solutions may be modified. Consequently, the outgoing wave $h_l^{(1)}(r)$ can be assigned the same amplitude as the incoming wave, but with its phase altered by the phase shift δ_l. Thus the wave functions outside the muffin-tin sphere can be written as

$$f_l(\rho) = \beta_l[\exp(2i\delta_l)h_l^{(1)} + h_l^{(1)})] \tag{3.28}$$

If the ion-core potential were also zero, so that no scattering occurs, this solution should also be valid down to $r = 0$. However, both $h_l^{(1)}$ and $h_l^{(2)}$ have magnitudes proportional to $1/r$, resulting in a singularity in the wave function at the origin. This is not permissible for a physical description of an electron, so the phase shift δ_l must equal zero when there is no scattering so that the two singularities cancel each other out.

When scattering occurs, one may separate out terms from equations that describe an unscattered wave—namely

$$u_l(\rho) = \beta_l[h_l^{(1)} + h_l^{(2)}] \tag{3.29}$$

leaving the scattered wave components

$$s_l(\rho) = \beta_l[\exp(2i\delta_l) - 1]h_l^{(1)} \tag{3.30}$$

such that

$$f_l(\rho) = u_l(\rho) + s_l(\rho) \tag{3.31}$$

Equation (3.30) is known as an 'S-matrix', and it shows distinctly the relationship between phase shifts δ_l and the scattering amplitude $s(\rho)$. It can be called a matrix because it is a function of two variables, ρ and l.

For any given ion-core potential and incident electron energy, the scattering process may be described fully by a set of phase shifts, one for each possible value of the angular momentum quantum number, l. Fortunately, the number of values of l required to characterize the scattering process accurately is relatively few at typical LEED energies. This is clearly demonstrated in Figure 3.2 in which the values of phase shifts for successive values of l are given for Mo, which is a moderately strong scatterer. In practice, up to 10 phase shifts are sufficient to describe accurately the scattering from Mo and W for incident energies up to 250 eV, but only 6 or 7 are necessary for a 'weak' scatterer such as Al over a similar energy range. The physical reason for this is quite simple.[105] The effective potential in the radial equation (3.20) is $V_{\text{eff}}(r) = V(r) + l(l+1)/r^2$. Consequently as l is increased so the centrifugal barrier term $l(l+1)/r^2$ becomes increasingly important. The kinetic energy required for an electron to overcome this repulsive force increases with l, and thus the electron is more strongly excluded from the main scattering centre of the potential. Roughly speaking, if the potential extends over a range a and the electron has a wave number k, the only important partial waves will be those for which $l < ka$.

3.3.2 t-Matrices and the scattering cross-section

A graphical way of illustrating the scattering effect of a potential is to consider the asymptotic form of a scattering wave function including the angular solutions in addition to the radial solution. The governing equation is

$$\psi(k,\mathbf{r}) \xrightarrow[r\to\infty]{} A(k)[\exp(i\mathbf{k}\cdot\mathbf{r}) + \frac{f(k,\theta,\phi)}{2\pi r}\exp(i\mathbf{k}\cdot\mathbf{r})] \qquad (3.32)$$

The potential is spherical, so the scattering amplitude is independent of ϕ and thus $f(k,\theta,\phi)$ reduces to the product $R(k,r)\Theta(\theta)$, which can be written

$$f(k,\theta) = \frac{\pi}{ik}\sum_l (2l+1)\{\exp[2i\delta_l(k)] - 1\}\ \rho_l(\cos\theta)$$

$$= \frac{2\pi}{k}\sum_l (2l+1)\sin\delta_l(k)\exp[i\delta_l(k)]\ \rho_l(\cos\theta) \qquad (3.33)$$

Far from the ion core, the scattered wave can be envisaged as a spherical wave with amplitude modulated as a function of θ and determined by the energy $E(=\hbar^2k^2/2m)$.

This scattering amplitude is sometimes referred to as the 't-matrix' of the ion core. It is a function of l and m, and relates the scattering of a wave $\psi_{l,m}$ into a wave $\psi_{l',m'}$. It is therefore written as $t_{lml'm'}$. In practice, the assumption that the potential is spherical prevents changes in angular momentum (hence the fact that the scattering can be described simply in terms of phase shifts δ_l), and this in turn means that the resulting t-matrices are diagonal (i.e. all off-diagonal matrix elements are zero). t-Matrices can be given graphical form by plotting them in Cartesian coordinates f (or t) vs. θ for any particular energy E, but are perhaps more easily visualized in polar coordinates. In Figure 3.3 t-matrices for Mo are illustrated in polar coordinates for a number of different incident beam energies. These energies are expressed with respect to the muffin-tin zero of the crystal, and were calculated assuming Slater's non-localized approximation to the exchange and correlation term. Only 180° of the complete 360° range of θ is given, because of the symmetry of the function about the incident beam axis. The most interesting features are the comparative strength of the back-reflected portion at low energies and the increasing angular complexity, in terms of lobes, that occur as the energy is increased.

The total cross-section can be shown to be

$$\sigma_T(k) = \frac{4\pi}{k^2} \sum_l (2l + 1) \sin^2\delta_l(k) = \sum_l \sigma_l(k) \tag{3.34}$$

(see question 12) where each partial wave cross-section $\sigma_l(k)$ is given by

$$\sigma_l(k) = \frac{4\pi}{k^2} (2l + 1) \sin^2\delta_l(k) \tag{3.35}$$

The maximum scattering contribution from each partial wave occurs whenever $\delta_l(k) = (n + \frac{1}{2})\pi$ with $n = 0, \pm 1, \pm 2, \ldots$ (all integers) and decreases inversely with energy ($1/k^2$). On the other hand, there is no scattering from that particular partial wave whenever $\delta_l(k) = \mathrm{n}\,(n = 0, \pm 1, \pm 2, \ldots$ all integers).

In the Mo example given in Figure 3.3, the $l = 0$ partial wave (S-wave) passes through π at about 5.5 Hartress (150 eV) at which energy a significant number of other partial waves have a moderately large value. In special cases such as in low-energy collision of electrons with rare-gas atoms, the potential may be strong enough to make the $l = 0$ partial wave pass through at an energy sufficiently low that no other phase shifts contribute significantly to the scattering. Consequently, essentially no scattering occurs at that energy; this phenomenon is known as the Ramsauer–Townsend effect. Although this occurs outside the field of typical LEED studies it is an interesting demonstration of the relevance of the partial-wave description of scattering.

Multiple-scattering theories (see Chapter 4) can be readily modified to include spin-polarization effects by replacing the t-matrix $t_{lml'm'}$ by a spin-dependent t-matrix $t_{lmsl'm's'}$ (see section 4.). The phase shifts are related explicitly to the total angular momentum, j, rather than s, and can be used to define the following t-matrix:

$$t_{lmjl'm'j'} = \delta_{ll'}\delta_{jj'}\delta_{mm'}\, i \sin(\delta_l^{\pm})\exp(i\delta_l^{\pm}) \tag{3.36}$$

The spin-dependent t-matrix can then be derived from this, via the following transformation:[159]

$$t_{lmsl'm's'} = \sum_{LMJL'M'J'} U_{lmsLmj}\, t_{LMJL'M'J'}\, U_{L'M'J'L'm's'} \tag{3.37}$$

where $U_{lms,\,LMJ} = 0$, unless $l = L$, $M = m + s$, $J = l \pm \frac{1}{2}$. Otherwise,

$$U_{l,\,(M - \frac{1}{2}),\,\frac{1}{2};\,l,\,M,\,(l + \frac{1}{2})} = +\left(\frac{l + M + \frac{1}{2}}{2l + 1}\right)^{1/2}$$

$$U_{l,\,(M + \frac{1}{2}),\,-\frac{1}{2};\,l,\,M,\,(l + \frac{1}{2})} = +\left(\frac{l - M + \frac{1}{2}}{2l + 1}\right)^{1/2}$$

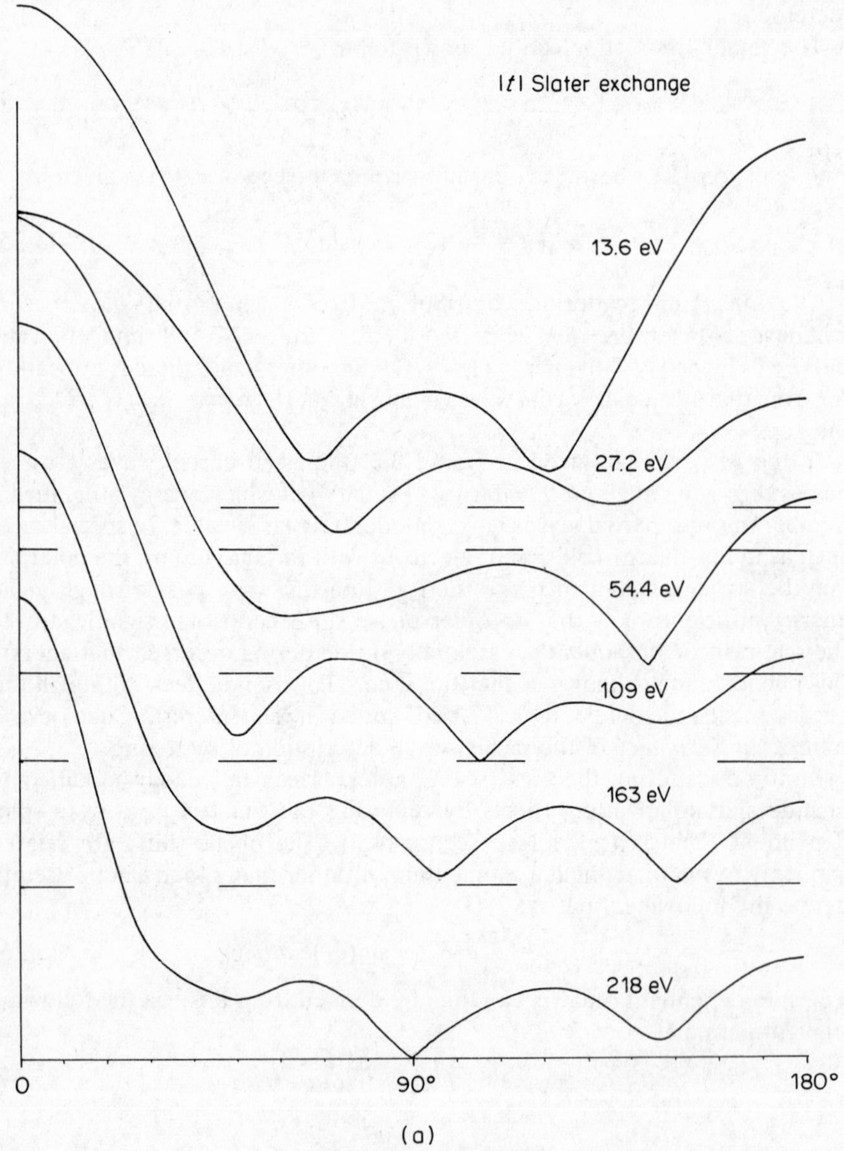

Figure 3.3. *t*-matrices for Mo: (a) using Cartesian coordinates; (b) using polar coordinates

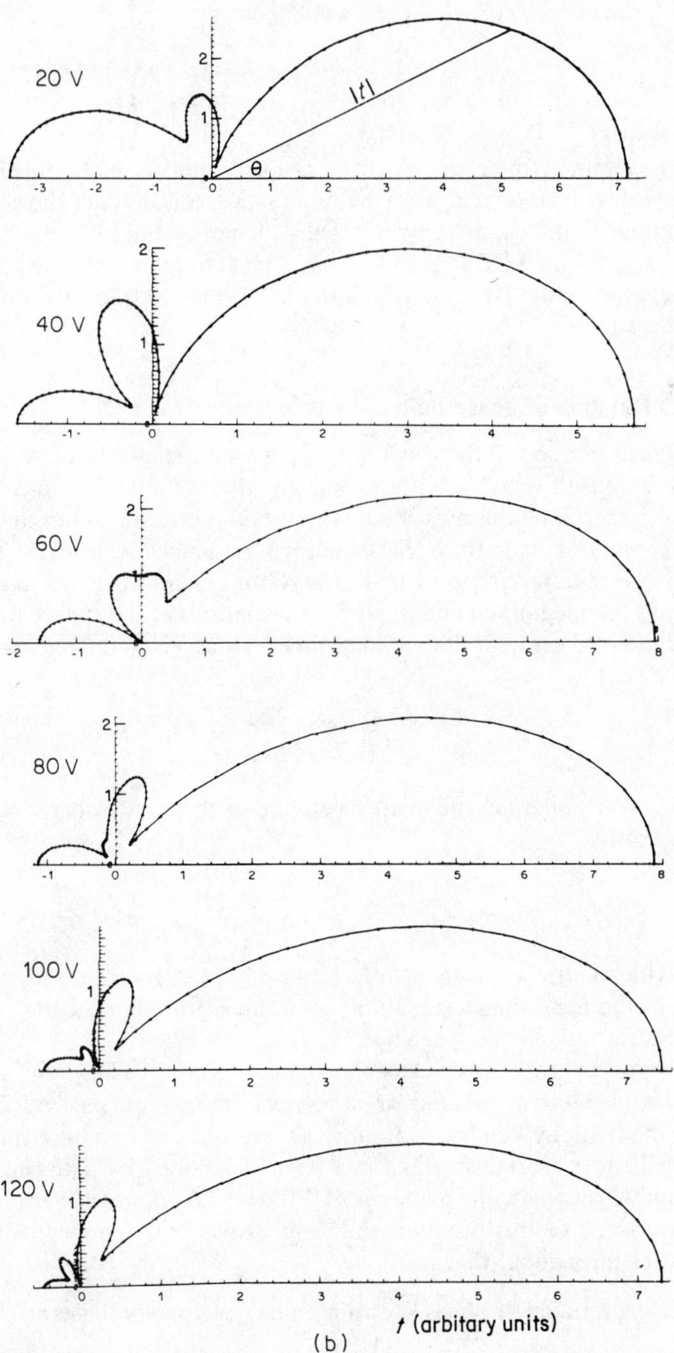

(b)

$$U_{l,\,(M-\frac{1}{2}),\,\frac{1}{2};\,l,\,M,\,(l-\frac{1}{2})} = +\left(\frac{l-M+\frac{1}{2}}{2l+1}\right)^{1/2}$$

$$U_{l,\,(M+\frac{1}{2}),\,-\frac{1}{2};\,l,\,M,\,(l-\frac{1}{2})} = -\left(\frac{l+M+\frac{1}{2}}{2l+1}\right)^{1/2}$$

The extra spin terms s and s' in the t-matrix introduce the possibility of spin-flip during the scattering process. In practice, this doubles the size of the corresponding matrices, thereby increasing the computational time necessary to make relativistic LEED calculations. Apart from this, the basis of the computational method is not substantially altered by the introduction of relativistic effects.

3.3.3 Calculation of phase shifts

An analytical solution to the wave function within the muffin-tin sphere does not exist, because $V(r)$ is itself not a simple analytical function. Consequently, the Schrödinger equation must be integrated numerically within the muffin-tin sphere, and at its surface, the solutions are matched in value with the outgoing waves appropriate to that energy. In general, it is not possible to match both the magnitude and slope of the solutions at the sphere boundary, and the ratio of these quantities at the sphere radius S is called the logarithmic derivative (L_1)

$$L_1 = \frac{f_l'(S)}{f_l(S)} \tag{3.38}$$

Substitution of equation and rearrangement of terms provides a value for the phase shifts

$$\delta_l = \frac{1}{2i}\ln\left[\frac{L_1 h_l^{(2)} - h_l'^{\,(2)}}{h_l'^{\,(1)} - L_1 h_l^{(1)}}\right] \tag{3.39}$$

This provides a direct means of calculating phase shifts from the numerical solution of the Schrödinger equation, once the scattering potential $V(r)$ and the muffin-tin radius S are specified.

A computer program to calculate phase shifts, using the standard Runge–Kutta–Merson method of numerical integration, is described and listed in the book by Pendry.[105] A more general suite of computer programs, based on Pendry's original, has since been developed by Titterington and others, and is known by the name of MUFPOT.[160] Although never published, various versions of this program suite are in use in a number of research laboratories throughout the world.

3.3.4 Interpretation of phase shift diagrams (Levinson's theorem)

The solution to the spherical Bessel differential equation (3.24) in terms of spherical Hankel functions as given in equation (3.25) is only one of a number of possible solutions. Another can be written[161]

$$f_l(\rho) = \rho[D_l^{(1)}(k)j_l(\rho) + D_l^{(2)}(k)n_l(\rho)] \tag{3.40}$$

where $j_l(\rho)$ is the spherical Bessel function and $n_l(\rho)$ is the spherical Neumann function. In the asymptotic limit, these take the form

$$j_l(\rho) \xrightarrow[\rho\to\infty]{} \frac{1}{\rho}\sin\left(\rho - \frac{l\pi}{2}\right) \tag{3.41}$$

$$n_l(\rho) \xrightarrow[\rho\to\infty]{} -\frac{1}{\rho}\cos\left(\rho - \frac{l\pi}{2}\right) \tag{3.42}$$

As we noted earlier, these solutions may differ in phase through some angle $\delta_l(k)$, but not in amplitude, therefore, in this asymptotic limit we may express (3.40) as [161]

$$f_l(\rho) \xrightarrow[\rho\to\infty]{} A_l(k)\sin\left[\rho - \frac{l\pi}{2} + \delta_l(k)\right] \tag{3.43}$$

where

$$\tan\delta_l(k) = -D_l^{(2)}(k)/D_l^{(1)}(k) \tag{3.44}$$

This enables us to give significance to the sign of $\delta_l(k)$. The nodes (zero-value solutions) of equation (3.43) occur wherever

$$\rho = n\pi + (l\pi/2) - \delta_l(k) \tag{3.45}$$

whereas the nodes of the free solution (zero-scattering potential) are simply

$$\rho = n\pi + (l\pi/2) \tag{3.46}$$

But

$$\rho = kr \tag{3.47}$$

so the effect of an attractive potential (negative at $r = 0$) is to pull-in the resulting solution, corresponding to a negative shift in the nodes with respect to r. This requires the phase shifts δ_l to be positive in value (see equation 3.45). Conversely, a repulsive potential (positive at $r = 0$) requires $\delta_l<0$. The resulting solutions are sketched in Figure 3.4.

It is very interesting to apply this method to an easily characterizable potential such as an attractive square well. If $V(r) = -V$ for $r < a$, then the radial equation can be written, for any given value of k,

$$\left[\frac{d^2}{dr^2} + \kappa^2 - l(l+1)/r^2\right]u_l(r) = 0 \qquad (r < a) \tag{3.48}$$

where $\kappa = \sqrt{(k^2 + V)}$

The regular solution of equation (3.48) (valid for the region $r \to 0$) is the spherical Bessel function

$$u_l(r) = c_l r j_l(\kappa r) \qquad (r < a) \tag{3.49}$$

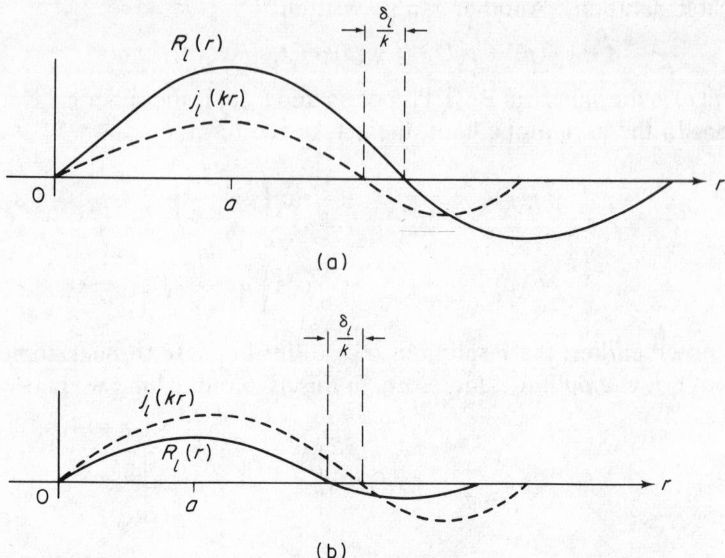

Figure 3.4. Schematic representation of the effect on the free radial wave $j_l(kr)$ of: (a) a repulsive (positive) potential; (b) an attractive (negative) potential.

from which the radial wave function is simply

$$R_l(r) = c_l j_l(\kappa r) \qquad (r < a) \tag{3.50}$$

The solution outside the attractive well is

$$R_l(r) = c_l[j_l(kr) - \tan(\delta_l)n_l(kr)] \qquad r > a \tag{3.51}$$

At $r = a$, these two solutions and their first derivatives must match. It is left as an exercise for the reader to determine the general solution for tan (δ_l). The phase shifts for each value of l can be deduced by substituting explicit polynomial solutions for the spherical Bessel and Neuman functions. These may be expressed as a sum of products of sin ρ and cos ρ with polynomials of (where $\rho = ka$ or κa as appropriate). The general expression and first three particular solutions are given in Table 3.1. Using the appropriate values for $l = 0$ it can be readily determined that

$$\delta_0 = -ka + \tan^{-1}\left[\frac{k}{\kappa}\tan \kappa a\right] \tag{3.52}$$

If the potential is weak, then at low energies $\kappa a \ll 1$ and

$$\delta_0 \simeq ka\left[\frac{\tan \kappa a}{\kappa a} - 1\right] \tag{3.53}$$

If the potential were steadily increased until $\kappa a = \pi/2$, δ_0 would increase

Table 3.1 Polynomial expansions of the spherical Bessel and Neuman functions (or spherical Bessel functions of the first and second kind)

$$j_n(\rho) = f_n(\rho)\sin\rho + (-1)^{n+1}f_{-n-1}(\rho)\cos\rho$$

where

$$f_0(\rho) = \rho^{-1} \qquad f_1(\rho) = \rho^{-2}$$

$$f_{n-1}(\rho) + f_{n+1}(\rho) = \frac{(2n+1)}{\rho} f_n(\rho) \qquad (n = 0, \pm 1, \pm 2, \ldots)$$

so that

$l = 0$:

$$j_0(\rho) = \frac{\sin\rho}{\rho}$$

$l = 1$:

$$J_1(\rho) = \frac{\sin\rho}{\rho^2} - \frac{\cos\rho}{\rho}$$

$l = 2$:

$$j_2(\rho) = \left(\frac{3}{\rho^2} - \frac{1}{\rho}\right)\sin\rho - \frac{3}{\rho^2}\cos\rho$$

and

$l = 0$:

$$n_0(\rho) = -j_{-1}(\rho) = -\frac{\cos\rho}{\rho}$$

$l = 2$:

$$n_1(\rho) = \ j_{-2}(\rho) = -\frac{\cos\rho}{\rho^2} - \frac{\sin\rho}{\rho}$$

$l = 3$:

$$n_2(\rho) = -j_{-3}(\rho) = \left(-\frac{3}{\rho} + \frac{1}{\rho}\right)\cos\rho - \frac{3}{\rho^2}\sin\rho$$

$l = n$:

$$n_n(\rho) = (-1)^{n+1}j_{-n-1}(\rho)$$

from zero to about $\pi/2$. From equation (3.25), the s-wave ($l = 0$) scattering cross-section at low energies is seen to be

$$\sigma_{\text{tot}k \to 0} \simeq 4\pi\left(\frac{\sin\delta_0(k)}{k}\right)^2 \tag{3.54}$$

and so if $\delta_0 \to \pi/2$, this diverges as k^{-2} and produces an anomalously large scattering cross-section known as a 'zero energy resonance'. This can be physically equated with the existence of an s-wave bound state in the potential well. This condition cannot be satisfied if $\kappa a < \pi/2$. However, if the potential is increased substantially, it may be able to support a series of bound s-states. Although it is common to plot phase shifts modulo $\pi/2$, strictly speaking, the zero-energy s-wave phase shift for a potential that can support n-bound s-states should be

$$\lim_{k \to 0} \delta_0(k) = n\pi \tag{3.55}$$

This will enable the phase shift to be a continuous function that can pass through modulo $(\pi/2)$ n times as k is increased, finally converging towards zero as $k \to \infty$ (as necessary to satisfy equation 3.52). This relationship between the absolute value of the phase shift and the number of bound states is valid for all types of potential, and is known as Levinson's theorem.[162] In its general form, Levinson's theorem may be expressed as follows:[163] the zero-energy value of the phase shift δ_l is π times the number N_l of bound states with angular momentum l, unless a zero-energy s-wave resonance occurs, in which case $\delta_0(k \to 0) = (N_0 + \frac{1}{2})\pi$

As a practical example, consider the phase shifts for Mo, given in Figure 3.2. The $l = 0$ phase shift changes magnitude very rapidly at low energies,

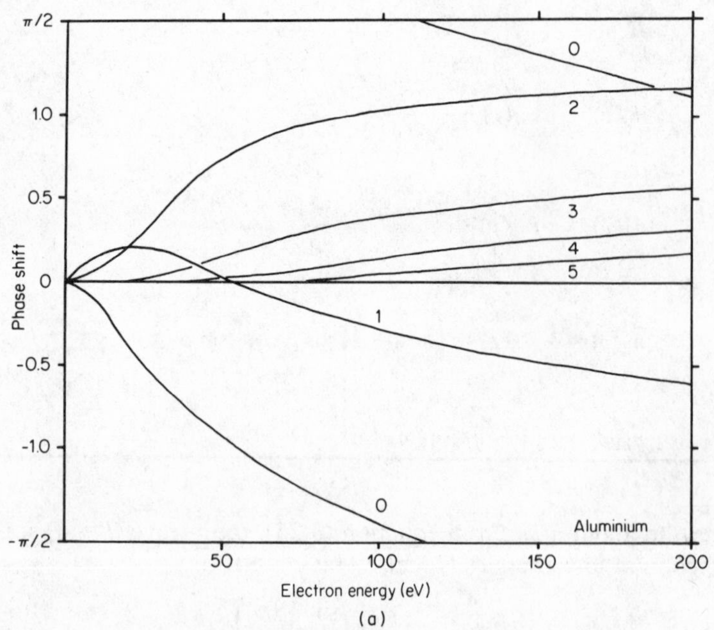

Figure 3.5. Phase shifts calculated for (a) aluminium, (b) nickel, (c) sulphur, in the range 0–200 eV above the muffin-tin zero. Deviations from these curves may result from the use of different potential construction methods and approximations, but the curves here are representative of the distinct types that may occur

passing through modulo $\pi/2$ at an energy of roughly 25 eV. This indicates the existence of a fairly tightly bound s-state. Above 100 eV, this phase shift passes through zero and this indicates the existence of at least one more weakly bound state. If the absolute value of $\delta_0(O)$ were set to 2π, then the whole phase shift would maintain a positive value at all energies, complying with the general relationship that attractive potentials lead to positively valued phase shifts.

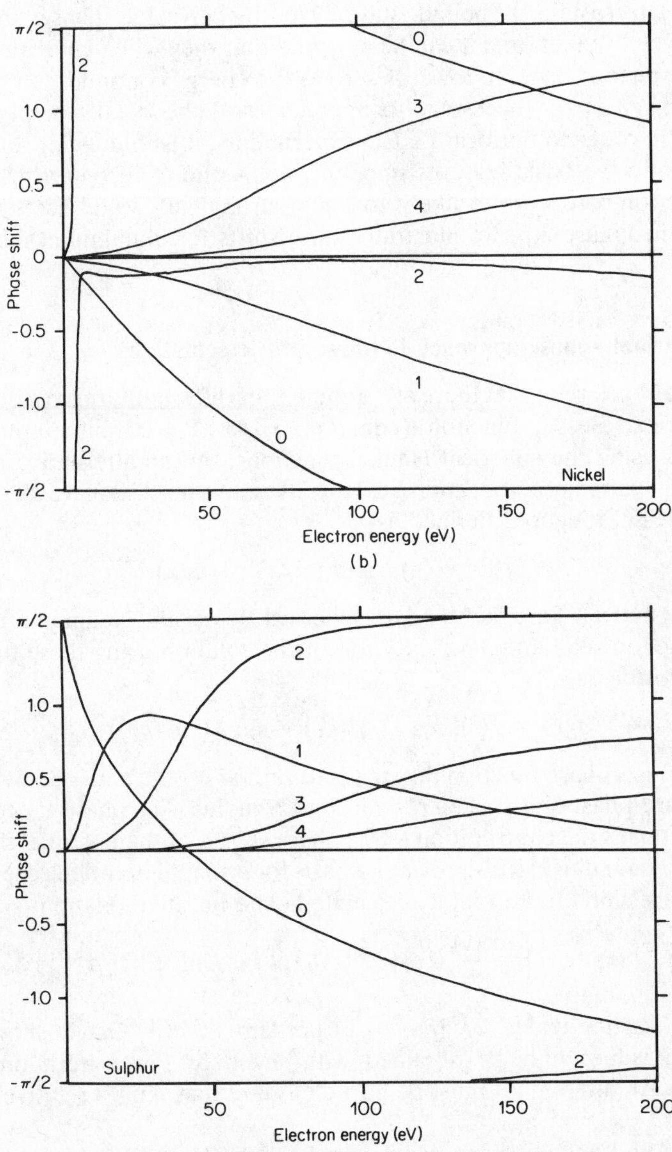

The behaviour of higher angular momentum phase shifts requires more complex arithmetic, but is in essence very similar to that discussed above for the $l = 0$ phase shift. Spectroscopic notation is customarily used to describe the first few partial waves so that the $l = 1$ phase shift corresponds to a p-wave, $1 = 2$ to a d-wave, $1 = 3$ to an f-wave, and so on. If we refer to the Mo phase shifts in Figure 3.2, we see that the $l = 1$ phase shift can sustain a rather loosely bound state or p-wave. The $l = 2$ phase shift, despite a rapid climb at low energies, turns over after an excursion of less than π, and so is not sufficient to sustain a bound state. The higher-value phase shifts are progressively weaker, and it can be assumed that they will all eventually turn over and start to converge towards zero as the energy continues to increase.

Although it is not necessary to appreciate all the details of phase-shift diagrams in order to perform LEED experiments, it is interesting and useful to appreciate the basic features since this helps one to recognize whether a particular ion-core type is likely to be a 'strong' or 'weak' scatterer. For example, in Figure 3.5, we illustrate phase shifts for aluminium, nickel and sulphur.

3.3.5 Variable-phase approach to phase shift calculations

We have already seen that there are a number of different forms of solution to the spherical Bessel function (equation 3.24). Phase shifts are usually calculated using the spherical Hankel functions, but an alternative method, called the 'variable-phase approach' by its proposer Calogero[163] uses the Riccati–Bessel functions defined as

$$\hat{j}_l(x) = xj_l(x) \quad \text{and} \quad \hat{n}_l(x) = xn_l(x) \tag{3.56}$$

where $j_l(x)$ and $n_l(x)$ are the standard spherical Bessel and Neumann functions as described in section 3.3.4. The form of the solution using these functions may be written as

$$f_l(r) = \alpha_l(r)[\cos \delta_l(r)\hat{j}_l(kr) - \sin \delta_l(r)\hat{n}_l(kr)] \tag{3.57}$$

where $\delta_l(r)$ is a phase function that tends to zero as $r \rightarrow 0$, and is equivalent to the standard phase shifts when r = muffin-tin radius. The phase function can be thought of as the contribution to the phase shift δ_l of the potential if it were truncated at a radius r. This provides a basis for a simple iteration scheme that allows phase shifts to be rapidly calculated. The iterative relation is

$$\delta_l^{(n+1)}(r) = -\int_0^r \frac{V(s)}{k}\{\cos[\delta_l^{(n)}(s)]\hat{j}_l(ks) - \sin[\delta_l^{(n)}(\mathrm{s})]\hat{n}_l^{(ks)}\}^2 \, \mathrm{d}s \tag{3.58}$$

with the starting value $\delta_l^{(0)}(r) = 0$. Aberdam *et al.*[137] have shown that converged values may be obtained with seven or fewer iterations when applied to Al, although it must be borne in mind that Al is a relatively weak scatterer.

This method has not been widely used in LEED studies, but is included here to illustrate that various alternative schemes do exist, and could undoubtedly be exploited further either in situations where they exhibit clear advantages over usual schemes, or simply to provide a useful check on the values obtained routinely.

3.4 The inner potential

In the foregoing sections, it has been shown that a crystal potential may be constructed mathematically by the overlap of atomic potentials, taking into consideration any interaction that may occur between electrons from neighbouring ion cores. The treatment is considerably facilitated by the muffin-tin approximation in which the potential is assumed to comprise an array of spherically symmetric potentials embedded in a constant background potential. This background is used as a reference level for the description of the potentials, and is called the muffin-tin zero. By analogy with the uniform potential that causes the refraction of light when it passes from one medium to another, this uniform background is sometimes called the 'optical potential'. More often, it is called the 'inner potential'. In practice, the inner potential is used to accommodate all the additional contributions to the scattering potential that can be given a uniform value throughout the crystal.

The muffin-tin zero is an important contribution to the inner potential, but alone would not result in good agreement with experimental results. This is evident from the fact that the work function (the energy necessary for an electron to leave the crystal) is different for different crystallographic surfaces. At first, this may seem strange—it implies that the difference in energy between the highest-filled energy levels within the crystal (the Fermi energy) and the vacuum level (the energy asymptotically reached well away from the crystal) depends on the face through which the electron is extracted. The difference lies in the relative polarizabilities of the electrons that spill out somewhat from the surface—the more open or corrugated the surface structure, the greater the effect. Consequently the work function for bcc crystals is greatest for (110) surfaces, less for (100) surfaces and least for (111) surfaces. For fcc crystals the reverse sequence applies. The accuracy of measurement is not substantially better than the differences themselves, but typical values for these three bcc surfaces would be 5.00, 4.6 and 4.4 eV respectively. An excellent review with many tabulated values has been recently written by Hölzl and Schulte,[164] and it is only necessary here to point out that such variations and uncertainties exist and cannot be directly related to the muffin-tin zero since this is a single-value bulk crystal parameter. For reference, mean work-function values for a range of different crystals are listed in Appendix 1 (p. 311).

Another important feature of the inner potential is that it varies slightly as the incident beam energy changes. It is found to be greater at low incident energies than at high incident energies. The greater the inner potential, the deeper the ion-core well, and hence the more attractive. This energy

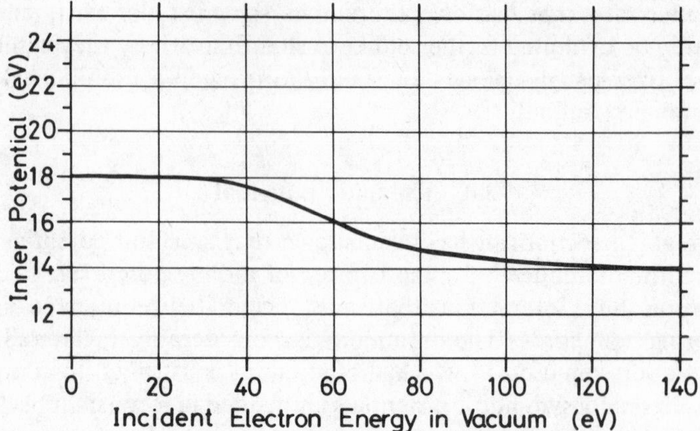

Figure 3.6. Variations in inner potential with incident beam energy, for Mo. (Reproduced by permission of North Holland Physics Publishing from Clarke and Morales de la Garza.[32])

dependence occurs because the ion-core potential is not perfectly formulated to deal with relatively high-energy electrons incident on a surface. At high energies, the effects of correlation are weakened, reducing screening and thus increasing the net repulsion experienced from electrons in the crystal. Detailed theoretical considerations indicate that this variation is not a linear function of energy, but instead of a form similar to that given in Figure 3.6, the inner potential varying by as much as 3 or 4 eV as the incident beam energy increases from 20 to 120 eV.[15,165]

In addition to the elastic scattering processes described so far in this chapter, the incident beam may also cause real excitations which thus remove energy. LEED experiments are devised so that only those electrons that do not lose energy can be detected, so energy loss processes can be equated with a loss of electron flux. The principal real excitations in a crystal that affect the incident beam are the collective vibrational modes of electrons, known as plasmons, and the collective vibrational modes of the ion cores, known as phonons. Phonon losses are treated separately in Chapter 5 because they are associated with ion-core vibrations and are temperature dependent; they are best dealt with through modifications to the ion-core scattering phase shifts. For the present we can ignore phonon effects if we assume that the crystal is at zero absolute temperature. There is a gap of roughly 15 eV between the Fermi energy and the lower threshold for plasmon excitation. Below this, an incident electron would encounter virtually no energy loss processes in a perfect crystal and could thus penetrate deeply into the bulk. However, the energy zero in the vacuum is between about 10 and 15 eV above the Fermi energy, so this situation could only be met by electrons with a few electron-volts incident energy. Above this threshold, the incident electrons will readily excite plasmons and so suffer loss of energy or elastic beam flux.

To a reasonable approximation, such energy loss processes can be represented by a single value throughout the crystal, and can therefore be incorporated as an imaginary component of the inner potential. The inner potential is thus expressible as a complex number

$$V_0 = V_r + iV_i \tag{3.59}$$

where both V_r and V_i may vary with incident beam energy.

Above the plasmon threshold V_i will, in general, increase to a roughly uniform level before decreasing again at very high energies due to the increased speed of the electrons. Although one may attempt to calculate the inner potential using many-body theory,[105] the assumption of uniform electron polarizability throughout the unit cell is only approximate and thus prevents the calculation of accurate values. The existence of an imaginary potential is a crucial element in the theory of LEED because it determines the minimum energy widths of the diffraction peaks, and restricts intensities to finite values; yet an accurate value is generally not essential because peak widths do not play an important role in the determination of structures, and their influence on intensities has little effect on the intensity–energy structure.

Experimentally, V_i can be determined by fitting theoretical beam widths to those obtained experimentally. Above the plasmon threshold, variations in V_i are slow and an experimental determination is often quite adequate, provided there are several isolated peaks for comparison. Near the plasmon threshold, V_i may vary more rapidly and experimental determination may be more difficult—but fortunately this energy is very low and is rarely approached in surface crystallographic studies.

It is quite usual to use a single value for V_i between 3 and 5 eV throughout the whole energy range. Simple algorithms designed to provide a reasonable simulation of the energy variation of V_i include[166]

$$V_i = -1.1 - 2.9\{1 + \exp[-5(E - 35.5)]\}^{-1} \text{ eV} \tag{3.60}$$

and

$$V_i = 3.8\left[\frac{E + V_r}{104}\right]^{1/3} \text{eV} \tag{3.61}$$

real and imaginary parts that to a first approximation can be given constant values independent of the incident beam energy. For a closer match with experiment it is recommended that suitable variations with incident beam energies be included. The inner potential is assumed to be uniform throughout the crystal. The one question remaining is how does it vary in the boundary region between crystal and vacuum? Both theoretical calculations, for example by Inkson, and experimental studies using very low energy electron diffraction[167] indicate that, in general, the potential varies relatively smoothly over a distance of 1 Å, between the vacuum level potential and the muffin-tin zero within the crystal. The point at which the energy is midway between the two extremes is positioned approximately half an interlayer

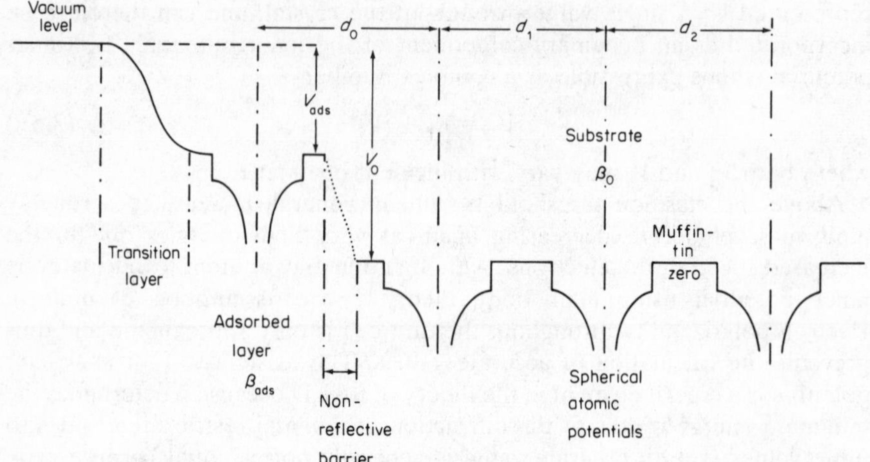

Figure 3.7. Potential profile at the surface of a crystal, assuming the presence of an adsorbate with smaller inner potential than the substrate

spacing outside the surface ion-core plane. It is difficult to be more precise than this, other than in cases where special studies have been made, because the surface potential does not form a perfect plane bounding the crystal. The surface is slightly corrugated, revealling the sites of the discrete ion-core potentials within the surface layer. Atom-scattering is a technique well suited to the study of such corrugations, as described in Chapter 8. To a first approximation, however, such corrugations can be ignored, so that the surface barrier may be considered to be a function of distance normal to the surface plane only.

For the purposes of LEED intensity calculations, the first important feature to note is that the variation in potential is smooth and relatively slow-varying (see Figure 3.7). Consequently, partial reflection of the incident electron beam is very weak. To approximate the reflective properties of the barrier, an abrupt potential step could be used, but this considerably overestimates the extent of its reflective properties. It is better to ignore reflections completely than use such a step. In the CAVLEED computer program package, the option of including a representative potential variation, based on the mathematical function 'tanh' is provided. This is sufficiently accurate for most LEED work, except perhaps for special extremely low-energy studies.

The second important feature of the surface barrier is that it provides the site for refraction of the incident electron beams as they cross from the vacuum region to the crystalline environment. This is entirely analagous to the refraction of light as it enters a refractive medium and it is for this reason that the inner potential is often called the 'optical potential'.

The position of the barrier is obscured both by the variation of inner potential with distance away from the crystal surface and also by the periodic rumpling of the profile across this two-dimensional plane. However, a mean value must be chosen in order to calculate its refractive effect. The mean position of the barrier in space does not affect the refraction, which can be associated with the variations of the real part of the inner potential, but it does affect the extent to which the electrons entering or leaving the crystal are subject to the imaginary, adsorptive potential.

Although this has an almost imperceptible effect on individual beam intensity–energy data, it has a clear effect on the relative intensities of the different beams at specific energies. Clearly, a beam leaving the surface near grazing emergence (almost parallel to the crystal surface) will experience much greater attenuation than one that emerges close to the surface normal.

It has been found[168] that the comparison of relative beam intensities at discrete beam energies can be used to determine the mean position of the surface barrier within about ±0.05 Å (see section 7.3.4).

CHAPTER 4

Multiple-scattering calculations

4.1 Introduction

Without a doubt it is the complexity of multiple-scattering theory rather than the intricacies of the experiment which have most daunted prospective users of the technique for surface crystallography. Reliable packages of computer programs have now been developed, and a fairly wide range of possible surface structures may be examined without much more than the manipulation of some input data statements. Any laboratory equipped with a modern large-capacity computer is potentially capable of engaging in the calculation of LEED beam intensities. Moreover, improvements in both the speed and capacity of new computers continually increase the range of surface structures which may be modelled. Indeed, since reliable computer programs currently exist which may treat many surface structures yet unresolved, it appears that the onus has now swung back to the experimentalist to produce the quantity and quality of data necessary to make the comparison of data as informative and reliable as possible.

Nevertheless, it is necessary to understand certain elementary facts about the various theories and available computer programs if the capabilities of the technique are to be exploited and extended, and if fundamental errors are to be avoided. This chapter is devoted to a summary of the salient features of multiple-scattering theory and the way in which it appears in the commoner packages of programs presently available. In addition, certain theoretical treatments are outlined which have been developed or are presently being developed for certain specialized tasks.

4.2 Representations in scattering theory

Two different types of mathematical description are commonly used in scattering theory, and this permits subsequent calculational methods to be divided likewise into two categories. In one method, scattering is described in terms of spherical waves—this is a natural basis with which to describe scattering from ion cores since, as we have already shown in Chapter 3, they are assumed to be spherically symmetric. This is often called the angular-momentum representation. Another name is l-space representation, since, for spherical waves the natural series of orthogonal terms is given by waves with ascending quantum angular momentum number (1), i.e. by 'partial waves'. As shown in Chapter 3, only 5–10 such terms are necessary to give an accurate description of the scattering of an electron at typical LEED energies.

The other method describes scattering in terms of plane waves. Since plane waves may be characterized by their wave-vectors **k** which are directly related to their momentum $h\mathbf{k}$, this is commonly known as momentum- or **k**-space representation. Of the two approaches this is perhaps the easier to visualize since scattering is described directly in terms of the diffraction beams which are experimentally observed. Typically up to 100 beams may be propagating at any particular energy so the description will include correspondingly more terms than the l-space representation. Despite this, **k**-space methods may often be more rapid than l-space methods, as we shall shown below.

The mathematical link between the plane and spherical waves at a distance r from a scattering centre at R (see Figure 4.1) is the expansion

$$a\exp(i\mathbf{k}\cdot\mathbf{r}) = a\sum_{lM}\exp(i\mathbf{k}\cdot\mathbf{R})4\pi i^{l}(-1)^{m}Y_{l\text{-}m}[\Omega(\mathbf{k})]Y_{lm}[\Omega(\mathbf{k})]j_{l}[\mathbf{k}\cdot\mathbf{r}] \quad (4.1)$$

where $Y_{l\text{-}m}$ and Y_{lm} are spherical harmonics and j_l is a Bessel function (see Chapter 3). Although mathematically this is an exact transformation when an infinite number of terms are included, in practical scattering calculations only a limited number of terms may be used. Consequently, even though certain calculational methods using the two different representations may be mathematically equivalent, in practice small differences in the results may occur due to the restrictions imposed on the number of beams or spherical wave terms.

The limitation on the number of beams using the **k**-space representation is, of LEED intensities will be determined by the 'strength' of the ion-core scattering which will in turn be dependent on the composition of the material and the incident electron energy. The temperature of the crystal and consequent vibrational amplitudes of the ion cores also have a minor influence on the number of partial waves required (if there is no vibration, fewer terms may be required). This is discussed in detail in Chapter 3 so need not be considered further here.

The limitation on the number of beams using the **k**-space representation is, on the other hand, primarily set by the geometrical arrangement of the surface atoms. The ideal surface of a crystal is a two-dimensional plane of atoms, which may be considered to scatter the incident beam into a set of diffraction beams. Scattering from such a plane will occur in both a forward direction, into the crystal, and in the reverse direction, back out of the crystal. Because the two-dimensional periodicity of the crystal is the cause of the diffraction beams, it is convenient to consider all the atoms in the crystal as though they were in planes parallel to the crystal surface. In fact there are very good mathematical reasons for conceptually splitting the crystal up in this way, as will be shown later in this chapter.

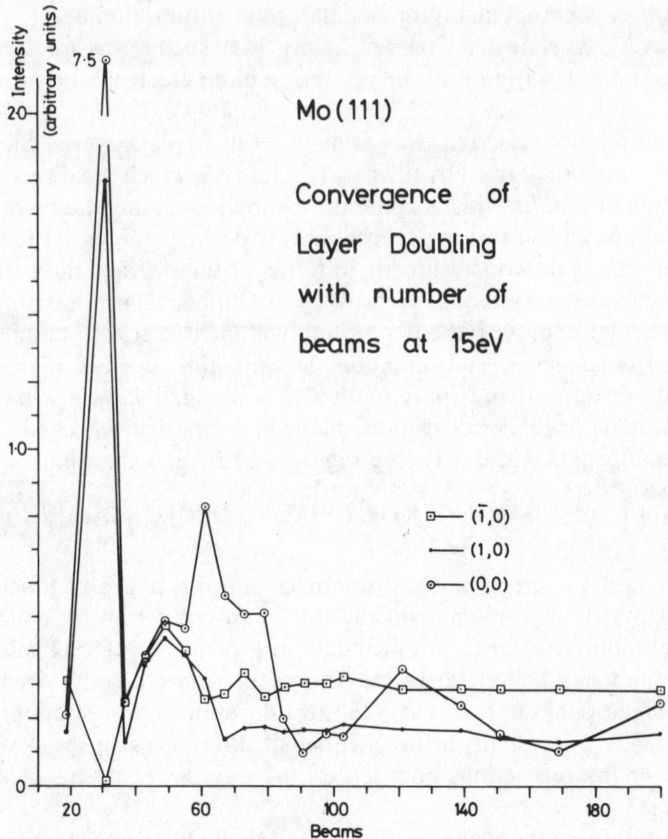

Figure 4.1 Variation in intensity of various beams, as the total number of beams used in the layer doubling model is increased, for Mo(111) with a 15 eV beam energy

For a given incident beam energy and direction, each diffraction beam may be characterized by its wave vector parallel to the surface plane. If $\mathbf{k}_{\|}^{\circ}$ is the in-plane component of the incident beam and $\boldsymbol{g}_i$ a reciprocal lattice vector then a diffraction beam i will have an in-plane wave vector

$$\mathbf{k}_{i\|} = \mathbf{k}_{\|}^{\circ} + \mathbf{g}_i \tag{4.2}$$

Because we are only considering electrons which have the same energy as the incident beam (see Chapter 2), the magnitude of the corresponding wave vector perpendicular to the surface plane is directly obtainable as

$$\mathbf{k}_{i\perp} = \pm \sqrt{(E - |\mathbf{k}_{\|}^{\circ} + \mathbf{g}_i|^2)} \tag{4.3}$$

where the ± sign may be used to describe forward or reverse scattering from a diffraction plane. As $\boldsymbol{g}_i$ increases, so $\mathbf{k}_{i\perp}$ decreases until eventually it becomes an imaginary quantity. This sets the limit to the number of propagating beams at that energy. In fact, because $\mathbf{g}_i$ is a discrete quantity, the number of propagating beams will increase in 'steps' as the energy increases, as may be seen from Table 4.1 in which the emergence energy of different beams are given in the case of normal incidence (i.e. with the incident beam perpendicular to the surface plane) upon a Mo(100) surface.

The number of beams which must be included is, however, rather greater than the number of propagating beams. This is because the amplitude of beams which have an imaginary wave vector perpendicular to the plane will decrease exponentially with distance from the plane, but, if they encounter the next plane before their amplitude has become negligible then they are able to 'couple-in' to corresponding imaginary beams from the next layer and survive to be included in the next scattering event. Such beams are called 'evanescent beams'; they may carry scattered flux from layer to layer. The first few evanescent beams (in order of ascending $\mathbf{k}^\circ_{/\!/} + \mathbf{g}_i$) must always be included in a calculation since their amplitudes may not decrease significantly between planes. The inclusion of additional beams may be necessary if the planes are close together or the ion cores are strong scatterers (so that the initially scattered amplitudes may be relatively large). The choice of the number of beams to be included cannot be predicted reliably and a series of calculations should be tried with increasing numbers of beams to check that convergence has been reached. One way of automatically including a certain proportion of evanescent beams at all energies is to calculate the number of beams propagating at an energy x times higher than the actual incident energy being considered, and to use those beams which propagate at that higher energy. This is essentially the method provided in the CAVLEED package of programs.[169] The parameter used in practice in the CAVLEED package has been selected to maintain a similar level of convergence at all energies, minimizing the wastage of computer time which would result from incorporating more beams than necessary.

If the interplanar spacing becomes very small (less than, say, 0.8 Å) and the ion cores are strong scatterers, the number of beams required to maintain a convergent solution may become impractically large. The (111) face of certain bcc crystals provides just such a situation. In particular, Mo(111) has an interlayer spacing of about 0.7 Å. At 15 eV, the number of propagating beams at normal incidence is only 13, yet the number of beams required to obtain convergence is much greater. The intensity of the (0, 0) beam oscillates wildly until 100 beams have been included, but even with 200 beams the intensity has still not stabilized (see Figure 4.1). By comparison higher-order beams were found to converge satisfactorily after 100 beams had been included. For higher incident beam energies, a convergent solution would involve several hundred beams, and be beyond the capacity of most present-day computers.

To give an idea of the limiting interplanar distance, Figure 4.2 shows how the intensity of the (0, 0) beam varies with the number of evanescent beams

Table 4.1 The emergence energy, in electron volts with respect to the Muffin-tin zero, of the first 69 beams from a Mo (001) surface. In this special case of normal incidence, the total set of beams may be reduced to 13 symmetrically equivalent beams

	Beam indices		Emergence Energy (eV)
1	0.0	0.0	0.0
2	−1.0	0.0	15.18
3	0.0	−1.0	15.18
4	0.0	1.0	15.18
5	1.0	0.0	15.18
6	−1.0	−1.0	30.36
7	−1.0	1.0	30.36
8	1.0	−1.0	30.36
9	1.0	1.0	30.36
10	−2.0	0.0	60.72
11	0.0	−2.0	60.72
12	0.0	2.0	60.72
13	2.0	0.0	60.72
14	−2.0	−1.0	75.91
15	−2.0	1.0	75.91
16	−1.0	−2.0	75.91
17	−1.0	2.0	75.91
18	1.0	−2.0	75.91
19	1.0	2.0	75.91
20	2.0	−1.0	75.91
21	2.0	1.0	75.91
22	−2.0	−2.0	121.45
23	−2.0	2.0	121.45
24	2.0	−2.0	121.45
25	2.0	2.0	121.45
26	−3.0	0.0	136.63
27	0.0	−3.0	136.63
28	0.0	3.0	136.63
29	3.0	0.0	136.63
31	−3.0	−1.0	151.81
31	−3.0	1.0	151.81
32	−1.0	−3.0	151.81
33	−1.0	3.0	151.81
34	1.0	−3.0	151.81
35	1.0	3.0	151.81
36	3.0	−1.0	151.81
37	3.0	1.0	151.81
38	−3.0	−2.0	197.35
39	−3.0	2.0	197.35
40	−2.0	−3.0	197.35
41	−2.0	3.0	197.35
42	2.0	−3.0	197.35
43	2.0	3.0	197.35
44	3.0	−2.0	197.35
45	3.0	2.0	197.35
46	−4.0	0.0	242.90

Table 4.1 *continued*

	Beam indices		Emergence Energy (eV)
47	0.0	−4.0	242.90
48	0.0	4.0	242.90
49	4.0	0.0	242.90
50	−4.0	−1.0	258.08
51	−4.0	1.0	258.08
52	−1.0	−4.0	258.08
53	−1.0	4.0	258.08
54	1.0	−4.0	258.08
55	1.0	4.0	258.08
56	4.0	−1.0	258.08
57	4.0	1.0	258.08
58	−3.0	−3.0	273.26
59	−3.0	3.0	273.26
60	3.0	−3.0	273.26
61	3.0	3.0	273.26
62	−4.0	−2.0	303.62
63	−4.0	2.0	303.52
64	−2.0	−4.0	303.62
65	−2.0	4.0	303.62
66	2.0	−4.0	303.62
67	2.0	4.0	303.62
68	4.0	−2.0	303.62
69	4.0	2.0	303.62

for the same two-dimensional planes of Mo atoms, but with the interplanar distance artificially ‘stretched’. In this case, convergence was attained with a reasonable number of beams only when the interplanar spacing was increased to about 1 Å.

The majority of surface structures investigated to date have involved interplanar distances greater than 1 Å within the crystal, but when light gas atoms are adsorbed on metal surfaces, the interplanar distance between the absorbed gas atoms and surface metal plane may be rather less than this. Although certain computational schemes (which are described later in this chapter) may be stable down to shorter interplanar distances than others, there will always be some minimum distance, about 1 Å, less than with the **k**-space approach will be unsuitable. By comparison, *l*-space methods are relatively insensitive to such geometrical factors, and can often treat interplanar separations down to zero. This should not be taken as a hard and fast rule, however: perturbation schemes based on the *l*-space representation may also be unable to converge in the presence of strong scatterers when the interplanar separation is small. Nevertheless, apart from certain exceptions, the ability to treat small interplanar separations must be considered to be a major advantage of *l*-space methods. Why then even bother with *k*-space methods? The answer is very simple: under favourable conditions, they may

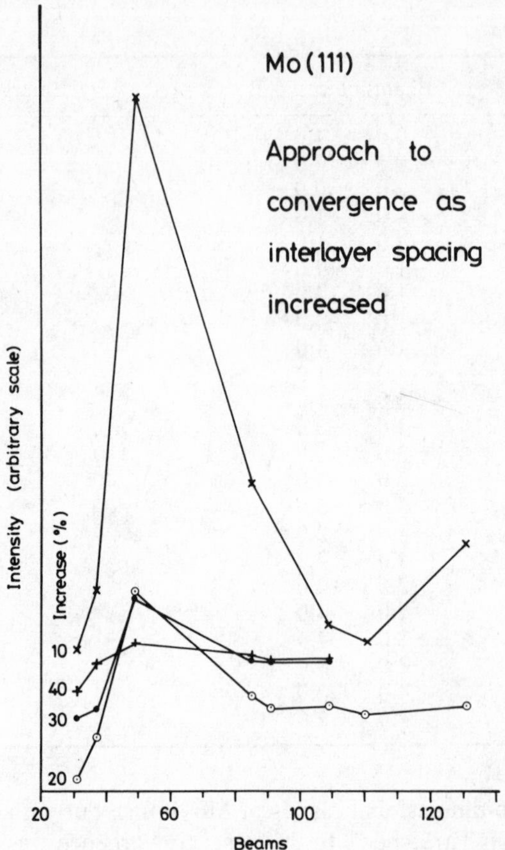

Figure 4.2. Intensity of the (0. 0) beam with increasing number of beams included in the model, as the bulk interlayer spacing for Mo(111) is artificially increased through steps of 10 per-cent.

be much faster, take us less computer memory space and therefore provide a more economical approach to solving a surface structure.

Let us now consider the most important computational methods presently available, to help us choose which will be the best for any particular structure in which we are interested. We shall begin with a brief outline of the development of the subject, to put into perspective some of the methods which have been used in the past and which may still have role to play in particular cases today.

4.3 The development of multiple-scattering theories

The origins of LEED multiple-scattering calculations may be traced to band-structure theories. A self-consistent formalism to describe the backscattering of electrons from a crystal using the *l*-space representation was developed by Beeby and published in 1968.[170] Likewise the Bloch-wave method, which uses the **k**-space representation as its basis was first used to describe scattering of low-energy electrons from the surface of a three-dimensional crystal by McRae.[171] Neither method provided a satisfactory description in its original form, due to the neglect of an important factor which distinguishes electrons propagating in a solid at typical LEED energies from electrons with energies near the Fermi level. This factor is the rate of energy loss which, although virtually zero for Fermi-level electrons, may be quite substantial when electrons exceed the plasmon excitation threshold (around 10–20 eV above the Fermi level).

Experimentally, electrons which have lost more than a few electron-volts of energy are filtered out by the suppressor grid, so do not contribute to the observed beam intensities. Therefore, electrons which lose energy may be treated mathematically as a loss of flux from the elastically scattered beams in which we are interested. A more important consequence from the point of view of LEED theory is the associated reduction in the mean free path. This is typically about 4 Å for electrons at 100 eV. After travelling this distance the electron flux will be reduced to $1/e$ of its original value. Since the decay is exponential, the distance which an electron beam may travel through the crystal before its intensity may be considered negligible will of course be much greater than this. Although it is possible to make theoretical calculations of the path length from details of the scattering potential,[105] the effective penetration depth is also influenced by multiple scattering, and so it is much easier to make use of data obtained empirically from the observation of Auger electron extinction lengths. The accumulation of data from many experiments on many different materials shows that there is an overall pattern to the energy dependence of the penetration depth, as illustrated in Figure 1.1. It is clear that, provided the electron beam is restricted in energy to the range approximately 40–300 eV, then the depth to which electron scattering may significantly affect the diffraction pattern will be quite limited. In fact, these values for electron penetration depths overestimate the depth from which electron scattering may play a significant role in the diffraction process. This data measures the extinction lengths of electrons generated from within the crystal, but in the LEED case, the initial electron beam comes from outside the crystal, must penetrate, be reflected and then return to the surface. Not only is the total distance to be travelled within the crystal double that of the Auger emission case, but at least one reflection is involved, and reflected flux is consequently much weaker than that scattered in a forward direction.

The gradual reduction in the reflected component compared with the transmitted component as the electron beam energy increases was described in Chapter 3. The implication of this is that the surface sensitivity of LEED may be extended to rather higher incident energies than might appear from the penetration depth data as described above. The mean free path parameter that enters LEED calculations is defined in terms of the reduction in propagation amplitude with distance between one scattering event and the next. In practice, multiple scattering will occur between scatterers and so the effective rate of flux loss will be greater than that which would occur if only single-scattering processes were present. If an approximate theory is used in which multiple scattering is ignored, it is justifiable to increase the mean free path parameter, to restore the penetration depth to a value close to that found in reality. It is not necessary to attempt to match closely the penetration depths predicted by LEED theory with the empirical values, since the calculated value will be closely linked with the numerical convergence of the computer program itself. However, it is usual to assume that the penetration depth cannot realistically be greater than some predefined distance (often chosen to be 32 atomic layers). If the propagating electron amplitudes after this distance are not all close to zero, this may be used as a criterion to indicate that the computational method is non-convergent for that particular case.

Neglect of energy-loss mechanisms in the original Bloch-wave and Beeby programs means that they lack surface sensitivity; they are also rather unwieldy and expensive to run. Despite this, their ability to treat infinite penetration may mean that they yet have a special role to play in extremely low-energy studies. (Flux attenuation is extremely small for electrons propagating with energies below the plasmon threshold region.)

Incorporation of inelastic damping is a very important ingredient in LEED theories because it limits the penetration depth and permits the calculation to be made on a layer-by-layer basis. Jepsen *et al.*[172] took a form of Block-wave method as used previously in band-structure calculations, incorporated inelastic damping and produced what is generally known as the 'layer-KKR method'. The name refers to Kramers, Kronig and Rostoker (KKR) who were instigators of the original band-structure formalism from which this method is derived. Often, it is simply called the Bloch-wave method for LEED, and a detailed description with associated computer programs is presented in Pendry.[105] All intraplanar and interplanar multiple-scattering events are included and so it is generally referred to as an *exact* method.

Layer-KKR has been used to determine surface structures on numerous occasions, and is able to match experimental intensity data closely. However, its precision is obtained only as a result of considerable computational effort. Despite successful demonstrations that it can model electron scattering from clean surfaces and simple overlayer structures, it would be excessively slow and cumbersome if more complex overlayers or superstructures were to be examined.

Inelastic damping was first incorporated into Beeby's T-matrix method by Tong and Rhodin,[173] also in 1971. Whereas earlier calculations using the T-matrix method had to be limited to just one partial wave (s-wave) the savings in computer effort concomitant with the inclusion of inelastic damping permitted several partial waves to be included. This was shown to reproduce relatively accurately experimental results obtained from the Al(100) surface by Jona.[174] This method was used to model a number of surface structures from 1971 to about 1975. As with the layer-KKR method, the realistic T-matrix method takes all possible scattering events into account via the multiple-scattering formalism, and so may be described as an 'exact method'. If N subplanes are included, there will be N inhomogeneous equations to be solved, each involving complex matrices of dimension l, where l is the number of partial waves. Consequently, as with the layer-KKR method, it is both time-consuming and extravagant in computer memory space. Indeed, it generally requires greater computer resources than the layer-KKR, but has the inherent advantage of l-space methods in that it can cope with very small interplanar spacings.

4.3.1 Terminology

The term *subplane* is often used to describe a two-dimensional plane of atoms with one atom per unit cell. A *plane* with more than one atom per unit cell can be considered as the combination of several subplanes, each associated with one of the atoms within the unit cell. In this case, the subplanes must be coplanar. A *layer* refers to a convenient grouping of subplanes which may or may not be coplanar. For most clean surfaces of elemental crystals, a layer may comprise nothing more than a single subplane: hence these terms are often used interchangeably.

As an example, consider the (100) surface of an alkali halide crystal such as NaCl. A 'rumpling' of the surface may occur which involves the outward shift of all the anions in the surface layer and an inward shift of al the cations. The surface may be modelled by the combination of two subplanes, one comprising only Na ions in a square lattice, the other comprising only Cl ions in a similar lattice. At the rumpled surface the subplanes are not coplanar, but together may nevertheless be described as the surface layer. Within the bulk, each layer may comprise the coplanar combination of two subplanes. It is possible to calculate the scattering from coplanar combinations using l-space methods, but with **k**-space methods, scattering from the whole layer must be explicitely calculated prior to the incorporation of interlayer scattering events. Consequently, although the terms *interplanar* and *interlayer* scattering are frequently used interchangeably in the literature of LEED theories, in general they apply to l-space and **k**-space representational methods respectively.

The word *dynamical* in the context of LEED calculations refers rather loosely to the fact that the complete scattering process has been described from an understanding of the detailed scattering events which comprise the whole. It generally means that multiple scattering has been included. By contrast the word *kinematical* is normally reserved for single-scattering theory.

4.4 Treatment of infinite orders of multiple scattering

The presence of inelastic damping limits not only the penetration depth, but the total number of scattering events possible. Consequently the 'exact' methods are inefficient because they include the possibility of many more scattering events than could actually make a significant contribution to the observed electron flux. The other extreme is to start by taking one scattering event at a time and summing them, one by one, until a convergent solution is obtained. Of course, the inclusion of an infinite number of events as used in the 'exact methods' does not take infinitely longer than the calculation of one event, because of the mathematical simplicitly regained by taking the asymptotic limit. If **G** is a vector which describes the propagation of the wavefield from one scattering centre to another, and V is a matrix which describes the actual scattering event, then the contributions from successive orders of scattering events may be expressed mathematically by a seres T with the following form:

$$T = V + V\mathbf{G}V + V\mathbf{G}V\mathbf{G}V + V\mathbf{G}V\mathbf{G}V\mathbf{G}V + \ldots \tag{4.4}$$

$$T = V\,[1 + \mathbf{G}V + (\mathbf{G}V)^2 + (\mathbf{G}V)^3 + \ldots] \tag{4.5}$$

$$T = V\,(1 - \mathbf{G}V)^{-1} \tag{4.6}$$

Consequently, *the summation of an infinite series of similar scattering events may be calculated via a single matrix inversion.* This is an important result which crops up again and again in the treatment of multiple scattering in LEED.

In **k**-space representation, a similar but more complex expression is necessary to describe the scattering matrix of a plane of atoms. The scattering of each ion core may be described in terms of a diagonal t matrix, and a mathematical transformation (in terms of spherical harmonics Y_{lm}) can be used to translate the description of scattering about one site to another. A consistent description of the total scattering effect can be made with respect to a chosen origin by making a 'lattice sum' over successive sites, combined with an appropriate transformation of axes and propagation factors. In CAVLEED[169] the lattice sums are given the notation F_{lm} and are evaluated within the subroutine *FMAT*. These then enter the equivalent to the product **G**V which appears in the above equation, but in CAVLEED this product is given the notation X, and is evaluated by the subroutine *XMAT*. Finally, the equivalent to the planar scattering matrix T above is denoted by $\boldsymbol{M}_{g'g}$ in CAVLEED and is evaluated within the subroutine *MSMAT*.

In the case of an infinitely periodic two-dimensional plane, it is possible to produce a relatively efficient computational model which can accommadate multiple scattering to an infinite order. If the electron energies are not too low, then adsorption between scattering events will ensure that the summation of terms converges fairly rapidly, and a scheme described by Pendry[105] based on Beeby's T-matrix method will be suitable. If very low energies are required then inelastic damping may be very small and the summation will not converge properly. This problem can be overcome by splitting the summation into two separate regions, one of which is transformed into reciprocal space. A method using this latter approach was formulated for LEED by Kambe in 1967.[175] Details of these two methods may also be found in Pendry.[105] If X is defined as above, δ_l^1 are the phase shifts which describe the ion-core scattering, A is the area of the unit cell and $\mathbf{k}$ and $\mathbf{k}(\mathbf{g}')$ are the total wave vectors of the incident beam and the component of the relevant diffracted beam wave vector normal to the surface respectively, then it can be shown[176] that the planar scattering matrix $M^{\pm\pm}$ is given by the expression.

$$\frac{8\pi^2 i}{A}\,\frac{1}{k_0 k_\perp^{\pm}(\mathbf{g}')}\sum_{LL'} Y_L(\mathbf{k}^{\pm}(\mathbf{g}'))[1 - X]^{-1}_{LL'} Y_{L'}(\mathbf{k}(\mathbf{g}))\, \mathrm{e}^{i\delta_l} \sin\delta_{l'} \tag{4.7}$$

or, in terms of t-matrices rather than phase shifts,

$$\frac{8\pi^2 i}{|\mathbf{k}_{\mathbf{g}}^{\pm}| A \mathbf{k}_{\mathbf{g}'z}^{+}} \sum_{\substack{l'm'\\ lm\\ LM}} [i^{-l'} Y_{l'm'}(\hat{\mathbf{k}}_{\mathbf{g}}^{\pm})] i^{-l} t_{l'm'LM} (1 - Ht)^{-1}_{LMlm} [i^{l}(-1)^m Y_{l\text{-}m}(\mathbf{k}_{\mathbf{g}}^{+})] \tag{4.8}$$

where

$$H_{l''m''lm} = \sum_{l'm'(j\neq 0)} 4\pi(-1)^{\frac{1}{2}(l-l-l'')}(-1)^{m'+m''} h_l^{(1)}(|\mathbf{k}^+|R_j) Y_{l'-m'}(\hat{\mathbf{R}}_j) \exp(ik_{\parallel}\cdot\mathbf{R}_j)$$
$$\times \int Y_{lm}(\Omega) Y_{l'm'}(\Omega) Y_{l''-m''}(\Omega)\,\mathrm{d}\Omega$$

The simplicity to be gained by dealing with infinitely periodic two-dimensional planes means that there is little room for improvements which may speed up the computational process, except in the case of high incident electron energies and grazing incidence, when the full two-dimensional symmetry of the plane is not exploited, or when the unit cell of the surface lattice is so large that scattering between cells is less important than scattering within the cell. Consequently, the scattering from planes or subplanes of atoms have been adopted as the basic building block for almost all subsequent schemes for dynamical LEED calculations. The cases of high incident energies and large unit cells are discussed separately in sections 4.8 and 4.9.2 respectively.

4.5 Direct perturbation methods

A direct perturbation scheme may be constructed in which successive terms include scattering contributions from one, two and three or more planes of atoms. Direct perturbation methods which use the subplane scattering matrix 'τ' as a basis are generally known as τ-matrix pertubation methods. First-order refers to scattering models in which scattering from each subplane is included once, but not multiple scattering between planes. Note that this is not the same as 'single scattering' or kinematical theory since multiple scattering within each subplane is treated exactly. The omission of multiple interlayer scattering is quite serious: only in cases where the scattering centres are very weak or the lattice spacing so large that significant attenuation of the beams occurs between layers is this approximation capable of reproducing experimentally obtained results. Solid Xe is one of the few materials that has been shown to sutisfy these criteria[177] (see Figure 6.2).

A second-order scheme was derived by McRae[178] in 1968, which, when adapted to incorporate realistic phase shifts by Tong and Rhodin,[173] was found to be capable of reproducing to a fair approximation experimental results for Al. However, Al is a weak scatterer; also, being a free-electron metal, its inelastic damping coefficient is relatively large. Beryllium is similar in this respect and Strozier and Jones[179] were able to obtain reasonable agreement with experiment using a variation of this second-order perturbation method. For most materials, however, second-order perturbation is also quite inadequate. Fortunately, the inclusion of third-order contributions is sufficient to permit quite reasonable agreement with experimental results for many scattering materials.[180] However, there is a problem with direct perturbation methods of this type, which is that the number of terms to be included increases rapidly with each successive order. The number of permutations of three successive layers from which scattering may occur is considerably greater than from two and, although third-order perturbation is quite acceptable when applied to weak or intermediate scatterers such as Al or Ni, it is still inadequate when strong scatterers such as W or Mo are to be treated. The various contributions which must be included in a third-order perturbation scheme are given in Figure 4.3. A fourth-order scheme would require so many terms that it would be quite unwieldy, and probably not particularly more efficient than one of the exact methods: consequently such a scheme has never been devised.

Although three orders of scattering are the practical limit of direct perturbation methods, there are several ways in which the τ-matrix scheme may be modified to improve efficiency or convergence. For example, it is possible to use the ion-core t-matrix as an alternative basis for a perturbation scheme. This approach has been shown to save time compared with the τ-matrix method, but does not improve on its quality of convergence. A more promising modification to the τ-matrix method is to treat separately the forward scattering of a beam into itself. This is because, at typical LEED energies, this 'zeroth-order forward scattering' component is significantly larger than forward scattering into any other beam or any reflected component. The matrices describing forward scattering may be separated into the sum of a diagonal matrix describing zeroth-order contribution, and matrices which include all the other (smaller value) elements. The strong forward-scattering contribution is included exactly, but does not involve considerable additional computational effort because of the simplifications resulting from the diagonal nature of the matrix. The weaker elements are treated in a perturbation scheme as before, but convergence is now improved due to the absence of the stronger contributions. Using such a method, Jennings[181] managed to obtain results which agreed quite well with

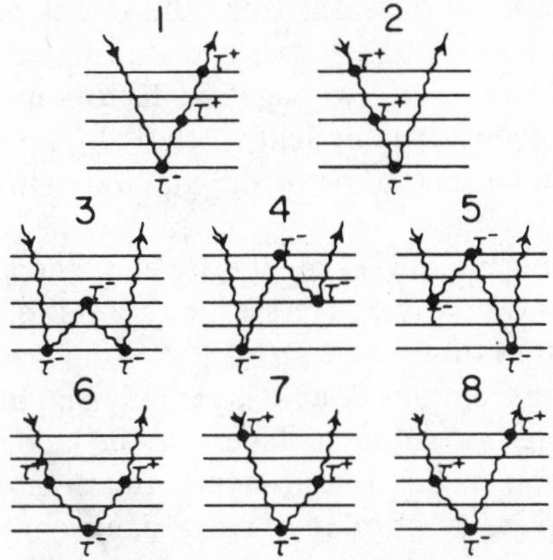

Figure 4.3. Diagrammatic presentation of all third-order τ-matrix scattering events. The number of planes between any two scattering events is arbitrarily chosen. (Reprinted by permission from *Progress in Surface Science*, **7**, Tong.[180] Copyright (1975) Pergamon Press Ltd.)

experiment for the moderately strong scatterer Cu, and the Group d'Etude des Surfaces met with similar success with a related scheme called the NIS (*N*th order interlayer scattering) method.[182] Despite all this additional sophistication, all these third-order perturbation schemes lack the ability to cope with strong scatterers for the simple reason that significant multiple scattering beyond the third order occurs in such cases. Fourth order or beyond, although in theory possible, is not practical as already noted above, and it is necessary to consider alternative schemes for perturbation calculations.

4.6 Iterative methods

For a perturbation method to be generally applicable, it must be capable of having its order increased without a dramatic increase in the associated number of terms. The requirement is to produce an iterative perturbation method, i.e. one in which each order of scattering can be expressed in exactly the same form as the previous order, so that the computational times involved would increase linearly with increasing order. There are two major types of approach to this problem. One is called the renormalized forward scattering (RFS) method, the other, reverse scattering perturbation (RSP). In the section 4.5 we saw that a reasonable pertubation scheme could be devised by assuming that the forward scattering of a beam into itself is the dominant contribution. In the RFS method, the restriction to zeroth-order forward scattering is lifted, and all forward scattering contributions are included exactly, but the reflected components are included one order at a time. This approach is fundamentally different from the direct perturbation methods discussed previously because the forward scattering components are added via modified (or renormalized) propagation factors in which the effects of successive scattering events are implicitly included. This permits considerable savings in computational time to be made, and a detailed description will be given in section 4.6.1.

It is in fact possible to devise an iterative method capable of treating multiple scattering between two layers exactly to all orders of forward and reverse scattering. From this, a new set of scattering matrices which describe the effect of scattering from this double layer may be obtained. Repetition of this procedure enables two double layers to be added together, to give scattering matrices from a slab of four layers. Iteration of the procedure will then permit the complete description of scattering from 8, then 16 layers and so on. This method, called layer (or matrix) doubling may be grouped together with layer-KKR as an exact method applicable to crystals of arbitrary but finite thickness; it is, however, more rapid in execution because it makes explicit use of the fact that successive bulk layers have identical structures and interlayer spacings.

4.6.1 Layer (or matrix) doubling

Let us see how the layer-doubling method can be affected mathematically. Scattering from a plane is most conveniently described in terms of the electron wavefield at a certain distance from the plane itself, since this description will include not only the scattering from the plane of ion cores but the effect of propagation through a certain distance between planes. Even if the incident wave were not scattered by the ion-core plane, the translocation of a propagating beam from one side of the plane to the other will involve a phase change, and we may define propagation of a beam with wave vector $\mathbf{K} + \mathbf{g}_i$ through a distance $\mathbf{a}$ where the $+$ and $-$ signs refer to propagation of a beam into and out of the crystal respectively. If the ion-core scatterers lie in a plane, we may assume that scattering from either side is symmetric, and we may define $\tau_\alpha^+(\mathbf{K})$ and $\tau_\alpha^-(\mathbf{K})$ as the transmission and reflection matrices for a layer α. If I is the unit matrix which takes into account the unscattered waves, then it is clear that the scattering wavefields propagating from planes on either side of the ion-core plane may be described fully by the four matrices

$$
\begin{aligned}
[T_\alpha^{++}] &= [P^+(I + \tau_\alpha^+(\mathbf{K}))P^+] \\
[T_\alpha^{--}] &= [P^-(I + \tau_\alpha^+(\mathbf{K}))P^-] \\
[R_\alpha^{+-}] &= [P^+\tau_\alpha^-(\mathbf{K})P^-] \\
[R_\alpha^{-+}] &=]P^-\tau_\alpha^+(\mathbf{K})P^+]
\end{aligned} \tag{4.9}
$$

These are illustrated in Figure 4.4(a). Note the order in which the matrices appear: the first event is that corresponding to the term furthest to the right of

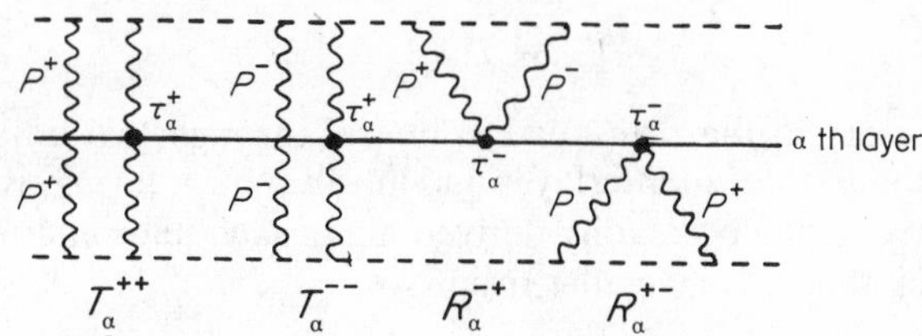

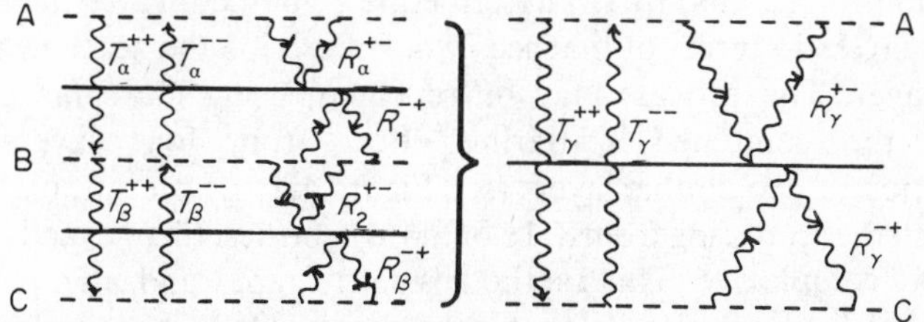

Figure 4.4. Diagrammatic representation of: (a) transmission and reflection matrices, in terms of propagation vectors and scattering matrices; (b) the mathematical process of layer-doubling (the representation of all multiple-scattering events between planes A and C by transmission and reflection matrices corresponding to an effective, single-scattering layer).

the equation, the last is on the left; this conforms with the usual mathematical convention for the operation of matrices. If all the ion-core planes are equally separated then the planes at which the wavefields are described may be chosen to lie midway between planes. In that case, all the propagation factors will be identical apart from a change of sign, and the description can be reduced even further to a single transmission and reflection matrix. When two such layers are adjacent we may consider the combined effect in terms of two new matrices T_γ and R_γ. Referring to Figure 4.4(b) we see that, if a wavefield at A is transmitted to B (T_α^{++}) it may then either pass through from B to C (T_β^{++}) or be reflected back from β to B (R_β^{-+}). This in turn may be reflected back from α to B (R_α^{+-}). As before, this remaining contribution may be either reflected yet again, in which case the process will continue to repeat itself, or otherwise be transmitted through β to C. Summing all possible scattering paths from A to C we obtain, for the total transmitted effect,

$$\begin{aligned} Y^{++} &= T_\beta^{++}T_\alpha^{++} + T_\beta^{++}R_\alpha^{+-}R_\beta^{-+}T_\alpha^{++} + T_\beta^{++}R_\alpha^{+-}R_\beta^{-+}R_\alpha^{+-}R_\beta^{-+}T_\beta^{++} + \ldots \\ &= T_\beta^{++}[1 + R_\alpha^{+-}R_\beta^{-+} + (R_\alpha^{+-}R_\beta^{+})^2 + \ldots]T_\alpha^{++} \end{aligned} \tag{4.10}$$

which has a mathematical form similar to equation (4.5). It may therefore be rewritten as

$$T^{++} = T_\beta^{++}[1 - R_\alpha^{+-}R_\beta^{-+}]^{-1}T_\alpha^{++} \tag{4.11}$$

If the planes, and the distances between are symmetric, this may be written simply as

$$T_{(2)} = T[1 - R^2]^{-1}T \qquad (4.12)$$

where $T_{(2)}$ refers to the transmission matrix through two adjacent identical planes to infinite order in interlayer and intralayer scattering. Reflection from these two layers may be readily derived along such lines and the reader may wish to check that this gives the result

$$R_{(2)} = R + TR(1 - R^2)^{-1}T \qquad (4.13)$$

Equations (4.12) and (4.13) describe the combined scattering effect of two adjacent layers in terms of matrices which describe the scattering effect of a double layer. The process may be repeated, using the same equations to describe the combined scattering effect from four layers from the combination of two double layers (see Figure 4.4(b)). Iteration of the process will permit the scattering from 8, 16 or more to be described exactly without any additional complexity. This method is fairly rapid and always numerically convergent. Its limitation is that common to all **k**-space methods—if the planes are too close together, the number of evanescent beams which must be included will become excessively large (see section 4.2). Because it treats all scattering events exactly, it is capable of the same degree of accuracy as may be obtained from the slower layer-KKR 'exact' method.

4.6.2 Renormalized forward scattering (RFS)

Iterative schemes which treat the reflected flux as a perturbation may be devised using both **k**-space and *l*-space representations. Pendry first introduced a method which used the **k**-space representation in 1971.[183] He called it the renormalized forward scattering perturbation method (RFS). It was found to be significantly faster than all previous methods, yet convergent when dealing with quite strong scatterers.

The first stage of RFS is to define a new propagation vector in the forward direction that includes all forward scattering events exactly. Between the *i*th and *j*th layers (where the *j*th layer is the deeper of the two into the crystal) there are $(j - i)$ layers at which forward scattering could occur. The sum of all possible permutations of propagation and forward scattering events (i.e. excluding paths which would involve one or more reflections) can be shown to reduce to

$$[P^+(I + \tau^+)P^+]^{j-i} = [P_{\mathrm{RFS}}]^{j-i} \qquad (4.14)$$

so that an original wave amplitude a_i^+ at i will become $[P_{\mathrm{rfs}}^+]^{j-i}a_i^+$ which may be compared with a simple propagation without scattering between these layers $[P^+]^{2j-2i}a^+_i$ where P^+ is the propagation factor from a plane to the midpoint between layers. Consequently, by replacing the forward scattering propagation between layers $(P^+)^2$ by P_{RFS}, a *renormalized forward scattering propagator* can be defined which takes all possible forward scattering events into account.

Starting with the incident electron beam arriving at the surface layer, the incident wavefield on each successive layer into the crystal may be described, to a first approximation, by the application of the *renormalized forward scattering propagator*. As transmission from one layer to the next into the crystal is calculated using these new propagators, the reflected component for each layer may be derived by applying the reflection matrix to each impinging wavefield. However, subsequent multiple scattering of this reflected component is not included at this first level of approximation. Once the forward-travelling wavefield has become sufficiently small to satisfy some suitable criterion (e.g. $|\Sigma(a_i)^2| < 10^{-6}$), attention is switched to the outward-travelling flux, starting with reflected flux from the next layer in. If the layer at which the forward-travelling flux has fallen below the minimum limit is the nth layer, then the outward flux from this nth layer will comprise the reflected component from this nth layer plus the transmitted flux which is reflected from the $n + 1$)th layer. An outward 'pass' is now conducted in which the outward-travelling flux from each successive layer is given as the sum of the transmitted flux from all deeper layers, using the *renormalized forward scattering propagator,* plus the reflected contribution calculated during the previous 'inward pass'. On reaching the surface layer again, the total flux will comprise the sum of all possible paths which involve no more than one reflection. The flux reflected back into the crystal from the surface layer provides the starting-point for a second 'inward pass' which will involve the summation of forward-scattered components and reflected components from the previous 'outward pass'. This process is illustrated in Figure 4.5.

Each successive pass will increase the order of reflected contributions by one. The observed diffraction beams necessarily involve scattered flux with a normal component away from the crystal surface, that is, in a direction opposite that of the incident beam, and this can only result from paths that involve one, three, five or other odd numbers of reflections. Consequently, the 'order' of perturbation usually refers not to the total number of reflections, but to the series of reflection types which can yield observable results. Consequently, the first order of perturbation in RFS is given by the net result of the first two passes, one 'inward' and one 'outward', the second order is given by the net result of including a further two passes and so on. Iteration of this process continues until the total wavefield leaving the crystal differs negligibly from one iteration to the next. Under favourable conditions (e.g. a weak scatterer, energy above 50 eV or large interlayer spacing) a converged result may be achieved after only two orders. Commonly three orders are necessary, and in less favourable conditions four or more orders will be required. The great advantage of this scheme is that the addition of successive orders increases the computational time only linearly, so the inclusion of higher orders is easily accomplished and not excessively

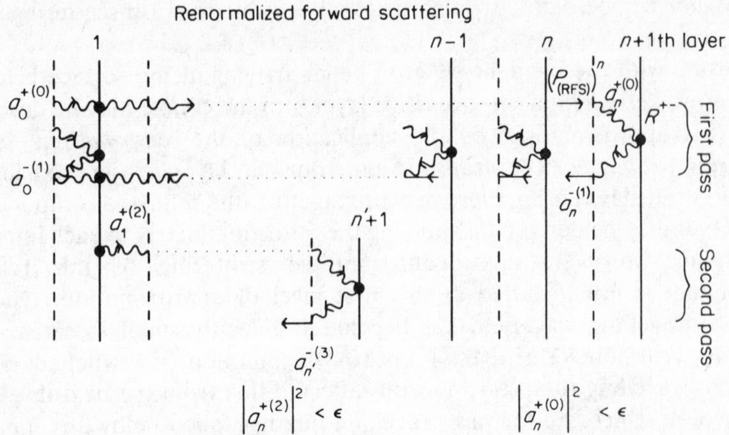

Figure 4.5. Schematic representation of the scattering processes included in the renormalized forward scattering (RFS) perturbation scheme

expensive. If convergence has not been achieved after the inclusion of six orders, however, then there is good reason to assume that a meaningfully converged result will not be possible with the method, and an error message is printed.

Although the basic assumption underlying the RFS method is that the reflected component should be small enough to be well represented as a perturbative contribution to the total flux, it has been shown that it may provide well-converged results even when the reverse scattering is comparably strong to the forward scattering, due to the influence of inelastic damping.[184]

4.6.3 Reverse scattering perturbation (RSP)

It is possible to construct an iterative perturbation scheme which works along similar lines to RFS, but in l-space representation. Such a method, called reverse scattering perturbation by its originators Zimmer and Holland,[185] combines the benefits of l-space representation with the speed and economy of RFS. Layer doubling fails to give meaningful results if the interlayer spacing falls below about 0.7 Å, and RFS is even more sensitive, breaking down for spacings less than about 1 Å. By comparison, RSP can deal with spacings down to zero provided the scatterer is not to strong.

The iterative mechanism of RSP is not unlike that described above for RFS, except that in practice the ordering of the subplanes may be less clearly defined. In RFS the subplanes must be well separated, and their natural ordering from the surface determines the order of inclusion of scattering events in the iterative series. RSP can be applied to complex layers which may include coplanar subplanes, and consequently any form of natural ordering may be lost. In this scheme, the ordering of the subplanes is assumed to be determined not by the depth of the subplanes within the bulk, but rather by the number of scattering events which must occur to produce the appropriate wave amplitudes. Nevertheless, for the sake of identification the subplanes must be ascribed some initial order, and they are therefore numbered sequentially from the surface, working inwards into the bulk. The perturbation order is determined by the number of 'reversals' which the electron wavefields experience, and this is defined in terms of the ascribed subplane order, even though in practice certain subplanes may be coplanar.

For example, if we are considering the nth-order scattering term from the ith layer, in a direction into the bulk, then we take the wavefield incident upon the ith subplane from 'above', i.e. from layers 1 to $(i - 1)$ which have already experienced n reversals, and consider the 'forward' scattering of this wavefield through subplane i into the bulk. This new, scattered wavefield preserves the criterion that it has experienced n reversals. To this wavefield, we add the reverse-scattered contribution of the wavefield incident on subplane i from 'within' the crystal, i.e. from layers $(i + 1)$ and greater (remembering that in practice the sum or all of these subplanes could be coplanar with i or themselves) which had previously experienced only $(n - 1)$ reversals. Reflection of this outgoing wavefield back into the bulk from subplane i brings the total reversals that it has experienced to n, and so the total 'inward'-travelling wavefield, which comprises both transmitted and reflected components, preserves the required order of perturbation. Boundary conditions are set by defining the total number of contributing layers outside of which no reflected flux can come, and an iteration scheme can be described which enables the process to be carried out to any required order of perturbation.

4.6.4 Bootstrap acceleration method

Adams has recently shown[186] that both RFS and RSP methods are directly related to the Gauss–Seidel–Aitken (GSA) double-sweep iterative method for the solution of systems of linear equations. This provides a precise mathematical analysis of the convergence behaviour of these methods. Here, RFS is found to be as efficient as possible for a given starting condition, but RSP is marginally less efficient in its original form for the simple reason that, on reaching the nth layer, instead of turning around to start the outward pass at once, the reflected values from the $(n + 1)$th layer should also be included (as is done in RFS).

An additional improvement may be made to both RFS and RSP methods. The convergence rate will obviously be related to the difference between the initial set of amplitudes and the final set. We have already noted that the process must be repeated in its entirety for each new model, and this reduces its advantage over, say, the layer-doubling approach when numerous models are to be tested. However, if the starting set of amplitudes is taken from the final results of a previously calculated model which is closely related in energy and structure, then the convergence rate may be greatly enhanced. The closer the starting set to the set to be eventually calculated, the greater the gain in speed. Called the 'bootstrap acceleration model' by Adams,[186] initial tests have shown that the average number of iterations may be halved compared with the standard RSP method, when using results for a surface interlayer spacing of 1.05 Å as input for a new calculation with interlayer spacing of 1.1 Å. The savings over standard RFS may, for normal conditions, be less great, but significant improvements may be obtained in cases where RFS would be otherwise struggling to converge. The bootstrap acceleration method cannot induce convergence where RFS would fail to converge, but provided the magnitude (spectral radius) of the iteration matrix is less than one, the bootstrap method may serve to improve the convergence rate. Trial results for V(110) indicate that advantages in convergence rate may be gained by using previous results separated by up to 4 eV in energy or 0.1 Å in surface interlayer spacing from the new model to be calculated.

4.6.5 Comparison of exact and perturbative iterative schemes

The iterative calculational schemes described in this section have significant advantages over previous schemes in terms of speed and computer memory space requirements. We have dealt in some detail with the relative merits of RFS and RSP, but it is important to recognize that the layer-doubling scheme is often comparably rapid when a large number of surface models are to be examined. This is because the iterative perturbative schemes in their standard forms must be rerun in their entirety for each new model. Layer doubling, on the other hand, is usually executed in such a way that scattering from the second and all equivalent layers within the bulk is calculated initially and the surface layer is only added at the end. If the various models only involve alterations or shifts of the surface layer, then the bulk scattering matrix may be retained from one model to the next, and only one layer combination process calculated per model. However, if the bootstrap acceleration procedure is applied to the perturbative methods, then this may restore their advantages in speed when dealing with multi-model calculations.

4.7 Combined space method (CSM)

Interlayer spacings associated with the adsorption of small atoms, e.g. H or one of the chalcogenates C, O or S, may be significantly less than 1 Å, and therefore unsuitable for calculation using **k**-space methods. On the other hand, treating the whole crystal by an *l*-space method is frequently very expensive in computer resources. The most efficient approach would be to treat the substrate with a relatively inexpensive **k**-space method and reserve the usage of the *l*-space representation for the closely spaced surface layers.

Van Hove and Tong developed a set of computer programs to achieve this marriage of methods, and called this approach the combined-space method (CSM). An outline and complete computer program listing is provided in their book.[15] It is a fair assertion that the CSM package provides the most flexible and useful set of LEED programs presently available. The reader is recommended to contact the authors or other established users of the programs before making use of the published programs, since in common with virtually all published program packages, they contain a few minor errors which have come to light since the original publication.

Quite complex surface structures can be modelled using the CSM. The only restriction is on the available computer memory space and time which restrict the number of subplanes in the surface layer to about six (depending on available computer resources and the number of phase shifts required to describe the individual ion-core scattering). Because subplanes can be coplanar in the l-space representation, this is equivalent to permitting unit cells of up to six atoms. A calculation involving six atoms per unit cell using the CSM package has recently been made by Gauthier *et al.* for a carbon adsorbate induced surface reconstruction of the Ni(100) surface.[187]

4.8 Chain method

All the conventional methods discussed so far have concentrated on the various approximations which may be used to model the interplanar scattering. It has been assumed that scattering within a plane can be efficiently and precisely calculated due to the infinite two-dimensional periodicity which exists in such a plane. Even when the interplanar scattering is described in **k**-space representation, the intraplanar contribution will involve calculations using l-space representation. (The concept of propagating diffraction beams loses its meaning within a plane.) Consequently, the most time-consuming steps in **k**-space models may actually be the evaluation of the two-dimensional lattice sums F_{lm} and the inversion of $(1 - X)_{lml'm'}$ where $X_{lml'm'}$ is the multiple-scattering matrix for a plane or a layer (see section 4.4). The dimension of $X_{lml'm'}$ is $2(l_{\max} + 1)^2$, so any increase in $l_{\max}$ will cause a dramatic increase in the required computational time.

Rather than treat the two-dimensional plane in a single step, it is possible to consider it as a series of parallel chains of atoms. Multiple scattering within a chain is solved and then the chains are assembled into a plane and multiple scattering between them also treated.[188] This has the advantage that the lattice sums F_{lm} involve a straightforward linear summation which can be executed very rapidly. Also, the matrix describing multiple scattering within a chain $X(m)$ has cylindrical symmetry and will therefore be diagonal in m. This matrix has dimensions $(2l_{\max} + 1)$ which is significantly less than that required in the two-dimensional case. However, there is an additional matrix inversion step required to describe multiple scattering between the chains and this may not be readily factorized. This offsets the savings gained by the reduction to one-dimensional stages and overall computational times will be similar to, or, in general, rather slower than, those associated with a direct two-dimensional calculation.

The chain method displays unique advantages in cases when the incident beam is at a large angle from the normal, i.e. approaching grazing incidence. The full multiplicity of the azimuthal quantum number, m, namely $(2l_{max} + 1)$, is no longer required. Pendry and Gard[188] showed that the maximum value of m necessary to preserve adequate convergence properties is given by $m_{max} = k_{\perp}r$, where $k_{\perp}$ is the perpendicular component of the incident momentum and r is the muffin-tin radius. This is clearly reduced by going to oblique angles of incidence. At $\theta = 80°$ over a typical range of energies from 50 to 250 eV, it could be shown that the $I(E)$ structure from typical surface is retained satisfactorily even for $m_{max} = 2$, compared with the full multiplicity of 7. In this case times could be reduced to less than one-quarter of that required for a full calculation.

The obvious advantages of this approach are in the extension of dynamical calculations to the higher-energy regions described as MEED (medium energy electron diffraction)[189] or RHEED (reflection high-energy electron diffraction).[190] It has been shown that incident electron energies of up to 5 kV can be treated by this method. Fully documented computer programs are available in the CAVLEED package.[169]

4.9 Specialized programs

4.9.1 Two-atom per unit cell program (CAVLEED)

In all the above discussions of computational methods, we have considered structures which can be divided into discrete planes, each of which comprise simple lattices with one atom per unit cell. If a gaseous species is adsorbed on to a surface, or surface reconstruction occurs, then a surface layer or selvedge may form that may involve two or more atoms per unit cell. In such cases, the constituent atoms of the surface unit cell may be divided conceptually into subplanes which may or may not be coplanar. It is frequently possible to treat such systems using l-space representation methods, as appear in the CSM package.[15] However, in Chapter 1 we noted that a wide variety of important binary compounds exist which involve two atoms per unit cell, not only in the surface region but throughout the bulk. So long as we are constrained to one atom per unit cell, **k**-space representation methods cannot normally be used to model such structures, and l-space methods would have to be applied throughout the bulk, in addition to the surface. This, as we have already noted, is not particularly efficient in its use of computer resources.

Specifically reformulating the intralayer scattering calculations to incorporate two atoms per unit cell (not necessarily coplanar) permits the redivision of the crystal into composite layers, each of which may be sufficiently well separated from its neighbours that **k**-space methods, with their concomitant improvements in program efficiency may be reintroduced.

The incorporation of more than one atom per unit cell into the standard intralayer scattering formalism necessitates extensive alterations and the only program available is that developed by Kinniburgh and published in the CAVLEED suite of programs.[169] These programs have been successfully employed to model a number of binary compound structures, including those with surfaces which 'rumple' due to different relaxations of the two atomic species.[191] No program has ever been developed to deal explicitly with more than two atoms per unit cell, but the classes of structures which would benefit from such a program are much less frequently encountered in LEED work, and CSM could be invoked if the need arose.

4.9.2 Reconstructed surfaces

'Reconstruction' is a term loosely used to describe any systematic change in the positions of the surface layer atoms away from the two-dimensional periodicity of corresponding bulk crystal planes. With alkali halides, surface reconstruction may involve the shift of alternate ionic species outwards or inwards with respect to the mean surface plane, in other cases, reconstruction may involve shifts of the atoms within the surface plane, resulting in atom pairing or zigzagged rows. If the resulting surface unit cell is directly related to the bulk unit cell then the structure can most likely be modelled using conventional programs described in sections 4.6 or 4.7. Often, however, reconstruction or weak adsorption may result in a surface layer in which the surface atoms do not all coincide with the same high-coordination sites provided by the substrate. If the surface layer has its own periodicity which differs form the substrate periodicity, but is nevertheless related in such a fashion that identical bonding sites are occupied at periodic intervals, then this is described as a 'coincidence structure' (see Chapter 1).

Well-known coincidence structures are those observed on clean Au, Pt and Ir(100) surfaces. Each of these surfaces possesses a periodicity which can be described as (5×1) or similar. (In fact that Au(100) pattern is really $c(68 \times 26)$.) Another example is the (7×7) structure found on the clean Si(111) surface. Noble gases adsorb very weakly (physisorb) on surfaces at low temperatures, and usually group into hexagonal close-packed structures which have periodicities incommensurate with the substrate structure. Again, these have been discussed in Chapter 1. Although it is not possible to deal with such structures rigorously using presently available theories, reasonable approximation schemes may be devised to deal at least with the simpler forms of reconstruction which occur.

If the reconstructed surface can be considered as a plane comprising one type of periodicity placed upon a substrate of a different periodicity, then we can use the analysis outlined in section 1.4.2 to find the scattering contributions which combine to produce each diffracted beam. The important conclusion drawn in that section was the multiple scattering within the bulk only intermixes beams whose wave vectors $\mathbf{k}'$ differ by integral multiples of the bulk reciprocal lattice vector **g**. If, for example, the coincidence structure has (5 × 1) periodicity, then five separate sets of diffracted beams may be considered to be propagating throughout the crystal. Recalculation of the bulk crystal planar scattering matrices five times for each possible diffracted beam set permits the bulk-scattering contribution to be described precisely. The only problem occurs at the surface layer because those beams emerging from the bulk could in principle be rescattered backwards into the bulk and cause further mixing of beam intensities. Fortunately, this does not provide a significant contribution to the total because backscattering is much weaker than forward scattering. Flux scattered back into the bulk would need to be reverse scattered a third time in order to be observed, and thus must be very weak. Admittedly, reverse-scattering contributions from five or more orders are commonly used in standard RFS calculations, but in that case the reverse scattered flux is the accumulation of contributions from a large number of periodically stacked planes, which may be significant even though any individual contribution alone could be considered to be negligible. In the case of a coincidence structure or incommensurate adlayer, we are only concerned with the reverse scattering from a single plane which is probably vibrating thermally with an amplitude rather greater than typical of the bulk (see Chapter 5). Neglect of mutliple scattering between bulk and surface is therefore to some extent justifiable. Experimental confirmation of the weakness of the surface–substrate scattering interaction has recently been made by Yang and Jona.[192] Using this assumption, a reasonably good estimate of the diffracted beam intensities of reconstructed surfaces may be made. Stoner *et al*. have applied this approach to the determination of the interlayer spacing between an irrationally related surface layer of Xe adsorbed on Ag(111),[193] and Van Hove *et al*. have also investigated (5 × 1) surface reconstructions on Ir, Pt and Au(001) surfaces with similar multiple-scattering theories.[194]

Although we have argued that single back-scattering may be acceptable to describe the scattering between a single surface layer and the bulk, the use of single-scattering (kinematic) theory to model the complete scattering process is much less justifiable, for reasons given above. Nevertheless, various attempts to use kinematic theory to model such structures do seem to have met with a fair degree of success. Ar and Kr graphite have been examined by Shaw *et al*. using kinematic theory,[195] and the (7 × 7) reconstructed surface of Si has been examined using kinematic analysis by Levine *et al*.[196] In both these cases, the substrate ion cores are relatively weak scatterers, and this undoubtedly improves the viability of kinematic analysis. A major constraint on using multiple scattering for reconstructed surfaces is the enormous demand made upon available computer resources. A potentially valuable approach to the analysis of such complex surfaces may be the $I(\mathbf{g})$ method described in Chapter 7.

4.10 Relativistic effects and spin-polarized LEED

Although the velocities of electrons in LEED experiments are very low compared with the velocity of light, relativistic effects may play a role in their interaction with heavy atom cores. Spin-orbit coupling terms in the scattering potential which can be ignored in the case of light atoms play an increasingly important role as the orbital angular momentum term increases. If the incident beam is not preferentially polarized, this effect can be accommodated through the modification of the ion-core scattering potential, since the spin orientation of the incident electrons will be random and any specific spin interactions will average out. However, if the experiment is divised to include either the emission or detection of spin-polarized electrons, then the scattering behaviour of electrons polarized in one direction may be distinguished from those polarized at 90° to them. At present, the primary advantage of using spin-polarized electrons in surface structural determinations is to provide an alternative means of obtaining results, thereby providing complementary data by which to check the reliability of the structural determination. Initial investigations using this approach with W(001) indicate that the diffraction features associated with spin-polarized LEED (SPLEED) can be more sensitive to details of the surface structure. Unfortunately, they are also more sensitive to values of non-structural parameters which enter the theory, and in particular to the scattering-core potential. If the non-structural parameters can be optimized, one may be hopeful that the structural results may be at least as accurate as those obtainable from conventional LEED techniques.

SPLEED has a unique advantage when it comes to the study of ordered magnetic structures, since it is sensitive to the exchange interaction between the propagating electrons of the LEED beam and the ground state electrons of the crystal. Such effects have been predicted theoretically[197] and demonstrated experimentally,[198] but the technique is still very new and its capabilities remain to be properly explored.

Modification of multiple-scattering programs to include to a good approximation of effects of spin polarization within the beams is in principle fairly straightforward. Relativistic effects only enter the calculation in the ion-core scattering process itself, and this appears in the planar multiple-scattering matrix $\boldsymbol{M}$ (equation 4.8) as the t-matrix $t_{lmsl'm's'}$ which replaces the non-relativistic $t_{lml'm'}$. It was shown in Chapter 3 that the restriction on spin quantum numbers limits the number of spin-dependent components to four—essentially, an incident electron may be polarized into either a spin up or spin down state, and when scattered from an ion core may either retain its original polarization, or be 'flipped-over'. The dimensions of the scattering matrices are consequently doubled, resulting in a fourfold increase in the necessary computer storage space. Matrix inversions scale as the cube of the dimensions, so the planar scattering matrices increase by a factor of 8 over the non-relativistic case. By comparison with non-polarized LEED, the computation of intensities is expensive, but not excessively so. In making an investigation it is clearly better to perform as many preliminary tests of structural and non-structural parameters as possible using non-polarized LEED, and then use SPLEED programs to test only a restricted number of possible models.

If standard non-polarized techniques are to be used in a study of the surface of a very heavy material, it is valuable to know the extent to which the neglect of specific SPLEED interactions will affect the quality of agreement between theory and experiment. When relativistic effects are ignored, the resulting intensity–voltage curves may differ significantly in peak heights and shapes when compared with calculations using spin-averaged relativistic phase shifts. The peak positions (energies) are hardly affected. However, if a full relativistic calculation is performed and the results averaged to simulate the non-polarized situation, the *I(E)* curves are almost indistinguishable from the standard LEED calculations which used spin-averaged phase shifts.[199]

This demonstrates the sensitivity of beam intensities to details of the scattering potential, but also shows that the peak positions are more representative of the geometrical configuration of the scattering atoms. One may conclude that *unless specific spin-polarized effects are being investigated experimentally, standard non-relativistic LEED programs are quite adequate for the treatment of heavy atoms provided the phase shifts include spin-averaged relativistic effects.*

4.11 Treatment of molecular adsorbates

Standard techniques as described so far in this chapter have been applied quite successfully to the adsorption of CO molecules on a number of different substrates, whilst acetylene (C_2H_2) and ethylene (C_2H_4) have been investigated by ignoring the extremely weak contribution to the scattering from the hydrogen atoms. Fortunately, not only are hydrogen atoms sufficiently weak scatterers that they can normally be ignored in LEED studies of this type, but typical constituents of adsorbed molecules, C, O and N for example, are relatively weak compared with metal substrate atoms. This means that their scattering can be fully described through the use of only a small number of phase shifts.

The combined space method (CSM)[15] was described in section 4.7 as the ideal compromise, in that it permits rapid **k**-space methods to be used within the regularly periodic bulk of the crystal, whilst the more cumbersome yet versatile *l*-space method can be employed to treat the small interlayer spacings which commonly result from the adsorption of small gas atoms on to the substrate surface.

This method has clear advantages when it comes to the treatment of molecular adsorption—if the atoms comprising the molecule can be described in terms of only a few phase shifts, then the *l*-space method is able to cope fairly easily with systems of up to six atoms, whilst **k**-space remains an efficient basis for dealing with the rather heavier but well-spaced atoms of the substrate.

Even if LEED theories are able to cope with the complexity of typical adsorbed molecules, is the information available from the resulting diffraction-beam intensities sufficient to provide adequate resolution of the principal geometrical features of the molecules? It is well known from clean surface studies that LEED is predominantly sensitive to atomic displacements normal to the surface. It is less sensitive to lateral displacements (parallel to the surface plane). However, it is very difficult to make quantitative estimates of the relative sensitivity of LEED to different types of displacement because the kinds of changes which occur in the $I(E)$ curves are not overtly systematic. Some beams will be affected more than others, but these variations are not very predictable, since they stem from a complex chain of relationships between the angular and energy features of the ion-core scattering, and the multiple scattering which occurs between neighbouring cores. Van Hove and Somorjai have recently published the results of an investigation of the sensitivity of LEED to various structural changes which could occur in a hypothetical benzene-type adsorbate.[200] They noted the changes in $I(E)$ curves which result from various distortions of the hexagonal ring of carbon atoms.

Expanding the rings away from the central axis induced changes in the $I(E)$ curves sufficiently large that displacements of about 0.2 Å could be clearly distinguished. This is comparable to the sensitivity to registry shifts or lateral displacements which may occur in the case of simple atomic adsorption. On the other hand, there was very little sensitivity to rotation of the rings about their central axis. In order to appreciate the significance of this, it is valuable to consider in more detail the actual structural model that Van Hove and Somorjai were investigating. The molecular adlayer was not sitting on an independent substrate as would normally occur, but on a stack of identical layers of carbon rings. Consequently, rotation had no signifant effect on the scattering between members of each ring or between corresponding rings in successive layers; rotation primarily affected the longer-range scattering between different rings within a layer, and this was shown to be a secondary effect. Expansion, on the other hand, had a marked effect on the scattering between nearest neighbours. It is reasonable to assume that significantly more sensitivity to rotation of molecular rings would have been obtained if they had been adsorbed on to a normal fixed substrate, since this would directly influence the scattering between atoms within the molecule and their nearest neighbours in the substrate surface. On buckling the carbon rings, so that alternate carbon atoms move in opposite directions perpendicular to the plane, numerous features of the $I(E)$ curves were found to change, leading to a reasonably high level of senstivity. This is consistent with the usual result that LEED $I(E)$ curves are predominantly sensitive to atomic displacements normal to the surface plane.

4.11.1 Approximation methods

Even if there is sufficient sensitivity to the required geometrical features of the adsorbed molecules, there remains the problem of actually performing the computing. Provided the molecules do not involve more than about six atoms, CSM is a viable method, as we have already noted. Nevertheless, there are many more permutations of geometrical configurations with six atoms per molecule than with just one or two, and the total computational effort necessary to consider all possible models exhaustively would pose a daunting task. Clearly some short-cut methods are necessary to 'screen' the major categories of molecular geometry so that a full detailed investigation need only be used to distinguish between a small number of likely candidates. In their study of the sensitivity of LEED to ring-like molecules, Van Hove and Somorjai considered the influence of increasing levels of approximation on the overall structural sensitivity of the *I(E)* curves.

The simplifications which may be introduced are linked to the mathematical treatment of the multiple scattering. Multiple scattering may be limited if the ion-core scattering amplitude is small, as is the case for the light gas atoms which commonly make up organic molecules. Low densities will also reduce the level of multiple scattering as a direct result of the inelastic scattering which limits the mean free path of the electrons. This effect is not as great as one might expect from previous studies of metal surfaces, because the light atoms provide fewer possibilities for single-electron or collective excitations and so the mean free path of the LEED energy electrons is greater. The results of the tests of Van Hove and Somorjai cited above indicated that the *I(E)* curves were not very sensitive to geometry changes between different carbon rings, and this confirms the presence of ample inelastic damping.

A first approximation to the multiple-scattering formalism would be to ignore multiple scattering between molecules. This avoids the matrix inversion which stems from this intralayer multiple scattering, thereby saving computer time, but not producing significant changes in the resulting *I(E)* curves provided inelastic damping is large. Using the RSP method to calculate the scattering within the molecular adsorbate region provides the possibility of deliberately restricting the order of scattering between neighbouring atoms. This is simply achieved by limiting the number of passes allowed within the scheme or alternatively by relaxing the convergence criterion. Since the number of passes is automatically terminated when convergence is achieved, this requires a deliberate deviation from normal levels of convergence, and so *I(E)* curves are more substantially affected. However, it is found that relative heights rather than peak positions are the principal features altered in the *I(E)* curves, so that sensitivity to geometry is not grossly degraded.

It is possible to go one stage further, by ignoring completely all intralayer scattering. Significant savings in computer time can be made with this approximation. For normal metal surfaces, structural sensitivity is so severely restricted at this level of approximation that little advantage is to be gained. However, we have already noted that multiple scattering is restricted by the weak scattering effect and low density of the molecular adsorbate species, and so the loss of structural information is much less severe.

An efficient method for solving the structures of molecularly adsorbed species would be to make a preliminary 'scan' of possible geometrical forms by ignoring intralayer scattering (the quasi-dynamical approach), then introduce a first level of intramolecular scattering to improve the quality for a more detailed survey of the more promising candidates and finally introduce full intralayer scattering to refine the model.

The deliberate approximations introduced into the theory for the preliminary investigations have been shown to affect peak heights rather than positions. If a numerical comparison scheme is used to compare theoretical and experimental curves it is clearly important to choose a simple scheme which is predominantly sensitive to peak positions, rather than a more comprehensive one (see Chapter 7).

Spurious peaks may appear in the theory when severe approximations are introduced, because dissipation of electron flux through multiple scattering is not permitted with the molecular layer. If the imaginary part of the inner potential, which determines the inelastic damping is increased, this will help dissipate the electron flux without introducing effects which significantly alter the structural conclusions. This is possible because the structural information is more closely linked to the phase of the electron wavefield between ion cores rather than their amplitudes.

4.11.2 Asymptotic cluster model

The second level of approximations given above involved the calculations of scattering between nearest neighbours within a molecule (i.e. within a surface unit cell) but not between molecules. This is an interesting step, because it overlaps with an alternative theoretical approach (the asymptotic cluster model[201]) in which attention is focused on the scattering properties of discrete clusters of atoms. Such clusters of atoms, which may or may not correspond to adsorbed molecules, may then be used as the basic scattering unit ready for assembly into a model of the surface.

Apart from the conceptual benefit of treating adsorbed molecules as integral units rather than as a collection of neighbouring atoms from separate subplanes (which, if the intralayer scattering is neglected, no longer form subplanes in any case), there is a more fundamental advantage to be gained from a cluster approach. This is to do with the thermal vibration of the molecule.

Modelling the effects of thermal vibrations would be a very complex task if it were not for certain simplifying assumptions, and these will be described in detail in Chapter 5. However, it is useful in the present section to anticipate the discussion of Chapter 5 and point out that thermal vibration are commonly introduced via a description of scattering from isotropically vibrating ion cores. Correlations of vibrations between ion cores is usually ignored. Thermal vibrations of an adsorbed molecule will tend to be greater than vibrations of substrate ion cores due to the weaker bonding between molecule and substrate, but, more importantly, the vibration of the constituent atoms is likely to be highly correlated. An obvious advantage of the cluster approach is that, if the scattering of a molecule is described in terms of a single origin, and the corrections for thermal vibrations applied retrospectively to the resulting 'molecular' scattering matrix, then the result will correspond to complete correlation of the molecular vibrations. Molecular vibrations do not really comprise a single correlated mode of vibration, but this is nevertheless much nearer actuality than a model based on the assumption of zero correlation. This is particularly important when the molecule comprises atoms stacked one above the other along an axis perpendicular to the surface, and when the incident electron beam is also along the normal to the surface. The separation between small atoms, for example C—C or C—O, will be small and scattering between these atoms may be significant in this perpendicular configuration—a good approximation to the thermal vibrations may be quite important. Indeed, in the absence of a specific cluster-based model the best approximation might well be to ignore completely the thermal vibrations within the molecule layers, since the introduction of non-correlated vibrations will very likely overcompensate for this effect.

The only calculation yet performed using a specifically calculated molecular scattering matrix is for CO on Ni(100) by Anderson and Pendry.[202] To achieve this they derived sets of multiple scattering equations for waves centred on each of the C and O atoms in the molecule, transforming to an expansion about a single origin only after this calculation. This is a relatively straightforward method, utilizing mathematical methods already extant from standard LEED programs. For small molecules this is quite appropriate as a course of action. For large molecules, however, a different approach is more suitable—atoms within the molecule are subdivided into shells about a chosen origin, and multiple scattering described in terms of intershell matrices, in a spherical analogue of more usual interplanar methods. This method has formed the basis of previous EXAFS calculations (see section 8.3.2), and has been recently proposed by Pendry as a potentially powerful method for extending LEED calculations to the solution of large molecules.[201]

Calculation in terms of spherical shells has the advantage that the full scattering matrix can be built up, shell by shell using a method rather like layer doubling (except, of course, that each shell will differ from the previous), and the complete wavefield is only fully described at the surface. If R_m is the radius of the molecule then the basis will scale as R_m^2. Clearly, the origin should be chosen to minimize this radius. A multicentre expansion requires the definition of the wavefield completely thrughout the volume of the sphere and consequently it scales roughly as R_m^3. For molecules which uniformly fill a three-dimensional volume, the shell expansion represents a considerably more efficient approach. For planar molecules there is little difference in efficiency between the two approaches whilst for linear or long-chain molecules, the multicentre expansion approach will be the more efficient. For linear molecules it is conceivable that an expansion in cylindrical coordinates could provide as efficient method, making use of basic computational elements from the chain method, but this has never been attempted.

The cluster approach not only provides an efficient basis for molecular adsorption calculations but also affords a novel approach to general structural calculations which involve large two-dimensional unit cells. If n is the number of atoms in the unit cell, then present methods scale as n^3 and the calculation becomes unmanageable if n is increased beyond 6 or so.

Consider any atom in the surface region of a crystal—the number of possible paths which a scattered electron beam may take between this atoms and the surface is limited by the inelastic mean free path. Now, if a scheme could be devised to calculate all the possible paths, then the total scattering from n atoms would only involve a linear summation of these n contributions. Consequently, the extension to n atoms would scale as n, rather than n^3. Existing methods have been devised to treat structures in which the unit cells are small, and scattering throughout the periodic lattice is very important. The cluster method would provide a superior scheme only when the unit cells reach such a complexity that scattering within the cell dominates scattering between cells, the crucial factor being the limiation on scattering length which results from inelastic damping. Nevertheless, once this cluster-dominated regime is reached, the linearity of the scaling will provide considerable benefits over standard methods which scale as n^3.

Pendry has outlined a possible method which would scale as n, which he calls the asymptotic cluster model.[201] The scattering cross-section of any particular ion core is necessarily finite so that the incident electron beam propagating into the crystal will be partly scattered by successive layers of atoms, but will also include an unscattered portion. Every atom may be considered as the starting-point of a series of scattering events, and the total wavefield eventually back-scattered from the surface can be described as the sum of contributions from each atom. Inelastic damping will limit the number of scattering events which can follow the scattering from the initial atom. This defines a cluster of atoms within a sphere described about the initial atom. A 'shell-type' expansion about this atom can then be performed, with the simplification over the general molecular approach previously described that the external wavefield only impinges on the initial, central atom. This method remains to be programmed and tested, but promises to provide a valuable contribution to future extensions of LEED theory.

4.12 Treatment of disordered surfaces

Diffraction results from the scattering of a wavefront by a periodic scatterer. Complete disorder would produce no diffraction. However, the adsorption of a gas upon a surface may occur in a partially ordered fashion, limited to specific bonding sites on the substrate surface, yet occupying available sites in a random manner. The resulting diffraction pattern would be intermediate between that produced by a clean surface and that of a perfectly ordered overlayer, but in detail would depend on the mean surface coverage and extent of ordering within the adlayer. If the adlayer is highly ordered in discrete regions but these regions do not match together to produce a single, perfect phase, then the resulting diffraction-beam intensities may be predicted using kinematic approximations, as will be described in Chapter 6. On the other hand, if the disorder occurs over a relatively short range, comparable to the lattice cell dimensions, then it is necessary to consider the scattering amplitudes from small groups of surface atoms, each describing a possible configuration of occupied and unoccupied sites which may occur.

A successful approach to this problem was devised and tested by Jagodzinski *et al.*[203] who used Beeby's T-matrix method as a basis for calculating scattering within a disordered surface layer. In their method, the short-range order surrounding a particular atom is modelled by defining a region which comprises nearest and next-nearest neighbours within which multiple scattering is precisely calculated. Scattering from outside this region is represented by an averaged T-matrix. This method therefore bears certain similarities to the asymptotic cluster model described in section 4.11.2. The calculation is repeated for each possible configuration of nearest neighbours. In practice, certain configurations may be more probable than others, due to interatomic forces, but for a given set of interaction parameters the statistically most probable combination of configurations may be derived. From this information, the resulting diffracted beam intensities may be derived by superposition.

A particular interest in such an approach is the possibly of analysing one-dimensionally disordered surfaces: these occur not infrequently whenever the substrate surface lattice is rectangular. Clean Au(110) appears to disorder in one direction, oxygen adsorbs into one-dimensionally disordered structures on W(112) and Au(110), and sodium on Ni(110) has also been observed to adsorb in this way.[204] Correlation functions for one-dimensional disorder can be calculated analytically, and consequently a direct relationship between interaction parameters and diffracted intensities can be found.

Visually sharp LEED patterns do not necessarily indicate perfect order in the surface.[205]

4.13 Summary of program packages available

Most of the recent computational schemes described in this chapter can be found in one or other of two computer program libraries, one written by Titterington and Kinniburgh, called CAVLEED,[169] the other by Van Hove and Tong[16] which contains the combined space method (CSM) as a major feature. CAVLEED is a compilation and generalization of programs developed in the Cavendish Laboratories, Cambridge, UK (hence the name) and owe much to the programs listed previously by Pendry.[105] Van Hove also worked originally on LEED under Pendry, so that a number of similarities also exist between the CAVLEED and CSM packages. A few other program libraries have also been published in the past. In 1974, Hoffstein published some programs[206] based on pseudo-potential scattering and Bloch-wave matching, but this approach has now been superseded by more recent methods. Rundgren and Salwen published separate versions of the layer-KKR and RFS perturbation schemes in 1974, but, perhaps more importantly, in 1975 they published a scheme to make use of the symmetry properties of diffracted beams,[207] and this may be adapted to the CAVLEED package to make significant savings in computer time and storage in appropriate cases.

The most recent set of programs is called CHANGE and has been developed by Jepsen for IBM in New York.[208] These have not been published nor extensively documented and, unlike all the other program libraries, are not restricted to FORTRAN language. A substantial part of the programs is written in IBM optimizing PL/1 and two subroutines use IBM 370 Assembly language, thus restricting the range of machines on which this library may be used. Despite these drawbacks, CHANGE is important due to its high degree of versatility. It is based on a scheme similar in approach to CSM, and can cope with as many as four atoms per unit cell in bulk layers and eight atoms per unit cell in the surface layer—sufficient to deal with an enormous variety of possible surface structures.

The existence of essentially independent libraries of programs is clearly very valuable, considering the complexity of the theory and the wide variety of assumptions and approximations which necessarily enter the calculations. In practice, it is a wise precaution to check that results obtained for a particular structure using one method or set of programs agree with those obtained from another. From the user's point of view, this provides a very useful check that he has not make any mistakes in the input data, or in handling the programs (such as setting dimensions). At a more fundamental level, the differences which result from the numerical schemes used by CAVLEED, CSM and CHANGE will produce small variations in results even when the structural and non-structural features of the model are identical. Shih has recently published a comparison which illustrates the extent of these discrepancies, which are usually, but not always, negligible.[209]

CHAPTER 5

Temperature effects

5.1 Introduction

So far, we have considered diffraction from rigid, perfectly ordered crystal surfaces. In practice, the real experimental surface will deviate from this ideal in two major ways. Firstly, defects and impurities will disrupt the perfection of the lattice; secondly, at finite temperatures the ion cores will vibrate about their mean positions. The latter is a relatively regular phenomenon which can be treated mathematically in reasonable detail. The former will generally reduce the efficiency of the diffraction process and reduce beam intensities, but not affect specific diffraction features. In certain cases, however, the defects may be ordered—for example, if there are 'steps' on the surface, in which case the diffraction pattern itself may be modified in a characteristic manner. A detailed discussion of these effects is given in Chapter 6. In this chapter we concern ourselves specifically with the effects of thermal vibrations.

The consequences of a finite temperature are threefold. Firstly, the lattice will expand, shifting Bragg peaks towards lower energies. Secondly, the motion of the ion cores in the form of phonon modes may result in an interchange of energy with the incident electrons, resulting in inelastically scattered components. Thirdly, the momentary displacements of the ion cores from perfect periodicity will lead to incoherent scattering, reducing the intensities concentrated in the diffracted beams.

5.2 Lattice vibration: phonon scattering and the Debye–Waller factor

Before getting involved in the modifications required to accommodate thermal vibrations into multiple-scattering theory, it is instructive to obtain a qualitative feel for surface thermal behaviour using the much simpler kinematic approach.

In kinematic theory the resulting intensity is produced by the interference of wave amplitudes scattered singly from the set of ion cores, which may be defined to be at positions $\mathbf{r}_i$ from some chosen origin. Propagation of an electron with wave vector $\mathbf{k}$ to $\mathbf{r}_i$, scattering into $\mathbf{k}'$ and subsequent propagation with wave vector $\mathbf{k}'$ describes the reflected amplitude

$$A(\mathbf{k}',\mathbf{k}) \propto \sum_i \exp(-i\mathbf{k}'\cdot\mathbf{r}_i)t_i(\mathbf{k}',\mathbf{k})\exp(i\mathbf{k}\cdot\mathbf{r}_i) \tag{5.1}$$

If the scatterers are identical, we may replace the set of individual ion-core t-matrices $t_i(\mathbf{k}', \mathbf{k})$ by a single representative t-matrix $t(\mathbf{k}', \mathbf{k})$. Expressing the atomic position $\mathbf{r}_i$ as an instantaneous thermal displacement $\Delta\mathbf{r}_i$ about the corresponding mean lattice site U_i such that

$$\mathbf{r}_i = \mathbf{U}_i + \Delta\mathbf{r}_i$$
$$\Delta\mathbf{k} = \mathbf{k} - \mathbf{k}_i'$$

we obtain, for the resulting intensity

$$I(\Delta\mathbf{k}) \propto |\, t(\mathbf{k}',\mathbf{k})|^2 \left|\frac{k_z'}{k_z}\right| \sum_{ij} \exp\,[i\ \Delta\mathbf{k}(\mathbf{U}_i - \mathbf{U}_j)]\exp[i\ \Delta\mathbf{k}(\Delta\mathbf{r}_i - \Delta\mathbf{r}_j)] \qquad (5.2)$$

Because the atomic vibrations are very rapid compared to the duration of measurement, it is the time-averaged intensity $\langle I \rangle$ rather than the instantaneous value that is observed. This affects the final term in equation (5.2). Assuming that there is no systematic correlation between any two atomic vibrations (see section 5.9), this final term may be re-expressed as

$$\langle \exp[i\ \Delta\mathbf{k}\cdot(\Delta\mathbf{r}_i - \Delta\mathbf{r}_j)]\rangle = \langle \exp[i\ \Delta\mathbf{k}\cdot\Delta\mathbf{r}_i]\rangle\langle \exp[i\ \Delta\mathbf{k}\cdot\Delta\mathbf{r}_j]\rangle \qquad (5.3)$$

It can be shown[210] that (see question 13)

$$\langle \exp(i\ \Delta\mathbf{k}\cdot\Delta\mathbf{r}_i)\rangle \simeq \exp[-\tfrac{1}{2}\langle(\Delta\mathbf{k}\cdot\Delta\mathbf{r}_i)^2\rangle] \qquad (5.4)$$

Assuming that the displacements are small, this may be expanded and then substituted into equation (5.2), giving, for the time-averaged intensity,[211]

$$\begin{aligned}\langle I \rangle \propto\ & |t(\mathbf{k}',\mathbf{k})|^2 \left|\frac{k_z}{k_z}\right| \exp[-\langle(\Delta\mathbf{k}\cdot\Delta\mathbf{r})^2\rangle]\sum_{ij}\exp[i\ \Delta\mathbf{k}\cdot(\mathbf{U}_i - \mathbf{U}_j)] \\ & \times\{1 + \langle(\Delta\mathbf{k}\cdot\Delta\mathbf{r}_j)\rangle + [\exp\langle(\Delta\mathbf{k}\cdot\Delta\mathbf{r}_i)(\Delta\mathbf{k}\cdot\Delta\mathbf{r}_j)\rangle - 1 \\ & - \langle(\Delta\mathbf{k}\cdot\Delta\mathbf{r}_i)(\Delta\mathbf{k}\cdot\Delta\mathbf{r}_j)\rangle]\}\end{aligned} \qquad (5.5)$$

The first term within the curly bracket represents the scattering from the rigid lattice, ignoring the fact that the ion cores are in motion. The second term describes the one-phonon scattering and the third, within square brackets, describes the multiphonon scattering. Taking the first term alone, we may write the intensity

$$\begin{aligned}\langle I \rangle &\propto |t(\mathbf{k}',\mathbf{k})|^2\exp(-2M)\sum_{ij}\exp i\ \Delta\mathbf{k}\cdot(\mathbf{U}_i - \mathbf{U}_j) \\ &= I_0\exp(-2M)\end{aligned} \qquad (5.6)$$

The beam intensity is reduced by the factor

$$\exp(-2M)$$

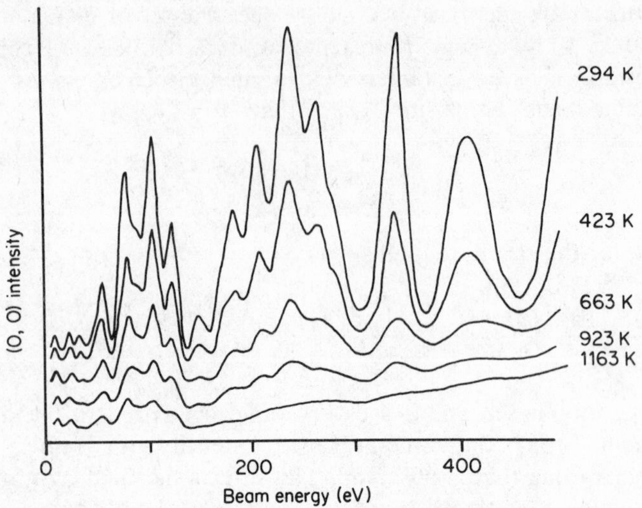

Figure 5.1. The (0, 0) beam from a Nb(001) surface, the incident beam being 8° off normal incidence. Curves for various temperatures are displaced relative to one another for clarity. (Reproduced with permission from Wilson and Bastow[215])

which is known as the Debye–Waller factor. This effect is clearly shown in Figure 5.1, in which diffraction-beam intensities from Nb(001) are plotted as a function of temperature.

The assumption that the atoms are displaced, yet motionless (zero-phonon scattering) is known as the Born–Oppenheimer approximation. It is not an unreasonable assumption since electrons at LEED energies may travel typically 10^4 times faster than the vibrating ion cores.

The factor M is thus given by

$$M = \frac{1}{2}\langle(\mathbf{\Delta k}\cdot\mathbf{\Delta r})^2\rangle = \frac{1}{6}|\mathbf{\Delta k}|^2\langle(\mathbf{\Delta r})^2\rangle \tag{5.7}$$

The mean-square amplitudes of vibration may be calculated from theories of lattice dynamics and we find relatively simple expressions in the limiting cases of high and low temperatures:[212]

$$\langle(\mathbf{\Delta r})^2\rangle_{T\to\infty} \simeq \frac{9}{mk_B\theta_D} \qquad (T \gg \theta_D) \tag{5.8}$$

$$\langle(\mathbf{\Delta r})^2\rangle_{T\to 0} \simeq \frac{9}{mk_B\theta_D}\left(\frac{1}{4} + 1.642\,\frac{T^2}{\theta_D^2}\right) \qquad (T\to 0) \tag{5.9}$$

where m is the atomic mass in electron mass units, k_B is Boltzmann's constant (3.17×10^{-6} Hartrees/kelvin) and T and θ_D are in kelvins; θ_D is a parameter which serves to describe the stiffness or compressibility of the lattice: it has the dimensions of temperature, and is known as the Debye temperature. The high-temperature limit is approached when $T \gg \theta_D$.

Frequently, the high-temperature limit is assumed, giving M as

$$M = \frac{3(\mathbf{\Delta k})^2 T}{2mk_B\theta_D^2} \tag{5.10}$$

Here, M is related to the beam intensity by the exponential expression given in equation (5.6). By substituting equation (5.10) and differentiating both sides of the equation by temperature, we see that the variation of the logarithm of intensity versus temperature should be linear in this approximation. Results from the Nb(001) data (see Figure 5.1) are given in Figure 5.2, and show that, in this energy and temperature range, this assumption is indeed valid. Deviations from linearity can, nevertheless, been detected at very low temperatures.[213,214]

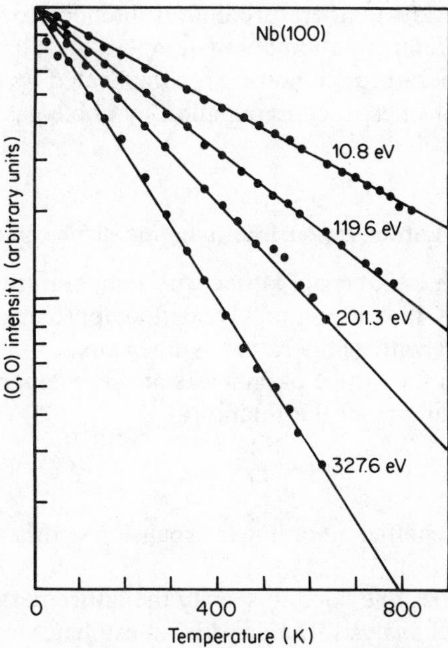

Figure 5.2. The (0. 0) beam intensity, plotted a logarithmic scale, as a function of temperature, for the Nb(001) data given in Figure 5.1. (Reproduced with permission from Wilson[234])

From the slopes of the curves, the Debye temperature may be calculated. If θ is the angle between $\mathbf{k}$ and $\mathbf{k}'$ (so that $\Delta\mathbf{k}|\mathbf{k}|\cos\theta$), and

$$\frac{2\pi}{|\mathbf{k}|} = \sqrt{\left(\frac{150.4}{E}\right)}$$

where E is the incident beam energy in eV, it is easy to show that

$$\Theta_D^2 = \frac{AE}{\left[\dfrac{d(\log I)}{dT}\right]} \tag{5.11}$$

where A is a material-dependent constant (see question 14).

In the LEED computer program suite CSM by Van Hove and Tong,[16] a facility has been provided for using an approximate mean-squared displacement formed by the mean of the high- and low-temperature limits

$$\langle(\Delta\mathbf{r})^2\rangle = \sqrt{\{[\langle(\Delta\mathbf{r})^2\rangle_{T=0}]^2 + [\langle(\Delta\mathbf{r})^2\rangle_{T\to\infty}]^2\}} \tag{5.12}$$

This inclusion of the low-temperature limit is intended to improve the model when applied to the interpretation of low-temperature experiments. Accurate analysis at low temperatures is, however, complicated by the high sensitivity of LEED beam intensities to contamination by noble-gas atoms.

5.3 Lattice expansion: dynamic displacements

The expansion of the bulk crystal lattice with temperature is quite accurately determinable from X-ray diffraction. To a good approximation, the variation in lattice parameters with temperature is linear over a limited temperature range, so that, given the lattice parameters at one temperature, the value at any other can be found from the relationship

$$l_t = l_0(1 + \alpha T) \tag{5.13}$$

Values of l_0 and the coefficient of linear expansion α for a variety of elements are given in Appendix 1 (p. 311).

How accurately does one need to specify the lattice parameters being used in a LEED structural analysis? To provide an estimate, we may consider the change in energy of the Bragg peaks associated with a change d in the interlaying spacing d. In the case of normal incidence, the Bragg equation simplifies to

$$E' = E_{\text{vac}} - v_r = \left(\frac{n\pi}{d}\right)^2 \tag{5.14}$$

From this it is easy to show that, assuming $\Delta d \ll d$,

$$\frac{\Delta E}{E} = \frac{2\Delta d}{d} \tag{5.15}$$

For example, if a crystal with lattice constant 4 Å has been annealed by heating prior to taking the experimental results, its temperature might be 50 °C or more above the assumed room-temperature value. If its coefficient of thermal expansion $\alpha = 2.5 \times 10^{-5}\,\mathrm{C}^{-1}$ then $\Delta d = \alpha d\,\Delta T = 5 \times 10^{-3}$ Å so that a peak at 200 eV would shift by 0.5 eV. This is at the limit of experimental detectability using a standard LEED system; however, it indicates that thermal expansion may significantly affect intensity peak energies if the crystal temperature is varied from, say liquid nitrogen to room temperature or above.

Expansion of the surface layer may be greater than that of the bulk, shifting further the energies of the Bragg peaks. Considering expansion perpendicular to the surface, we may define a coefficient of surface expansion α_{surf} by

$$\alpha_{surf} = \frac{1}{z}\frac{\mathrm{d}z}{\mathrm{d}T} \tag{5.16}$$

where z is the lattice spacing perpendicular to the surface. Observation of the Bragg peak shift ΔE will lead to an effective coefficient of expansion α_{eff} somewhere between surface and bulk values

$$\alpha_{eff} = -\frac{\Delta E}{2\Delta T(E - V_0)} \tag{5.17}$$

Assuming that the electron penetration depth increases with energy one might expect that α_{eff} will take values near α_{bulk} for high-energy Bragg peaks and near α_{surf} for low energies. In fact, the situation is considerably more complicated than this due to multiple scattering, the consequences of which are discussed more fully in section 5.6. If however we continue with this kinematic interpretation, the greater proportional shift at lower energies may be associated (via equation 5.17), with a coefficient of surface expansion about double that of the bulk. Figure 5.3 illustrates this for Mo(001),[215] and similar results have been found for Cr(001),[215] Ag,[216] Cu[217] and Ni.[216] Using for the moment the conclusion, derived later in section 5.6, that the kinematic interpretation may overestimate the influence of surface vibrations, we may assume that the surface coefficient of expansion is actually rather less than twice the bulk value. Indeed Unertl and Webb[218] found it difficult to isolate any surface enhancement effects, because of the inadequacies of the kinematical approach and uncertainties in actually extracting accurate energies from the LEED data.

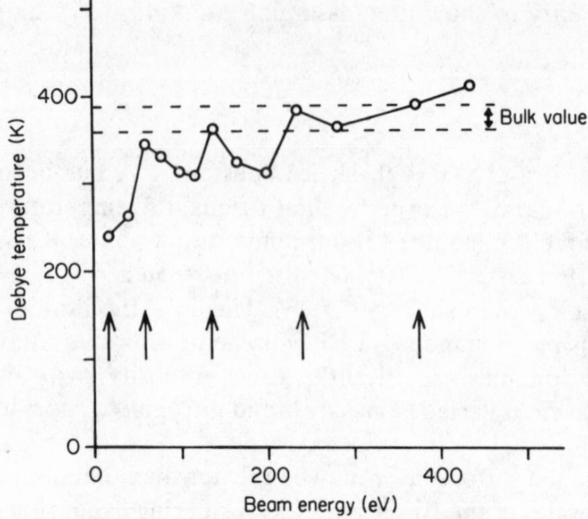

Figure 5.3. Variation of effective Debye temperature for Mo(001), as a function of beam energy, from kinematical analysis. The arrows indicate kinematic Bragg peak energies. (Reproduced with permission from Wilson[234])

Theoretical calculations of the coefficient of surface expansion have been made by Dobryzinski and Marududin[219] for α-Fe and by Kenner and Allen[220] for Ar, Kr and Xe. The latter authors derived the following relationship for the high-temperature limit:

$$\frac{\alpha_{\text{surf}}}{\alpha_{\text{bulk}}} = \frac{3}{4}\frac{\langle \Delta r_{\perp}^{2} \rangle_{\text{surf}}}{\langle \Delta r_{\perp}^{2} \rangle_{\text{bulk}}} \tag{5.18}$$

This may be compared with high-temperature limit derived by Wilson and Bastow[215]

$$\frac{\alpha_{\text{surf}}}{\alpha_{\text{bulk}}} \simeq \frac{\langle \Delta r_{\perp}^{2} \rangle_{\text{surf}}}{\langle \Delta r_{\perp}^{2} \rangle_{\text{bulk}}} = \frac{\theta_{\text{D bulk}}^{2}}{\theta_{\text{D surf}}^{2}} \tag{5.19}$$

which they derived from the Grüneisen relation (the expansion coefficient for cubic crystals is proportional to the specific heat) and the assumption that at high energies the frequency of vibration is equal to the Debye frequency. If we assume that the surface vibrational amplitudes are double that of the bulk, as estimated in section 5.4, and substitute this into equation (5.18), we obtain

$$\alpha_{\text{surf}} \simeq 1.5\alpha_{\text{bulk}} \tag{5.20}$$

Although work carried out to date leaves us with a rather uncertain knowledge of the surface thermal expansion, this is clearly an effect which should be borne in mind when interpreting bond lengths from complete multiple-scattering calculations. The observed surface displacement arrives as the sum of two separate effects—the static displacement resulting from alterations in the bonding arising from the presence of the surface itself and a thermal expansion. Dobrzynski and Marududin[219] termed this combination the 'dynamic displacement'. From a theoretical point of view it is this dynamic displacement—the sum of a static displacement and a thermal expansion—which would be calculated from a minimization of the total free energy of the surface with respect to interatomic distances.

5.4 The calculation of vibrational amplitudes

The simplest means of calculating vibrational amplitudes is to assume the atoms in the crystal are harmonic oscillators so that the problem reduces to that of determining the force constant σ. In this case, if σ is the force constant for a pairwise interaction, then

$$\langle |\Delta \mathbf{r}|^2 \rangle = \frac{kT}{\sigma} \tag{5.21}$$

As a first estimate of the effect of the surface, we may use the fact that, normal to the plane, the surface atoms have only half the number of nearest neighbours that surround bulk atoms, so $\sigma_{\text{surf}} \simeq \frac{1}{2}\sigma_{\text{bulk}}$ and, roughly, we would expect

$$\langle \Delta r_{\perp}^2 \rangle_{\text{surf}} \simeq 2\langle \Delta r_{\perp}^2 \rangle_{\text{bulk}} \tag{5.22}$$

These greater surface amplitudes at the same time increase the importance of anharmonic effects, a problem that will be considered later.

In the harmonic approximation, the equation of motion for an atom displaced through a distance Δr_i^α in the α-direction is

$$m\frac{\mathrm{d}^2}{\mathrm{d}t^2}(\Delta r_i^\alpha) = -\sum_{j\beta} \Phi_{\alpha\beta}(i,j)\Delta r_j^\beta \tag{5.23}$$

where m is the atomic mass and the force constant $\Phi_{\alpha\beta}\,(i,j)$ is, to a first approximation, the force exerted on atom i in the α-direction by a unit shift of atom j in the β-direction, all other atoms remaining fixed. Such force constants will take different values at the surface, due to the static relaxation of the surface atoms. As a further level of refinement, a 'quasi-harmonic' approximation may be introduced in which the 'dynamic equilibrium' positions of the atoms are incorporated in the calculations. This takes into account both static displacement and thermal expansion effects on the mean surface atom sites.

Allen and deWette[221] investigated the significance of various approximations based on the Lennard–Jones interaction potential

$$\Phi(r) = 4\varepsilon\left[\left(\frac{\sigma}{r}\right)^{12} - \left(\frac{\sigma}{r}\right)^{6}\right] \tag{5.24}$$

In the harmonic approximation, $\langle(\Delta r^{\alpha})^2\rangle$ was found to be linear with temperature down to about $\theta_D/3$, but in the quasi-harmonic approximation $\langle(\Delta r^{\alpha})^2\rangle$ was found to rise more rapidly than linear. This latter effect is to be expected since thermal expansion of the lattice will effectively reduce all the force constants describing interatomic interactions. They also studied the increasing significance of anharmonicity with temperature. At one-half the melting temperature, anharmonic effects in the (110) surface of a Fe crystal were found to be significant, increasing root-mean-square (rms) vibrational amplitudes by up to 50 per cent more than that deduced from the quasi-harmonic value (which in turn was about 50 per cent greater than the value calculated using the harmonic approximation). In practice, such effects cannot be observed easily in LEED experiments because, at the high temperatures where anharmonicity would be significant, the background resulting from thermal diffuse scattering tends to swamp the diffraction beams.

Most calculations of Debye–Waller factors use the assumption that the phonon spectrum at any temperature can be approximated by the phonon spectrum at room temperature. This, together with other necessary assumptions as outlined above, limit the accuracy of such calculations. Empirically, the bulk Debye–Waller factor can be obtained from X-ray diffraction, but even using this technique, differences of up to 50 °C are sometimes obtained between different investigations of the same crystal type. Representative values of the Debye temperature for a range of elements are given in Appendix 1 (p. 311).

As a rough guide to the surface Debye temperature, we may take our original estimate for the enhanced surface vibrational amplitudes when the number of nearest neighbours is halved, as given in equation (5.22). Substituting this into equation (5.19) we find, in the high-temperature limit,

$$\Theta_{D_{surf}} \simeq \sqrt{(2)}\Theta_{D_{bulk}} \tag{5.25}$$

Many theoretical studies have been carried out to improve the precision of calculated surface vibrational amplitudes. For example Clark *et al.*[222] studied a number of different surfaces of Ni, and determined the variation between crystallographic faces and successive layer depths, as well as the anisotropies between vibrations normal and parallel to the surface plane. Depending on the crystallographic face, rms vibrational amplitudes normal to the surface were estimated to be about 2.0 times the bulk value in the top layer, between

1.7 and 1.3 times greater in the second layer (the higher value corresponding to the open-packed (110) surface), and between 1.24 and 1.13 in the third layer from the surface. As we would expect from our simple model above, vibrational amplitudes parallel to the surface were found to be less strongly enhanced. The effect is greatest for the relatively open (110) surface, yielding values of 1.61, 1.19 and 1.10 for successive layers compared with 1.56, 1.15 and 1.07 for the (100) and 1.33, 1.11 and 1.06 for the (111) surfaces. Such values should be taken only as a qualitative guide, however, as the simple model used to calculate these values assumed central forces only, with force constants identical to those in the bulk. Although giving reasonably good agreement for bulk properties, it is not clear how accurate such assumptions would be for the surface region. For example, with Xe(111), the ratio of rms amplitudes normal to the surface relative to the bulk was found to be 2.0 by Clark *et al.*,[222] using a central force model, and 2.25 by Allen and deWette[223] using the harmonic approximation and 3.48 when anharmonic effects were included. The surface amplitude enhancement ratio for Xe(111) was estimated experimentally by Tong *et al.*[224] to be 2.0 ± 0.5. Increasing complexity in the theoretical model does not necessarily indicate that the result will always be better.

An additional effect which has not always been well characterized in the above calculations is the static displacement of the surface layer—clearly, in the case of Mo(100) where the surface displacement is in the form of a 10 per cent contraction from typical bulk spacing[15] or even in the case of Ni(110) where there is a 5–8 per cent contraction[37] the force constants may be substantially strengthened—a possibly significant factor in explaining the results of Wallis *et al.*[225] and Vail,[226] who found that stiffening of the force constants was required to improve agreement with experiment.

Recently, Bortalani *et al.*[227] set about a calculation of the surface lattice dynamics of (100) faces of bcc materials, in particular Fe(100). The bulk force constants were determined by fitting a phonon spectrum (calculated using a central force model up to second-nearest neighbours) to the experimentally determined bulk phonon spectrum. To treat the surface they used the 3 per cent contraction determined from LEED by Legg *et al.*[228] Previously, Allen and deWette[223] had suggested that bond length relaxations could affect the nearest-neighbour force constants with a variation r^{-p}, where r is the bond length and $p \geqslant 5$. Bortolani *et al.*[227] used these results to modify the second-order nearest-neighbour force constants which were increased by 20 per cent for the surface atoms. The resulting density of states $g_s(\omega)$ may be related to the surface Debye temperature by the following expression:

$$\int_0^{\omega_D^S} g_s(\omega)\mathrm{d}\omega = \leqq N^2 \tag{5.26}$$

where N is the number of surface atoms, each with three degrees of freedom. The conclusions are very interesting—if no relaxation is considered, $\Theta_{D1surf} \simeq \Theta_{Dbulk}/\sqrt{2}$ just as our simple estimate gave from counting nearest neighbours (equation 5.25). When the relaxation and consequent tightening of surface force constants is included, $\Theta_{D_{surf}} \simeq \Theta_{D_{bulk}}$; that is, the surface Debye temperature is restored to the bulk value. A similar conclusion was also drawn for W(001).

Using LEED intensity analysis, both Ignatiev *et al.*[33] using purely visual comparisons, and Clarke[15] using quantitative methods of comparison, found that the surface Debye temperature of Mo(001) was indeed lower than that of the bulk, and consistent with a specific investigation of the temperature dependence of the beam intensities by Tabor and Wilson[229] who found $\Theta_{D_{surf}} \simeq \Theta_{D_{bulk}}/\sqrt{2}$. For W(001), a number of LEED investigations were performed where $\Theta_{D_{surf}}$ was assumed to be equal to $\Theta_{D_{bulk}}$, but only Heilmann *et al.*[230] set out to determine specifically the surface Debye temperature with respect to the bulk. They found $\Theta_{D_{surf}} \simeq 210$ K, assuming $\Theta_{D_{bulk}} \simeq 380$ K. This ratio was successfully used by Debe *et al.*[231] and Clarke and Morales[32] in quantitative LEED analyses of the $I(E)$ data.

It appears that the theory overestimates the increase in the nearest-neighbour forces. For the moment, it seems that more extensive studies are required before the theory of surface lattice dynamics can be fully developed.

An alternative phenomenon which could be investigated to tie in with observed surface Debye temperatures is surface melting. Using Lindemann's melting criterion[232] and the harmonic approximation, the melting-point T_M and atomic weight m may be related to the Debye temperature as follows

$$\vartheta_D^2 = \frac{9\hbar^2 T_M}{k_B \langle \Delta \mathbf{r}^2 \rangle m} \tag{5.27}$$

Chatterjee[233] used this relation to relate surface and bulk parameters as follows, for any crystallographic face (hkl)

$$T_M(hkl) = \frac{\langle \Delta \mathbf{r}_\perp^2 \rangle_{Surf}}{\langle \Delta \mathbf{r}^2 \rangle_{bulk}} \left[\frac{\theta_{D_{surf}}(hkl)}{\theta_{D_{bulk}}} \right]^2 T_M^B \tag{5.28}$$

Knowing bulk properties and any two of the surface parameters the third may be estimated. To date, this approach does not appear to have been put to the test, probably due to a reluctance to melt a good single-crystal specimen!

5.5 Experimental measurements using kinematical analysis

Any attempt to determine the surface and bulk Debye temperatures from a kinematical analysis of LEED data faces a series of difficulties. It is found that

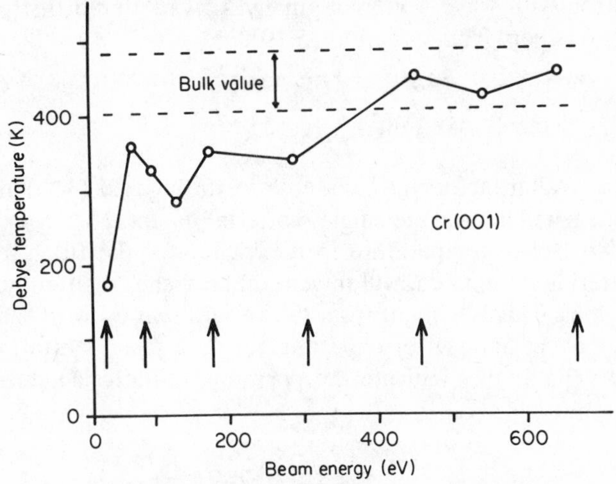

Figure 5.4. Variation in effective Debye temperature for Cr(001), as a function of beam energy, from kinematical analysis. The arrows indicate kinematic Bragg peak energies. (Reproduced with permission from Wilson[234])

the effective Debye temperature that can be deduced from the slope of the log intensity versus temperature relationship will depend strongly on the particular incident electron energy chosen. As an example, results from Cr(001) and Mo(001) taken from the work of Wilson[234] are illustrated in Figure 5.4. Three main factors lead to such behaviour—the existence of a lower Debye temperature in the surface, multiple scattering which limits the validity of the kinematical analysis and the angularly dependent term $\mathbf{\Delta k}$ $(=\mathbf{k}' - \mathbf{k})$ which occurs in the Debye–Waller factor and assumes particular significance in the presence of multiple scattering.

A lower surface Debye temperature will result in a greater damping factor associated with scattering from surface atoms. The extent to which this influences the observed effective Debye temperature will depend on the relative amount of scattering from surface and bulk atoms. This in turn will depend on the incident beam energy since the penetration depth decreases with energy.

Multiple scattering will tend to reduce intensities more strongly than predicted by single-scattering theory, since the scattered amplitude $a^{(i)}$ is reduced by a Debye–Waller factor by each successive scattering event i:

$$a^{(1)} = a^{(0)} \exp(-M^{(1)}) \tag{5.29}$$

$$a^{(2)} = a^{(0)} \exp(-M_1^{(2)} - M_2^{(2)}) \tag{5.30}$$

If an electron with wave vector originally $\mathbf{k}_0$ is scattered firstly into $\mathbf{k}_i$ then finally into $\mathbf{k}_f$, then, from equation (5.10),

$$M_1^{(2)} = M|\mathbf{k}_i - \mathbf{k}_0|^2 \tag{5.31}$$

$$M_2^{(2)} = M|\mathbf{k}_f - \mathbf{k}_i|^2 \tag{5.32}$$

The Debye–Waller factors in the double-scattering case ($M\binom{1}{1}$ and $M\binom{1}{2}$) will generally be smaller than the single-scattering factor, M. As a consequence, the effective Debye temperature in the vicinity of the Bragg peaks, where single scattering dominated, will in general be higher than in the intervening energy regions. This is evident from the variation in Θ_D with energy, derived using kinematical analysis of experimental data from Mo(001) as given in Figure 5.4. The arrows indicate the positions of the calculated Bragg peak energies.

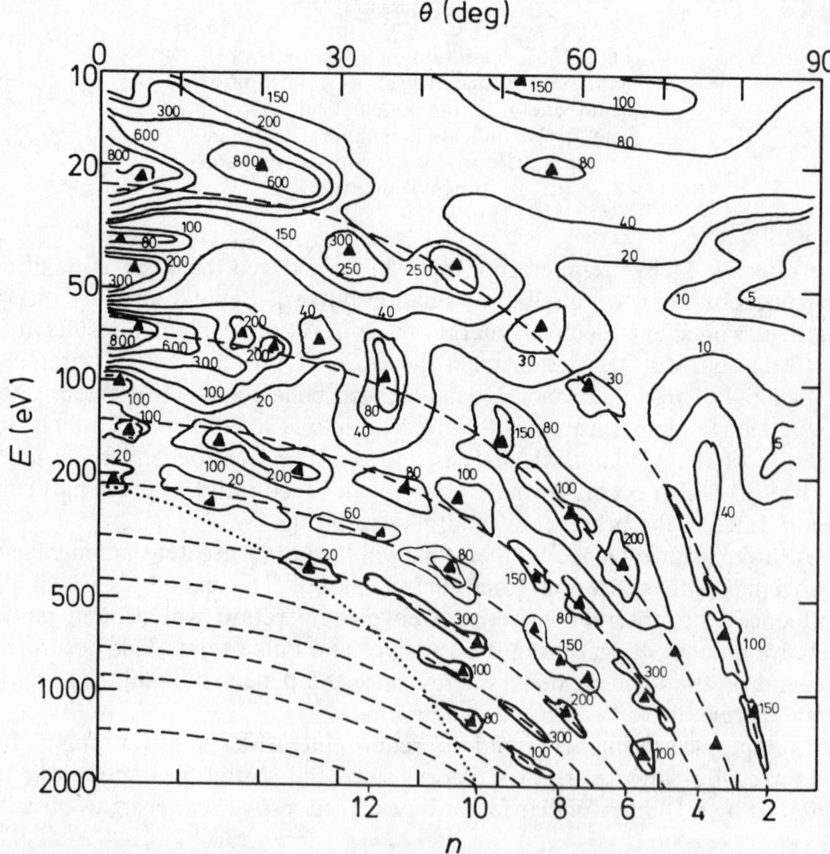

Figure 5.5. Variation of beam intensity with both energy and polar angle θ for the (0, 0) beam reflected from Al(001). Dashed lines marked 2, 3, . . . are Bragg lines. Figures indicate relative intensity on an arbitrary scale, solid triangles indicate local intensity maxima. (Reproduced by permission of the Institute of Physics from Aberdam[235])

As the energy of the incident beam decreases, so the proportionate amount of wide-angle scattering increases. For a given path length, this will permit a greater proportion of scattering to come from the surface region, and the Debye–Waller factors from such low-energy scattering events, which are proportional to $|\,\mathbf{\Delta k}\,|^2$ will also be correspondingly greater. Consequently, the effective Debye temperature at low energies may turn out to be lower than the true Debye temperature corresponding to single scattering from the surface atoms. Incidentally, this $|\,\mathbf{\Delta k}\,|$ dependence plays an important role not only at very low energies, but at very high ones (compared with typical LEED energies). Studies in the MEED energy range 500–3000 eV are only possible for near grazing incidence with scattering predominantly in the forward direction, so that $|\,\mathbf{\Delta k}\,| = |\,\mathbf{k}\,|\sin\theta$ remains small even for large $|\,\mathbf{k}\,|$. The rapid intensity drop with increasing polar angle θ is clearly visible for high energies in the iso-intensity contour map of Al(100) given in Figure 5.5, where the intensities of the (0, 0) beam is simultaneously plotted against θ and E^{235}. In this case, the lack of intensity variation in the area bounded by the dotted line results from the Debye–Waller factor together with the shape of the atomic scattering factor, which becomes increasingly peaked in the forward direction as the energy increases.

Thermal expansion will accompany any rise in crystal temperature, and, as found in section 5.3, this will cause the Bragg peaks to shift in energy. Measurement of the temperature dependence of peak intensities will be influenced quite strongly by this effect, if the measurement is carried out at fixed energies. When the temperature increases, the Bragg peaks will tend to shift downwards in energy. This will cause intensities on the high-energy side of the peak to drop with respect to the mean, but on the low-energy side, they will rise, as shown in Figure 5.6. In this figure, a peak at two different temperatures is drawn schematically, normalizing the maximum peak intensity. In practice, the intensity of the high-temperature peak could be much lower than the low-temperature peak, giving a mean Debye temperature measurement. The effect of thermal expansion is thus to cause a

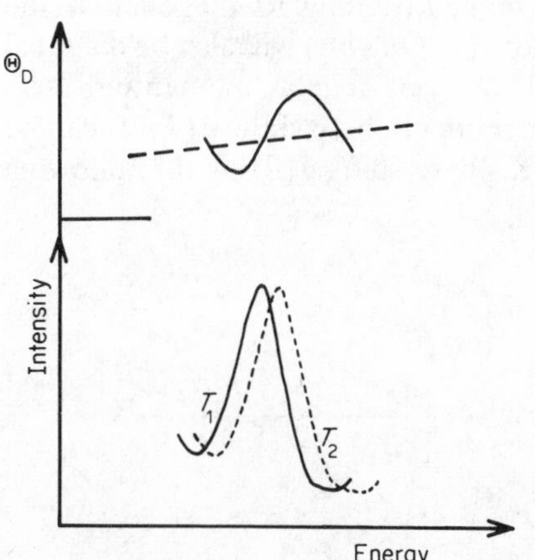

Figure 5.6. Variation in effective Debye temperature associated with a peak that shifts in energy as the temperature changes, if precise energies, rather than the peak energy, are used in the kinematic interpretation (the two peaks here are normalized to identical intensities, for clarity.)

systematic oscillation about this mean. It is clearly better to measure Debye temperatures at energies corresponding to the maximum peak position, rather than at a fixed energy. This effect has been measured in a study of Ni(111) by Gelatt *et al.*[216]

5.6 Multiple-scattering theory

In the preceding sections, we have seen that the surface Debye temperature may be derived approximately by observing the change in effective Debye temperature with decreasing energy, in the vicinity of the Bragg peaks that is, where multiple scattering may be assumed to be minimized. However, this kinematic approach has been shown to be insufficiently accurate to give more than a rather qualitative understanding of surface vibrations, and so multiple scattering should be considered. Indeed, the incorporation of thermal effects in multiple-scattering theory is essential if theoretical and experimental curves are to be accurately compared throughout the whole energy range of the data.

Temperature effects were first incorporated in the multiple-scattering formalism by Duke and Laramore,[236] and independently by Holland.[237] By re-expressing the ion-core scattering by a temperature-dependent factor, analogously with the Debye–Waller factor in kinematical theory, it is possible to include temperature effects directly into the multiple-scattering formalism as derived in Chapter 4. A temperature-dependent *t*-matrix may be defined as

$$\begin{aligned} t(T, \theta_{\mathbf{k}',\mathbf{k}}) &= t(0, \theta_{\mathbf{k}',\mathbf{k}}) \exp[-\tfrac{1}{2}\langle(\Delta\mathbf{k}\cdot\Delta\mathbf{r})^2\rangle] \\ &= 8\pi^2 \sum_{lm} \sum_{l'm'} t(T, l'm', lm) Y_{l'-m'}(\Omega(\mathbf{k}')) Y_{lm}(\Omega(\mathbf{k}))(-1)^{m'} \end{aligned} \tag{5.33}$$

which may be included in the layer diffraction matrix (equation 4.9) once the effects of non-spherical scattering have been included.

As a result of the sphericity imposed by the muffin-tin model, the 0 K scattering matrix, $t(0,\ l'm'\ lm)$ is only non-zero along its diagonal. If the vibrations are assumed to be isotropic then $t(T, l'm', lm)$ will also be diagonal and so the scattering may be described for each angular momentum term separately, in the form of temperature-dependent phase shifts $\delta_l(T)$. It can be shown[105] that these are related to the 0 K phase shifts $\delta_l(0)$ by the following expression:

$$\begin{aligned} \exp(i\delta_l(T))\sin(\delta_l(T)) &= \sum_{L'l''} i^{L'} \exp(-2\alpha\kappa^2) j_{L'}(-2i\alpha\kappa^2) \\ &\quad \times \exp(i\delta_{l''})\sin(\delta_{l''}) \left[\frac{4\pi(2L'+1)(2l''+1)}{(2l+1)}\right]^{1/2} \\ &\quad \int Y_{l''m''} Y_{L'0} Y_{l0} \mathrm{d}\Omega \end{aligned} \tag{5.34}$$

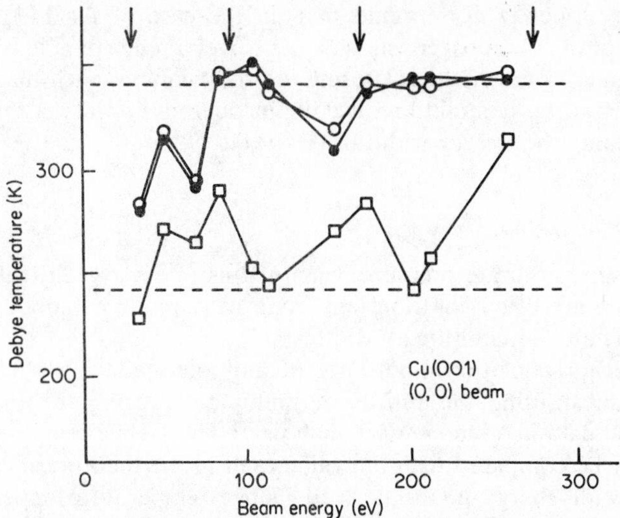

Figure 5.7. Effective Debye temperature of peaks in the (0, 0) beam from a Cu(001) surface at normal incidence, calculated when all atoms have a true Debye temperature of 343 K. (●) $T = 300$ K; (○) $T = 600$ K, and when the surface is given a different Debye temperature, $343/\sqrt{2}$ K; (□) $T = 300$ K. (Reproduced with permission from Pendry (1974), *Low Energy Electron Diffraction*. Copyright Academic Press Inc. (London) Ltd.)

The first thing to note about these new temperature-dependent t-matrices is that they are complex numbers. The imaginary part represents flux adsorption from the diffraction beams, thus including the effect of flux lost to incoherent scattering. Experimentally, McKinney *et al.*[238] found that the incoherent thermal diffuse scattering (TDS) background varies approximately as the reciprocal of the distance from the diffraction beam, with a total TDS intensity approximately $\exp(2M)$ times that of the discrete beam intensity. If thermal effects are large, then the background, which is simply ignored in the theory, should be subtracted from measurements of beam intensities.

The individual ion-core t-matrices $t(T,.\theta_{\mathbf{k}',\mathbf{k}})$ at temperature T are related to the 0 K t-matrices by a simple Debye–Waller factor

$$t(T, \theta_{\mathbf{k}',\mathbf{k}}) = t(0, \theta_{\mathbf{k}',\mathbf{k}})\exp\left(-\frac{3|\mathbf{k}' - \mathbf{k}|^2 T}{2m\mathbf{k}_B\theta_D^2}\right) \tag{5.35}$$

Hence, for forward scattering $\mathbf{k}' = \mathbf{k}$ and $t(T, 0) = t(0, 0)$. The total scattered flux, given by equation 3.34, is thus unchanged from the 0 K case; the difference being that at elevated temperatures the flux is redistributed into the incoherent background. This fact may be checked experimentally by integrating the total elastic and quasi-elastic scattered current as a function of temperature. Jones *et al.*[239] found that, in the case of Ag(111), the total scattered intensity is almost constant in value over a range of at least 500 °C.

The imaginary part of the scattering matrix combines with the imaginary part of the optical potential to reduce the incoherent lifetime of the electron, thus broadening the energy width of the peaks

$$\Delta E \geqslant \frac{1}{\tau} \tag{5.36}$$

Experimentally, accurate measurements of this effect are difficult to make because there are often contributions from overlapping neighbouring peaks which distort any apparent peak width.

An appreciation of the importance of multiple-scattering effects can be gained by substituting the results of multiple-scattering theory from the experimental data in a kinematical analysis.[105] The Debye temperatures thus derived may be compared with the value used in the theoretical calculation. Variations with energy in the derived Debye temperature appear, just as found in a comparison with experimental data, the bulk value being closely matched at high incident electron energies, and lower effective values occurring at low energies (see Figure 5.7). In this case, however, the surface layer can be given precisely the same vibrational amplitudes as the bulk, so that the effective decrease in Debye temperature derived for low incident beam energies can be attributed solely to the accumulation of Debye–Waller factors resulting from the increase in multiple scattering associated with wide-angle scattering. Inclusion of a lower surface Debye temperature in the theory reduced the effective Debye temperature derived from the kinematical analysis, but did not substantially alter the trend towards lower values at low incident energies. This is a clear demonstration of the fact that multiple scattering is at least as important as the enhanced surface vibrations in producing low effective Debye temperatures at low electron beam energies. Results for the surface Debye temperature derived from kinematical analysis alone should, therefore, be treated with caution.

5.7 Anisotropic vibrations

Although anisotropy plays a minor role in the treatment of bulk atomic vibrations, it may be significant for surface atomic vibrations, since, by definition, the surface environment is highly anisotropic. As discussed in section 5.4, the absence of neighbouring atoms on the vacuum side of the surface plane will lead to vibrational components in a direction normal to the surface which have amplitudes perhaps double that of atoms within the bulk, whilst vibrational amplitudes within the surface layer may be intermediate between the bulk value and surface normal value.

The theoretical methods discussed in section 5.4 may be applied to the calculation of in-plane amplitudes and thus the surface vibrational anisotropy. Again the same restrictions and uncertainties apply, and we may only take existing calculations as a qualitative guide. The ratio between these surface and bulk vibrational amplitudes was predicted by Wallis[240] as

$$\begin{array}{ccccc} \langle(\Delta r_{\perp})^2\rangle_{\text{surf}} & : & \langle(\Delta r_{\parallel})^2\rangle_{\text{surf}} & : & \langle(\Delta \mathbf{r})^2\rangle_{\text{bulk}} \\ 2.0 & : & 1.2 \sim 1.5 & : & 1.0 \end{array}$$

for (100) faces of fcc metals. For (110) faces there may be an additional anisotropy within the plane, reflecting the two-dimensional symmetry of the lattice. Calculations by Wallis *et al.* on Ni(110)[225] estimated the following ratios between vibrations in the various crystallographic directions:

	[110]	[1$\bar{1}$0]	[001]	bulk
(Theory:)	2.0	1.51	2.15	1.0

In principle such anisotropy could be detected by measuring the effective Debye temperature as the incident and emergent beam angle is varied, provided the kinematic hypothesis were valid. This approach is discussed by MacRae[241] who performed experiments on Ni(110) and deduced the following ratios and corresponding Debye temperatures:

	[110]	[1$\bar{1}$0]	[001]	bulk
(Expt:)	3.13	1.40	3.13	1.0
Θ_D (K)	220	310	220	390

These values for the [110] and [001] components are particularly large, and probably reflect the errors introduced into kinematical analysis by the neglect of multiple scattering, which itself will vary with the scattering angle and the changing ratio of backward to forward scattering. The trend, at least, is consistent with the theoretical predictions.

In a recent reassessment of the surface Debye temperature of Ni(110), by Mroz and Mroz,[242] anisotropic surface vibrations were identified, but the data were insufficient to provide any quantitative support for MacRae's findings. By contrast, Morabito *et al.*[243] observed no consistent anisotropy in the surface vibrations at various faces of Ag. Once again, this underlines the difficulties associated with trying to extract detailed information using only kinematic theory. For the time being, it seems that we must either be guided by theoretical calculations in order to estimate the degree of anisotropy in surface vibrations, or use some alternative technique to make suitable measurements.

The multiple-scattering theory discussed in section 5.6 was quite general to the point where the *t*-matrix (equation 4.47) was approximated by its diagonalized form *t*. The inclusion of anisotropy by retaining the non-diagonalized form of $t(T, l'\ m', lm)$ would considerably increase the

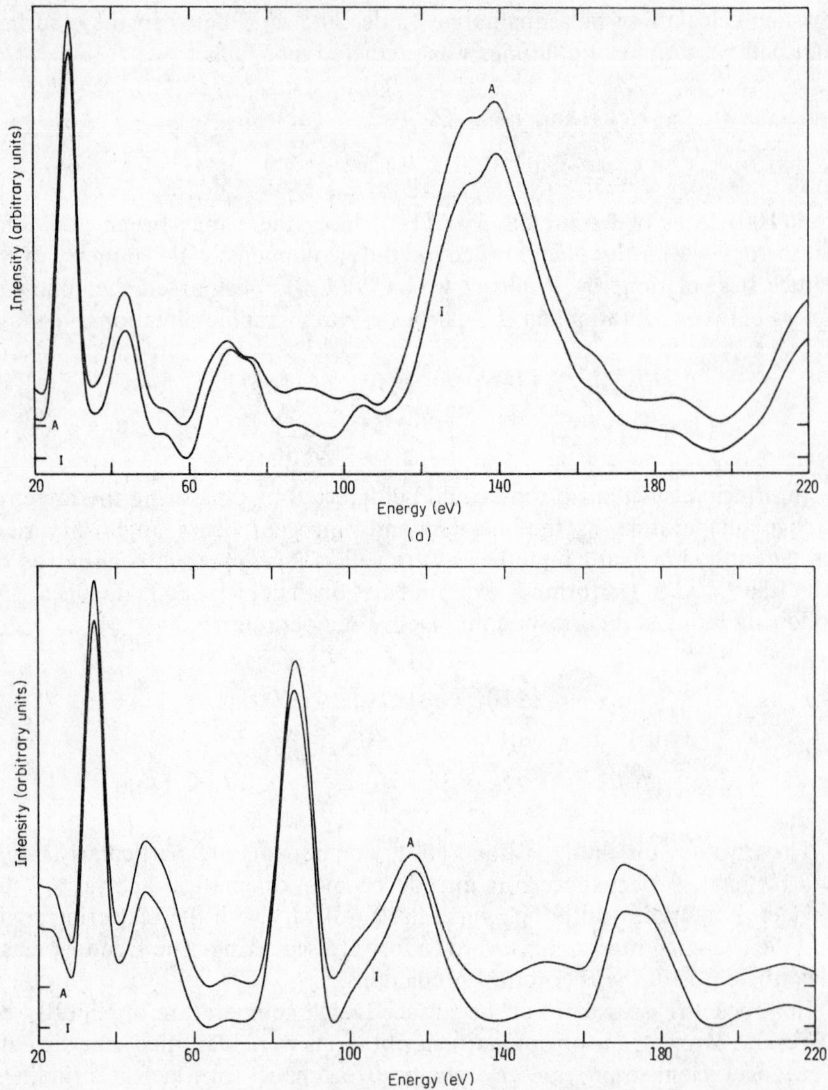

Figure 5.8. Theoretically calculated $I(E)$ curves for (a) the (0, 0) beam and (b) the (0, 1) beam, reflected from Cu(001) with normal incidence. The curves marked I result from the model with isotropic vibrations, the curves marked A result from the model with anisotropic vibrations. (Reproduced by permission of the American Institute of Physics from Ulehla and Davis[244])

amount of computing compared to the isotropic case, but Ulehla and Davis[244] have calculated $I(E)$ curves using a first-order approximation to the off-diagonal elements by including only those values of $t(T, l'\ m', lm)$ for which $l' = l, l \pm 2$. This is done by expressing $t(T, l'\ m', lm)$ as the sum of the isotropic contribution and the lowest non-trivial order of anisotropy

$$t(T, l'\,m', lm) \simeq t_0(T, l'\,m', lm) + t_1\,(T, l'\,m', lm) \tag{5.37}$$

This could then be substituted into the interlayer scattering matrix in order to perform multiple-scattering calculations. Using Cu(001) as a test case, $I(E)$ curves calculated assuming isotropy were compared with curves for which vibrational amplitudes perpendicular to the layer were assumed to be 50 per cent greater than those parallel to the layer. Results for two beams are given in Figure 5.8.

The differences between the two sets of curves are quite small. The specular beam at normal incidence hardly changes, but then it is probably the least sensitive to lateral vibration amplitudes. The (1, 0) beam is affected more and shifts in small peaks of up to 2 eV can be seen, in addition to small peak shape alterations. The results of further calculations for incident angles off-normal would be most interesting to provide us with a better appreciation of the possible extent of this effect, and the limitations on theory/experiment comparisons that may result from ignoring it.

A simple treatment permitting partial inclusion of anisotropic vibration in the surface layer has been provided as an option in the computer programs of Van Hove and Tong.[16] The intralayer scattering matrices are calculated using phase shifts as in the isotropic case, but the resulting matrix elements $M^{\pm}_{g'g}$ (see equation 4.9) are then modified by an anisotropic Debye–Waller factor e^{-M} where

$$\mathrm{M} = \frac{1}{6}\,[\,|\Delta k_{\parallel}|^2\langle(\Delta r_{\parallel})^2\rangle_T + |\Delta k_{\perp}|^2\langle(\Delta r_{\perp})^2\rangle_T] \tag{5.38}$$

5.8 Adsorbate atom vibration

Whereas bulk crystal parameters are generally well characterized from X-ray data and theories of crystal bonding, and corresponding surface parameters may be estimated theoretically or empirically with respect to the bulk as described in section 5.4, the force constants between an adsorbed gas and substrate are much less well understood.

Estimates or adsorbate vibrational amplitudes may be made empirically, using LEED results, as follows. The vibrational amplitude of an atom u ($\equiv\Delta r$) may be related to its Debye temperature by[211]

$$\langle u^2\rangle = \frac{3\hbar^2 T}{mk\,\Theta_D^2}\left\{\Phi(x) + \frac{x}{m}\right\} \tag{5.39}$$

Where m is the atomic weight, $x = \Theta_D/T$ and $[\phi(x) + x/m]$ is a function that takes values close to unity, as given in Table 5.1.

The relative amplitudes of vibration of adsorbate (A) and substrate (S) atoms may thus be related as follows:

$$\frac{\langle u^2\rangle_A}{\langle u^2\rangle_S} = \frac{m_S}{m_A}\,\frac{F(x_A)}{F(x_S)} \tag{5.40}$$

Table 5.1 Tabulation of the Debye function used in equation (5.39) (from) James[211]

x	$\Phi(x) + x/4$	x	$\Phi(x) + x/4$
0.0	1.000	1.2	1.040
0.2	1.001	1.4	1.054
0.4	1.004	1.6	1.069
0.6	1.010	1.8	1.087
0.8	1.018	2.0	1.107
1.0	1.028	2.5	1.164

In 1973, Theeten *et al.*[245] outlined a method for estimating theoretically the adsorbate force constants relative to the bulk values. They identified two relevant adsorbate parameters, α_{A-S} the force constant between adsorbate and substrate and α_{A-A} the interadsorbate force parameter. Concentrating on the simpler but more important case of vibrational modes normal (perpendicular) to the surface plane, α_{A-A} need not be considered. Adsorbate vibrations will have well-defined asymptotic values at the limits of high and low temperature, assuming that the vibrations remain harmonic.

At the high-temperature limit, equipartition indicates that the potential energy $\frac{1}{2}\alpha u^2$ is proportional to $k_B T$. Consequently,

$$\langle u_{\perp}^2 \rangle \propto \frac{1}{\alpha_{A-S}} \qquad (T \to \infty) \tag{5.41}$$

which is independent of the mass of the vibrating adatom. At temperatures sufficiently low that the quantum limit is reached, $\frac{1}{2}\alpha u^2$ is proportional to $\hbar\omega$. Thus

$$\langle u_{\perp}^2 \rangle \propto \frac{\omega}{\alpha_{A-S}} \frac{1}{\sqrt{(\alpha_{A-S} m)}} \qquad (T \to 0) \tag{5.42}$$

For a given force constant α_{A-S}, the temperature variation of $\langle u_{\perp}^2 \rangle$ would take mass-dependent values which have the form given in Figure 5.9.

In an attempt to determine adsorbate vibrational amplitudes from observable temperature dependence of LEED beam intensities, Theeten *et al.*[245] compared the temperature dependence of integral-order and fractional-order beams from various $c(2 \times 2)$ adsorbate structures. For $c(2 \times 2)$S on Pt(100) they found that the temperature dependence of the (0, 0) beam did not change significantly from the clean surface, whereas the $(\frac{1}{2}, \frac{1}{2})$ beam, an artefact of the adsorbate structure, had a substantially different temperature dependence. Attributing this difference to reductions in the adsorbate vibrational amplitude, they concluded that

$$\alpha_{S-Pt} > \alpha_{Pt-Pt}$$

from the high-temperature limit derived above.

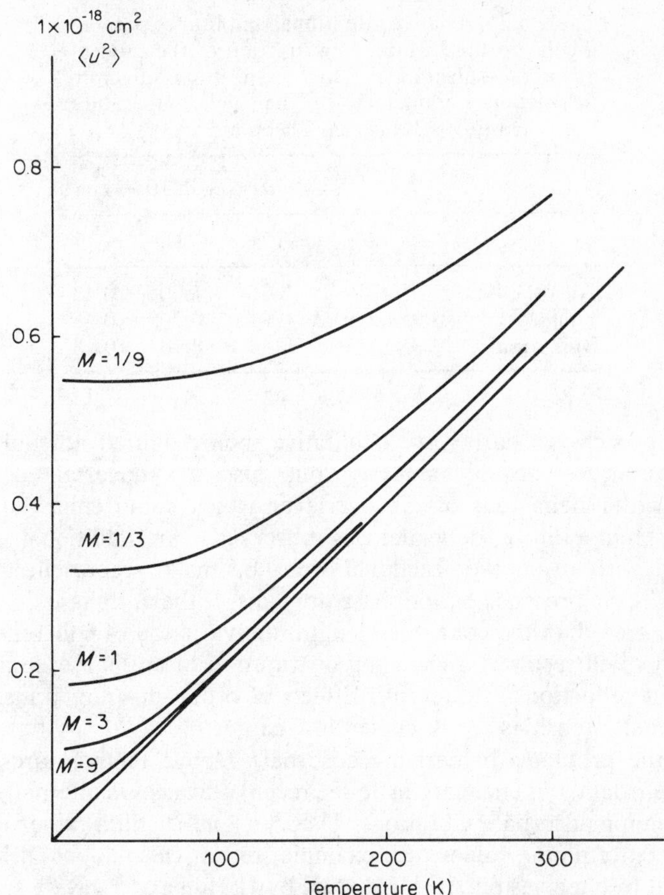

Figure 5.9. Schematic behaviour of $\langle \mu_{\perp}^2 \rangle$ (100) of an adsorbed atom versus its mass, with the force constants fixed. Here, m = (mass of an adsorbed atom)/(mass of a substrate atom). (Reproduced by permission from North Holland Physics Publishing from Theeten *et al.*[214])

A similar inequality was found for $c(2 \times 2)$S on Ni(001). By contrast, $c(2 \times 2)$Pb on Ni(001) showed similar thermal behaviour for (0, 0) and $(\frac{1}{2}, \frac{1}{2})$ beams, indicating that

$$\alpha_{\text{Pt-Ni}} \simeq \alpha_{\text{Ni-Ni}}$$

For the high-temperature limit, they calculated relative vibrational amplitudes as a function of $\alpha_{\text{A-S}}$. Using a scale in which the clean surface amplitude is 1.00, so that the bulk value is 0.48, they obtained the results given in Table 5.2. If, for example, $\alpha_{\text{A-S}} > \alpha_{\text{S-S}}$ as found for S on Pt(100) and Ni(100) then $\langle u_{\perp}^2 \rangle_{\text{A}}$ approaches $\langle u_{\perp}^2 \rangle_{\text{S}}$ whence, from equation (5.40), $F(x_{\text{B}})/m_{\text{A}} \simeq F(x_{\text{A}})/m_{\text{B}}$ from which the adsorbate Debye temperature may be found.

Table 5.2 Relative vibrational amplitudes perpendicular to the surface, as a function of the ratio of adsorbate–substrate to substrate–substrate interaction parameters, in the high-temperature limit (500 K) (From Theeten *et al.*[245])

	$\langle u_{\perp}^2 \rangle_{(100)}$ in 10^{-18} cm^2		
$\alpha_{A-S}/\alpha_{S-S}$:	1/3	1	3
Adsorbed atom	2.56	1.00	0.43
First layer substrate atom	0.91	0.70	0.48
Bulk atom	0.48	0.48	0.48

This approach is clearly very qualitative, being limited not only by the kinematic nature of the analysis, but also by uncertainties in the order–disorder behaviour of the overlayer which could contribute to the adsorbate temperature dependencies observed. An additional difficulty associated with using the fractional-order beams to represent adsorbate vibrations is the presence of antiphase domains. If the antiphase domains are smaller in area than the coherence length, the variation of coherence length with energy will result in a changing distribution of antiphase domains and consequent reductions in intensity. Effects of order–disorder transitions on LEED beam intensities are discussed in Chapter 6.

Given the problems of deriving adsorbate Debye temperatures directly from LEED data, it is encouraging to see recent advances in alternative means of determining adsorbate vibrations. Theory alone is still a rather uncertain means of determining values. For example, recent calculations of hydrogen vibrational frequencies on Al, Mg and Ni by Hjelmberg[246] gave a substantial spread in possible values, dependent on which of several plausible models were used.

With LEED, surface vibrations are measured by noting their effect on scattered beam intensities. An alternative is to measure vibration energies, but this requires an energy resolution down to about 10 meV, well below the resolution limit of typical LEED systems.

High-resolution electron energy loss spectrometry (HREELS) is a method capable of achieving such resolution, with good surface sensitivity. Infra-red spectroscopy is another possible technique which gives excellent resolution, but has been developed only recently as a method capable of providing adequate surface sensitivity. Both are described in more detail in Chapter 8. Even if specific surface measurements have not been made, it seems reasonable to extract force constants or vibrational frequencies from well-characterized molecules as input to calculations of adsorbate bonding. In this way Richardson and Bradshaw[247] were able to calculate mode frequencies for CO on Ni(001) using data from the nickel carbonyl Ni(CO_2), obtaining values in good agreement with HREELS experiments by Andersson.[248]

Once the vibrational frequency ν of a normal mode is known, it is possible to calculate the vibrational amplitude from the expression[249]

$$\langle u^2 \rangle = \frac{16.857}{\nu} \coth\left(\frac{0.7194\nu}{T}\right) \tag{5.43}$$

where $\langle u^2 \rangle$ is in Å^2, ν in cm^{-1} and T in K^0, assuming the harmonic approximation is valid.

When experiments are carried out with suitably high-energy resolution, they may provide considerably more information, concerning bonding sites and adatom interactions, than the simple identification of surface Debye temperatures. A detailed understanding of vibrational modes at a surface is most important, particularly when dealing with adsorbate structures. For example, Bader[250] has recently shown that hybridization between substrate and adsorbate vibrations in a Ni(001)–$p(2 \times 2)$O structure may substantially alter its apparent Debye temperature. Bader deduced a value of 480 K for the oxygen Debye temperature, in contrast with the value of 843 K used previously in a multiple-scattering LEED analysis of the same surface structural system by Van Hove and Tong.[251]

Atom-scattering is a technique that is particularly sensitive to the surface plane. Using very low energies, the atoms, typically helium, are scattered from the surface potential barrier, as described in Chapter 8. It is found that the specularly scattered intensity decreases linearly with increasing temperature, as with LEED, from which the surface Debye temperature may be deduced. In practice, however, it is found that the surface Debye temperature determined in this way may take a value even higher than the bulk value. Rieder and Wilsch,[252] for example, measured a surface Debye temperature of 500 K for Ni(100) using He scattering, which may be compared with the bulk value of 480 K. They suggested that the reason for this result is that the He scattering measures an effective surface Debye temperature appropriate to the measuring technique itself, rather than a true measurement of the surface atomic vibrational amplitudes. This occurs because the He atoms interact with several surface atoms simultaneously, and this makes the technique more sensitive to correlations in the vibrations of the surface atoms. Levi and Suhl,[253] and Meyer[254] have studied the theoretical basis of Debye–Waller effects in atom scattering, and have shown that the Debye–Waller theory is only appropriate to fast collisions from 'hard' lattices, as may occur, for example, with He scattering from alkali-halide surfaces. With slow collisions (for which the period of interaction is long compared with respect to the atomic vibrational frequency), as may occur, for example, with the scattering of Ne from LiF, it is found that high-frequency phonons average to zero, reducing their contribution to the inelastic scattering. Experimental verification of this model has been obtained recently by Mattera *et al.*[255]

Ion scattering, whether in the low-, medium- or high-energy range, appears to be more suitable than the atom-scattering method for studying surface atom vibrations. This is because the corresponding energy ranges used are much higher than in atom scattering, so that the scattering is dominated by the positions of the surface ion cores.

In a study of the Pt(111) surface using 0.2–2.0 MeV helium ion scattering (Rutherford back-scattering), Davies *et al.*[256] demonstrated the sensitivity of this technique to both normal and lateral vibrational modes of the surface atoms. They originally concluded that the surface Debye temperature corresponding to the normal vibrations was very close to the bulk value of 239 K, in contrast with the value of 111 K determined by Kesmodel and Somorjai[257] using LEED. However, as with atom scattering, correlations between vibrating atoms may have contributed to this discrepancy, and a theoretical treatment of this effect has been outlined by Jackson and Barrett.[258] Another source of error may be inaccuracies in the background correction.[259] Mindful of these considerations, subsequent investigations by Davies *et al.*[259] and, independently, by Bogh and Stensgaard,[260] concluded that a surface Debye temperature of Pt(111) is, indeed, close to that originally derived using LEED. This example illustrates how cross-checking results from essentially independent techniques can be very useful in highlighting potential problem areas.

Using medium-energy ion scattering (173 keV), Van der Veen *et al.*[261] also investigated thermal vibrations in a Pt(111) surface. They found that the best agreement between theory and experiment occurred when the surface normal vibration was set equivalent to a Debye temperature of 111 K, the same as determined using LEED. They were also able to deduce that the in-plane surface vibrations correspond to a Debye temperature of 230 K, close to the bulk value of 239 K.

Low-energy ion scattering has also been used to determine surface Debye temperatures. For example, Poelsma *et al.*[262] have determined the surface Debye temperature of Cu(001) using this method. Clearly, ion scattering can be exploited in all energy ranges for the investigation of crystal surface vibrations.

5.9 Correlation and adsorbed molecules

It has been assumed, both in the kinematical formulation of the Debye–Waller factor and in subsequent extensions of the theory to include multiple-scattering effects, that there is negligible correlation between the vibrations of atoms in the lattice. The correlation between displacement $\mathbf{r}_i$ and $\mathbf{r}_j$ of two atoms may be expressed by

$$C_{ij} = \frac{\langle (\Delta \mathbf{r}_i \cdot \Delta \mathbf{k})^2 \, (\Delta \mathbf{r}_j \cdot \Delta \mathbf{k}) \rangle_T}{\{ \langle (\Delta \mathbf{r}_i \cdot \Delta \mathbf{k})^2 \rangle_T \, \langle (\Delta \mathbf{r}_j \cdot \Delta \mathbf{k})^2 \rangle_T \}^{1/2}} \tag{5.44}$$

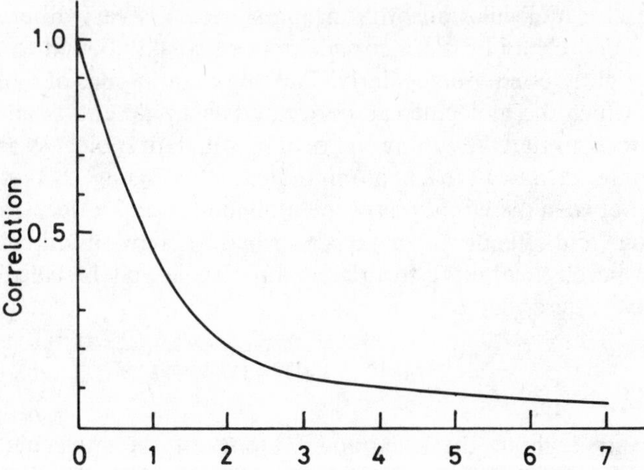

Figure 5.10. Correlation between displacements of atoms separated by units of the lattice vector evaluated in the high-temperature limit of the Debye model. (From Pendry[105])

which gives unity for perfect correlation and averages to zero when there is no correlation. This correlation will result from the collective modes of lattice vibration and may be calculated using the Debye model. In the high-temperature limit it may be shown that

$$c_{ij} = \frac{2}{3}\left(\frac{3}{4\pi}\right)^{2/3} \frac{a_0}{|\mathbf{r}_i - \mathbf{r}_j|} \int_0^{\chi} \frac{\sin(x)}{x}\, \mathrm{d}x \tag{5.45}$$

Where

$$\chi = 2\pi\left(\frac{3}{4\pi}\right)^{1/3} \frac{|\mathbf{r}_i - \mathbf{r}_j|}{a_0}$$

which is independent of θ_D and T. This correlation is plotted as a function of the atomic separation, in units of $|\mathbf{r}_i - \mathbf{r}_j|/a_0$, in Figure 5.10. Correlation is seen to fall rapidly with atomic separation, and will only have an appreciable effect on that small fraction of possible electron paths that pass through close neighbours (or the same atom more than once). Looped paths for which correlation could play an important role will involve several scattering events. However, the majority of scattering events that could involve vibrational correlations will involve forward scattering, for which $\mathbf{k}$, and therefore M, is relatively small. Consequently, correlation effects should be insignificant in LEED at high energies. At very low energies, however, the angular scattering becomes much broader, and correlation effects may assume a greater significance. Unfortunately, no practical formalism currently exists capable of incorporating this effect.

In the case of molecular adsorption, the situation is very different. As an example, CO adsorbed on a Cu surface may be weakly bound to the metal, but remain tightly bound molecularly. The dominant modes of vibration will be those in which the molecule can be treated as a rigid entity, and it is thus preferable to consider the vibrations of the complete molecule as a whole, rather than to deal with each atom individually. Using this assumption, correlation between the vibrations of the individual atoms is necessarily unity.

Andersson and Pendry[201] have described a new multiple-scattering formalism whereby scattering from a rigid molecule may be calculated. The incident waves, described as

$$\sum_{lm} A^{(0)} j_l(k|\mathbf{r} - \mathbf{R}_0|) Y_{lm}(\mathbf{r} - \mathbf{R}_0) \tag{5.46}$$

are re-expanded about the component atoms of the molecule. Multiple scattering between the component atoms is found using the basic form given by equation (4.6):

$$a = a^{(0)}(1 - \chi)^{-1} \tag{5.47}$$

from which the total wavefield scattered from the jth atom is found to be

$$\sum_{l''m''} a_{jl''m''} h^{(1)} (k|\mathbf{r} - \mathbf{R}_j|) Y_{l''m''}(\mathbf{r} - \mathbf{R}_j) t_{jl''} \tag{5.48}$$

The contribution from the various atoms may now be re-expanded about an origin R and summed together. The resulting asymmetrical scattering matrix is described by

$$A^{(s)}_{L''M''} = \sum_{lm} A^{(0)}_{lm} T_{lm,L''M''} \tag{5.49}$$

where

$$T_{lm,L''M''} = \sum G^{(k)}_{lm,l''m''} (1 - \chi)^{-1}_{kl''m'',jLM} t_{jL} G^{(j)}_{LM,L''M''}$$

The mathematical processes and meanings of the various symbols follows those described in Chapter 4.

This method has been found to be quite economical in use, and could be used, in principle, for molecules as complex as benzene adsorbed at a surface. An additional feature is that standard matrix theory permits the rotation of the molecular scattering matrix T through any angle, so that a scattering matrix may be described simply for any orientation of the molecule.

CHAPTER 6

Applications of kinematic theory

6.1 Kinematic theory

Low-energy electrons are scattered very strongly by the atoms or ion cores within a crystal and it is almost always necessary to use a detailed theory of electron scattering if all features of the diffraction beam intensity profiles are to be properly accounted for. Such a 'dynamical theory', which includes the effects of multiple scattering of electrons within the crystal is the subject of Chapter 4. However, there are numerous cases for which a simpler, single-scattering theory may be of value. For example, data averaging is a method devised to 'wash-out' the majority of multiple-scattering features and leave only those diffraction features dominated by single-scattering events. The methodology of data averaging is described and discussed in section 6.5. Single-scattering theory is also applicable in situations where the unit cell dimensions of the periodic lattice are large compared to the mean free path of the electrons, so that multiple scattering between equivalent scattering cells is negligible. For example, the presence of a regular array of steps provides a set of diffraction-intensity modulations that can be superimposed directly upon the diffraction features associated with the flat, unstepped surface. To calculate the intensity–voltage curves resulting from such a surface, the flat surface results would be calculated using multiple-scattering theory and the step-associated modulations from single-scattering theory as described in section 6.3.

Kinematics is the description of motion in which the causes of the motion are not investigated. In so far as it is an oversimplification in the interests of adaptability and ease of use, this explains the frequent use of the term 'kinematic theory' for the application of single-scattering theory to the problem of LEED. Another name sometimes used is 'first-order perturbation theory'. It is assumed that ion-core scattering within the crystal is weak, so that the probability of more than one scattering event per electron is negligible. Kinematic theory for LEED is similar to the scattering theory used in X-ray diffraction except that, in addition to scattering from the periodic lattice of ion cores, there is a constant background potential known as the 'inner' or 'optical' potential. As noted in Chapter 3, this inner potential, V_0, is represented by a complex number. The imaginary part (iV_i) accounts for the absorption of elastic flux resulting from energy-loss processes, and determined the mean free path of the electrons. The real part (V_r) leads to a change in the momentum component normal to the surface, with associated refraction and reflection effects at the surface barrier.

A detailed derivation of the mathematics of kinematic theory has been made previously by Pendry,[105] so we need only review the salient features here. Because multiple scattering is neglected, intralayer scattering need not be considered, and the principal contributions to the scattered electron flux come from the angular dependence of the scattering from individual ion cores and the interference between contributions scattered from successive atomic layers. The periodicity of the two-dimensional surface mesh determines the diffraction pattern, which can be derived directly and independently from the reciprocal lattice model (see Chapter 1), but the interference between contributions scattered from ion cores within an atomic layer will be important if there is more than one atom per unit cell, and for generality if is preferable to incorporate the single-layer scattering processes explicitly.

It was shown in Chapter 3 that an electron wave ψ_0 scattered from an ion core can be described in terms of angular momentum components, l, and phase-shifts, δ_l, about each ion-core centre. At a position r from the ion core, the scattered wave ψ_s is given in terms of the incident wave by the expression

$$\psi_s = \psi_0 X_s(\mathbf{r}) \tag{6.1}$$

where

$$X_2(\mathbf{r}) = \sum_l F(\delta_l, \mathbf{r}) i^{l+1} h_l^{(1)}(|\mathbf{k}|\,|\mathbf{r}|)$$

The scattered wavefield from a period array of ion cores will be simply the accumulation of such contributions summed over ion-core sites. If the origin in space is defined to lie within the atomic layer, and the lattice vectors are $\mathbf{a}_1$ and $\mathbf{a}_2$ respectively, then the position of any ion core within the layer may be specified in terms of its unit cell origin, $(m\mathbf{a}_1 + n\mathbf{a}_2)$ (where m and n are integers) and its position within that unit cell. In general, the origin of the pth unit cell may be defined as $\mathbf{R}_p$, and the position of the qth ion core within that unit cell as $\mathbf{u}_q$ (see Figure 6.1), where

$$\mathbf{R}_p = m\mathbf{a}_1 + n\mathbf{a}_2 \tag{6.2}$$

An incident electron that has amplitude a_0 and zero phase at the origin will have an amplitude and phase at this general ion-core position given by the expression

$$\psi_0 = a_0 \exp[i\mathbf{k}\cdot(\mathbf{R}_p + \mathbf{u}_q)] \tag{6.3}$$

The total wavefield resulting from the scattering of the incident electron wavefield from the whole periodic array of ion cores is therefore given by the summation

$$\Phi_s = \sum_{p,q} a_0 \exp[i\mathbf{k}\cdot(\mathbf{R}_p + \mathbf{u}_q)]\chi_s(\mathbf{r} - \mathbf{R}_p - \mathbf{u}_q) \tag{6.4}$$

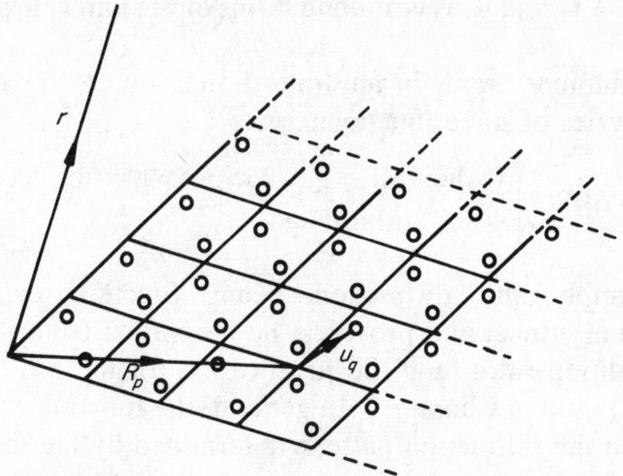

Figure 6.1. Notation used to describe the u_qth atom in the R_pth unit cell, and the position r in space, with respect to some origin within the surface layer

For the purposes of LEED beam intensity analysis, it is convenient to express this again in terms of a set of beams with wave vectors $\mathbf{k}'$:

$$\Phi_s = \sum_{\mathbf{k}'} b_{\mathbf{k}'} \exp(i\mathbf{k}'\cdot\mathbf{r}) \tag{6.5}$$

The coefficients $b_{\mathbf{k}'}$ may be determined by multiplying both sides of this equation by $\exp(-i\mathbf{k}\cdot\mathbf{r})$ and integrating over the whole area (A) of the scattering layer illuminated by the beam. By substituting Φ_s from equation (6.5), it can be shown that

$$b_{\mathbf{k}'} = \left(\frac{a_0}{AN^2}\right)\sum_p \exp[i(\mathbf{k}-\mathbf{k}')\cdot\mathbf{R}_p] \sum_q \exp[i(\mathbf{k}-\mathbf{k}')\cdot\mathbf{u}_q]\int_A \mathrm{X}_s(\boldsymbol{\rho}) \exp(-i\mathbf{k}'\cdot\boldsymbol{\rho})\mathrm{d}^2\rho \tag{6.6}$$

where $\boldsymbol{\rho} = \mathbf{r} - \mathbf{R}_p - \mathbf{u}_q$ and N^2 is the total number of unit cells contained within the area A.

The final integral contains the individual ion-core scattering function (see equation 6.1), and may be expressed in terms of phase shifts and momentum components:

$$\int_A \mathrm{X}_s(\boldsymbol{\rho}) \exp(-i\mathbf{k}'\cdot\boldsymbol{\rho})\mathrm{d}^2\rho = \frac{2\pi i}{|\mathbf{k}|\,|\mathbf{k}'_z|}\sum_l F(\delta_l,\mathbf{r}) \tag{6.7}$$

The two summation terms of atoms in the layer constitute the structure factor S, which may take the form

$$\begin{aligned} S(\Delta\mathbf{k}) &= \frac{1}{N^2}\left[\frac{1-\exp(iN\mathbf{a}_1\cdot\Delta\mathbf{k})}{1-\exp(i\mathbf{a}_1\cdot\Delta\mathbf{k})}\right]\cdot\left[\frac{1-\exp(iN\mathbf{a}_2\cdot\Delta\mathbf{k})}{1-\exp(i\mathbf{a}_2\cdot\Delta\mathbf{k})}\right] \\ &= \frac{1}{N^2}\;\frac{\sin^2(\frac{1}{2}N\mathbf{a}_1\cdot\Delta\mathbf{k})}{\sin^2(\frac{1}{2}\mathbf{a}_1\cdot\Delta\mathbf{k})}\cdot\frac{\sin^2(\frac{1}{2}N\mathbf{a}_2\cdot\Delta\mathbf{k})}{\sin^2(\frac{1}{2}\mathbf{a}_2\cdot\Delta\mathbf{k})} \end{aligned} \tag{6.8}$$

where $\Delta\mathbf{k} = \mathbf{k} - \mathbf{k}'$ and $\mathbf{R}_p$ is written in terms of the unit cell vectors (equation 6.2).

When the number atoms in any one dimension, N, is very large, $S(\Delta\mathbf{k})$ represents a series of spike functions, since

$$S(\Delta\mathbf{k}) = \begin{cases} 1 \text{ when } \Delta\mathbf{k} = \mathbf{g} \text{ (a reciprocal lattice vector)} \\ \sim 1/N^2 \text{ otherwise} \end{cases}$$

This is simply the diffraction beam function expected from a two-dimensional lattice, and provides the derivation from first principles of the Laue condition rules (and the justification for the use of the reciprocal lattice picture) used in Chapter 1. In general, the function of $\mathbf{u}_q$ does not add extra beams to the diffraction pattern determined by the lattice periodicity, although characteristic omissions from the anticipated set of diffraction beams may result if the extra atoms in the cell adopt certain special symmetries within the unit cell (see section 1.7). Thermal vibrations will slightly relax these stringent diffraction conditions, as discussed in Chapter 5.

Scattering from a layer, equation (6.5), may be written in the form

$$\Phi_s = \sum_{\mathbf{k}'} M(\mathbf{k}', \mathbf{k}) a_0 \exp(i\mathbf{k}' \cdot \mathbf{r}) \tag{6.9}$$

where

$$M(\mathbf{k}', \mathbf{k}) = \frac{i}{A|\mathbf{k}_z'|} s(\Delta\mathbf{k}) t(\mathbf{k}', \mathbf{k})$$

in which the t-matrix $t(\mathbf{k}', \mathbf{k})$ is given by

$$t(\mathbf{k}', \mathbf{k}) = -\frac{2\pi}{|\mathbf{k}|} \sum_{ls} (2l + 1) \sin[\delta_l(s)] \exp[i\delta_l(s)] P_l \frac{\mathbf{k} \cdot \mathbf{k}'}{|\mathbf{k}|\,|\mathbf{k}'|} \exp[i(\mathbf{k} - \mathbf{k}') \cdot \mathbf{u}_q] \tag{6.10}$$

The matrix $M(\mathbf{k}', \mathbf{k})$ is effectively the product of two functions, $S(\Delta\mathbf{k})$, which describes the two-dimensional pattern of beams, and $t(\mathbf{k}', \mathbf{k})$, which describes the influence of the individual ion-core scattering on the intensities of the beams.

Having expressed the way in which a layer will scatter an incident electron wave, we may now turn our attention to the contribution played by scattering from successive layers from the surface. Whereas the interferences between contributions scattered from the array of ion cores within an atomic plane produces the two-dimensional arrangement of diffraction beams, it is the interference between waves scattered from successive layers that provides the major contribution to the observed modulations in diffraction beam intensities.

So far, we have assumed that the electron beam incident from the vacuum will reach the surface layer of ion cores unimpeded. In fact, the surface potential barrier, which defines the boundary between the outer extent of the electron gas associated with the crystal and the effective vacuum region, provides a weakly reflective medium that, in the interests of generality, should be taken into account. If a_i is the amplitude of the incident electron wave in the vacuum, and transmission through the surface barrier into the crystal is given by the **k**-dependent transmission function $T^+(\mathbf{k})$, then the amplitude of the electron wave incident on the surface layer will be

$$a_0 = T^+(\mathbf{k})a_i \tag{6.11}$$

Another, and possible more significant, consequence of the surface barrier is that it causes refraction of the beam, so that the incident beam direction is altered. This phenomenon is dealt with in section 3.4, and for the present purposes we shall consider that the wave vectors **k** of the incident beam and **k**′ of the diffracted beams are those occurring within the crystal.

If each atomic layer is separated by the interlayer vector **c**, the incident wave at the nth layer will be

$$\exp(in\mathbf{k}\cdot\mathbf{c})T^+(\mathbf{k})a_i\exp[i\mathbf{k}\cdot(\mathbf{r} - n\mathbf{c})] \tag{6.12}$$

Substituting this into equation (6.8) and summing over n layers where n is large, the total scattered wave becomes

$$\Phi_s = a_i \sum_{\mathbf{k}} T^+(\mathbf{k})M(\mathbf{k}', \mathbf{k}) \frac{\exp(i\mathbf{k}'\cdot\mathbf{r})}{1 - \exp(i\Delta\mathbf{k}\cdot\mathbf{c})} T^-(\mathbf{k}') \tag{6.13}$$

where $T^-(\mathbf{k}')$ is the transmission function for diffracted waves passing out through the surface barrier.

If reflection R of the incident beam from the surface barrier is included (it can usually be ignored because it is very weak except at very low incident beam energies), the kinematic expression for the amplitudes of the diffraction beams becomes

$$\Phi_s = a_i f(\mathbf{k}, \mathbf{k}')\exp(i\mathbf{k}'\cdot\mathbf{r}) \tag{6.14}$$

where

$$f(\mathbf{k}, \mathbf{k}') = R\delta_{/\!/k_{/\!/}} + T^+(\mathbf{k})T^-(\mathbf{k})\left[\frac{M(\mathbf{k}', \mathbf{k})}{1 - \exp(i\,\Delta\mathbf{k}\cdot\mathbf{c})}\right]$$

At typical LEED energies (i.e. above the very low energy regime) the influence of scattering from the surface barrier will be small and usually negligible. In this case, the structure (or form) factor $f(\mathbf{k}, \mathbf{k}')$ may be approximated by

$$f(\mathbf{k}, \mathbf{k}') = M(\mathbf{k}, \mathbf{k}')D(\Delta\mathbf{k}\cdot\mathbf{c})$$

where

$$D(\Delta\mathbf{k}\cdot\mathbf{c}) = [1 - \exp(i\,\Delta\mathbf{k}\cdot\mathbf{c})]^{-1} \tag{6.15}$$

The factor $D(\Delta\mathbf{k}\cdot\mathbf{c})$ in this equation is a sharply peaked function with maxima occurring whenever $\Delta\mathbf{k}\cdot\mathbf{c} = 2\pi m$, where m is an integer. This is simply a more general expression of Bragg's law, and it is usual to refer to the intensity maxima predicted by this function as the Bragg peaks. If $\Delta\mathbf{k}$ were a real number, the value of $D(\Delta\mathbf{k}, \mathbf{c})$ would become infinite at each maximum. It is prevented from doing so by the absorptive part of the potential, which enters the equation as the imaginary part of the inner potential, iV_i.

Experimentally, one observes the intensities, not amplitudes of the scattered beams. The intensity of each beam is proportional to the squared modulus of the amplitude, the area illuminated and the component of momentum normal to the surface. Consequently, the reflected intensity of a beam, expressed in terms of a unit incident current, is given by

$$|f|^2 \left| \frac{k'_z}{k_z} \right| \tag{6.16}$$

Kinematic theory is applicable only in cases where elastic scattering is weak compared with the flux absorption, or the separation between scattering centres is large compared to the mean free path of the electrons. The (100) surface of Xe provides one of the few simple crystal surfaces for which close agreement has been found between untreated experimental diffraction beam

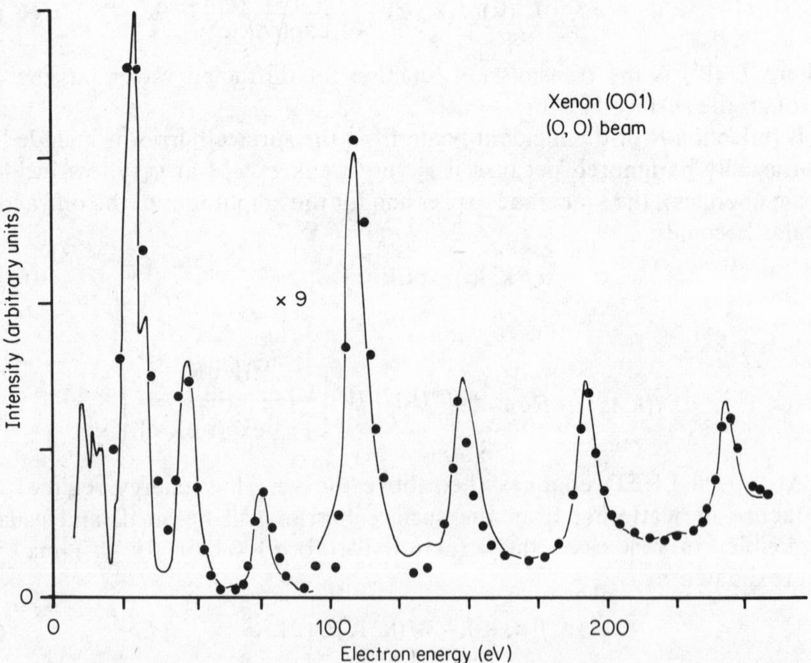

Figure 6.2. Intensity/energy spectrum taken near normal incidence from a Xe(001) surface. Points calculated using the kinematic approximation are indicated by (●). (Reproduced by permission of the American Physical Society from Ignatiev *et al.*[177])

intensity–voltage curves and predictions exclusively derived using kinematic theory (see Figure 6.2).[177] This can be attributed both to the large unit cell volume (232 Å^3) and relatively weak ion-core scattering.

6.2 Point defect structures: random mixing and vacancies

In section 6.1 we considered diffraction from perfect but finite crystal surface using the kinematical approximation. Two major factors that enter the analysis are the structure factor $S(\Delta\mathbf{k})$, which describes the two-dimensional pattern corresponding to the surface lattice, and $D(\Delta\mathbf{k}, \mathbf{c})$ which describes the modulation of diffraction beam intensities as a function of energy due to interference from successive layers. If the surface deviates from perfect two-dimensional periodicity, this will predominantly affect the structure factor S.

The best way of calculating or simulating the consequences of imperfect surface structures depends on the nature or complexity of the defect. If the irregularity can be characterized simply in mathematical terms, it may be possible to use an analytical solution to predict the resulting diffraction pattern. A number of examples of this approach are given in sections 6.4 and 6.5. Alternatively, a computer simulation may be carried out, by specifying the locations of all the atoms in the two-dimensional array (up to some practical limit, say 100×100), and summing the diffraction contributions directly assuming simple kinematical interference between scattered contributions. Ertl and Kuppers first introduced this computer-based simulation technique in 1970,[263] and it has been subsequently used on numerous occasions. As a recent example, Maurice *et al.*[264] used computer simulation to predict the consequences of varying antiphase domain shapes and sizes on the diffraction patterns obtained from the adsorption of oxygen on Cu(001).

If a computer simulation cannot be readily arranged, it is possible to simulate the pattern using optical diffraction from slides. If the structural model to be examined is plotted on a sheet of paper with points separated at about 5 mm intervals in an array comprising 25–50 scatterers in each dimension, and this is then photographically reduced to a total area of about 5 mm square, the pattern will fall into the range of optical light wavelengths. The consequences of varying structural models can then be examined by noting the resulting optical diffraction patterns obtainable from such slides.[265]

The simulation of defects provides a direct method identifying the relationship between any particular two-dimensional structures and the resulting diffraction pattern. However, it remains very useful to seek mathematical solutions where possible, since this may provide insights into the general consequences of surface irregularities, as well as providing a rapid and quantitative means of analysing the diffraction patterns that result from particular defect structures. This is necessary if a full dynamical scattering analysis of diffraction beam intensities is being carried out, and the consequences of imperfections are to be imposed on the final result.

For example, it may be that there is more than one atom type randomly distributed throughout some periodic surface lattice. For example, it has been observed that the dissociation of CO on Fe(001)[266] and Mo(001)[267] may result in a regular $c(2 \times 2)$ adlayer comprising either of the two atomic species in a random distribution. This could occur whenever the interatomic forces within the adlayer are too weak to impose the regular organization of the different atomic species within the two-dimensional structure. Further implications of order–disorder transitions are the subject of section 6.5. Such mixing may also occur at the surface of a crystal that is a substitutional alloy. The consequences of such random distributions can be calculated as follows.

If the form factor of one of the atom types i is f_i, then the intensity I of a group of N atoms is simply

$$I = \left| \sum_{i=1}^{N} f_i \, exp(i \, \Delta\mathbf{k}\cdot\mathbf{r}_i) \right|^2 \tag{6.17}$$

In general, if there are two atom types i and j, the equivalent expression is

$$I = \sum_{i,j}^{N} {}_i\cdot f_j \, exp[i \, \Delta\mathbf{k}\cdot (\mathbf{r}_j - \mathbf{r}_i)] \tag{6.18}$$

When the atoms are distributed randomly, $f_i \cdot f_j$ may be replaced by its average

$$\begin{aligned} \langle f_i \cdot f_j \rangle &= \langle f_i \rangle \cdot \langle f_j \rangle = \langle f \rangle^2 \qquad i \neq j \\ \langle f_i \cdot f_j \rangle &= \langle f^2 \rangle \qquad\qquad\quad\;\; i = j \end{aligned} \tag{6.19}$$

from which

$$i = N(\langle f^2 \rangle - \langle f \rangle^2) + \langle f \rangle^2 \sum_{i,j}^{N} \exp[i \, \Delta\mathbf{k}\cdot(\mathbf{r}_j - \mathbf{r}_i)] \tag{6.20}$$

The first term comprises the N terms for which $i = j$; it is independent of $\Delta\mathbf{k}$ and leads to a uniform background intensity. The second term is the normal diffraction term with a scattering factor that is simply the average of the different contributing atoms.

This result may be applied to the case of CO adsorbed on Fe(001). Through dynamical analysis, the $I(E)$ curves may be calculated for $c(2\times2)$C and $c(2\times2)$O separately, and the results averaged together: this would simulate the effect of diffraction from large separate domains of C and O on the surface. Another approach would be to use averaged scattering phase shifts, simulating a complete mixture of the two atom types, as treated in the above kinematical analysis. Jona *et al.*[268] have performed such calculations, and a typical result is illustrated in Figure 6.3. In this case, it is found that the results from either approach are very similar, although, if one looks closely at the results for the (1, 0) beam, the use of the averaged phase shift does

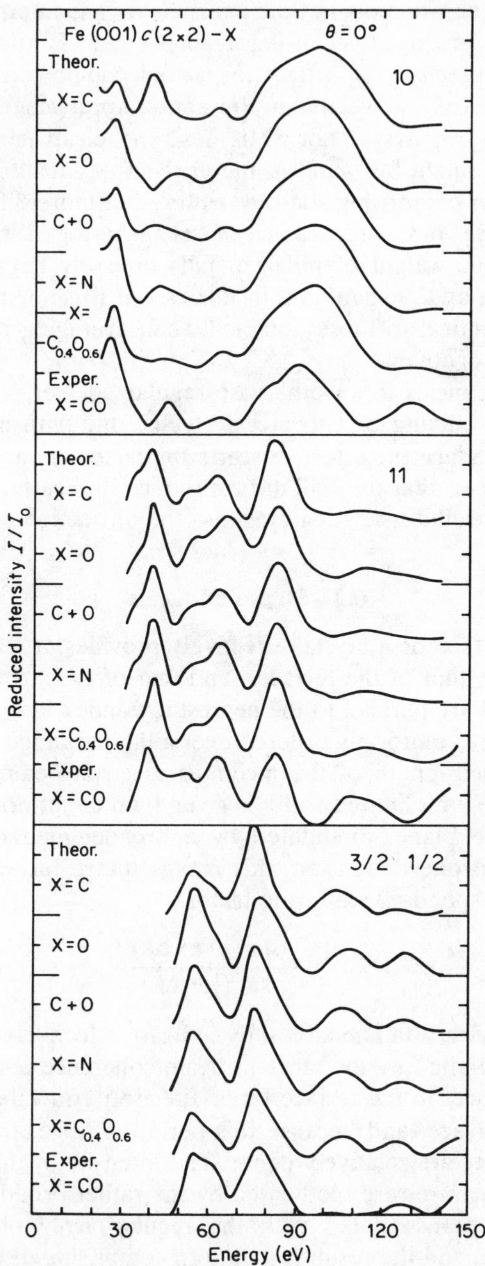

Figure 6.3. Calculated and observed 10, 11 and 3/2 1/2 LEED spectra at normal incidence. In each panel, starting from the top, the curves are as follows: calculated for Fe(001)$c(2 \times 2)$–C; calculated for Fe(001)$c(2 \times 2)$–O; 50–50 average of the above two; calculated for Fe(001)$c(2 \times 2)$–N, calculated for an 'averaged' atom ($C_{0.4}O_{0.6}$); experimental for Fe(001)$c(2 \times 2)$–CO. (Reproduced by permission of the American Physical Society from Jona *et al.*[268])

provide a more accurate model of the peak shapes and energies. In this case, the phase shifts were averaged using a 40 per cent C and 60 per cent O weighting, more precisely to reflect the actual composition of the surface layer. The differences between the alternative approaches is small in this example, because neither C nor O is a strong electron scatterer—more significant effects might be found in the analysis of substitutional alloys, or from a mixed layer comprising adatoms with rather more different scattering powers. In Figure 6.3, the results of using nitrogen phase shifts are shown—the atomic weight of nitrogen falls precisely between the atomic weights of carbon and oxygen, and in this case it is clear that this provides results almost identical to those obtained when averaging the $c(2\times2)$C and $c(2\times2)$O results together.

If there are vacancies in an otherwise regular adlayer, the effect will be equivalent to the mixing of two atomic types, one with a zero scattering factor. This will reduce the effective scattering factor $\langle f \rangle$ and from equation (6.20) it can be seen that this will in turn reduce the scattered intensity, but not necessarily introduce new features into the diffraction pattern.

6.3 Steps and facets

A step at the surface of a crystal effectively provides at least two separate diffracting planes: that of the terraces, and that of the macroscopic stepped face. The terraces are parallel to the nearest low-index face and provide sets of scattering centres uniformly ordered over a limited range, often much less than the coherence length of the incoming electron beam. The scattering factor of a terrace with limited width is given by the scattering factor for the effectively infinite plane, modulated by a broadening factor as given in equation (6.7). In one dimension, this means that a terrace with width N atoms has a corresponding scattering factor

$$F(N) \propto \frac{\sin^2(\frac{1}{2}N\mathbf{a}\cdot\,\Delta\mathbf{k})}{\sin^2(\frac{1}{2}\mathbf{a}\cdot\,\Delta\mathbf{k})} \tag{6.21}$$

The macroscopic crystalline face is, by contrast, effectively infinite in extent (experimentally limited by the incident beam coherence length) and slightly inclined with respect to the terrace faces. Precisely equivalent atoms appear only once per terrace, and because this period is large, the corresponding diffraction features are relatively dense. The intensity of these high-density, step-dependent features are modulated by the rather broad and low-density diffraction features associated with the regular array of atoms in each individual terrace, and the result is a pattern comprising discrete groups of a few beams each located in the vicinity of the beam positions associated with the perfect low index plane. A simulation in one dimension is illustrated in Figure 6.4, from the work of Wagner.[269]

The energy dependence of the resulting pattern can be derived likewise from the Ewald construction applied to the macroscopic stepped surface

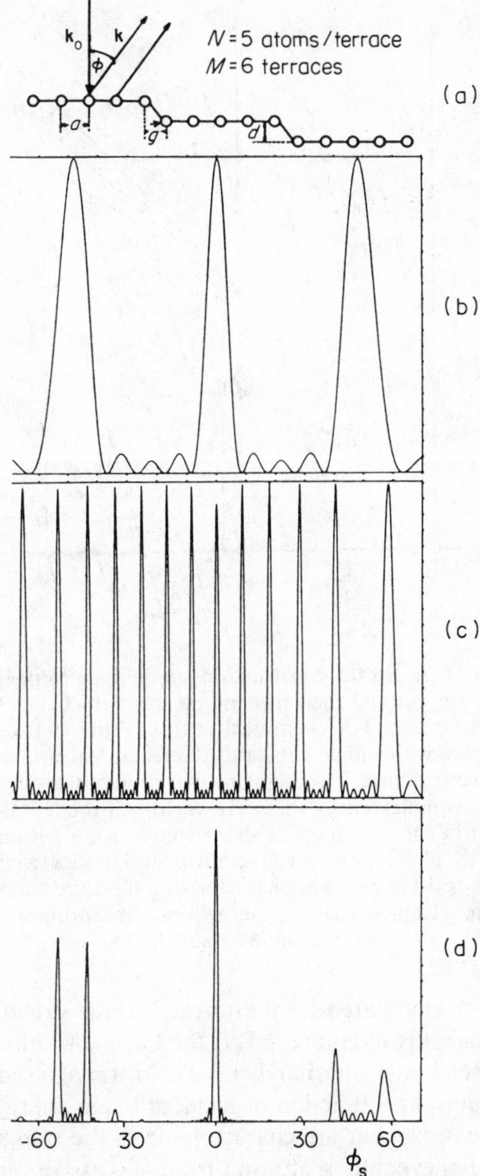

Figure 6.4. Two-dimensional representation of low-energy electron diffraction from a stepped surface with periodic step array. (a) Step structure in real space; lattice constant a. (b) Diffraction function from a single terrace with five atom rows as a function of the diffraction angle $\phi; a/\lambda = \sqrt{(9/5)}$ and $d/\lambda = 3/2$. (c) Diffraction function of step array with six terraces. Zero-order diffraction occurs at $\phi_s = -25.2°$. (d) Product of diffraction functions as shown in (b) and (c). For the wavelength chosen, the (0, 0) beam appears as a sharp single peak, whereas the $(\bar{1}, 0)$ beam is split into two beams of equal intensity. (Reproduced by permission of Springer-Verlag from Wagner[269])

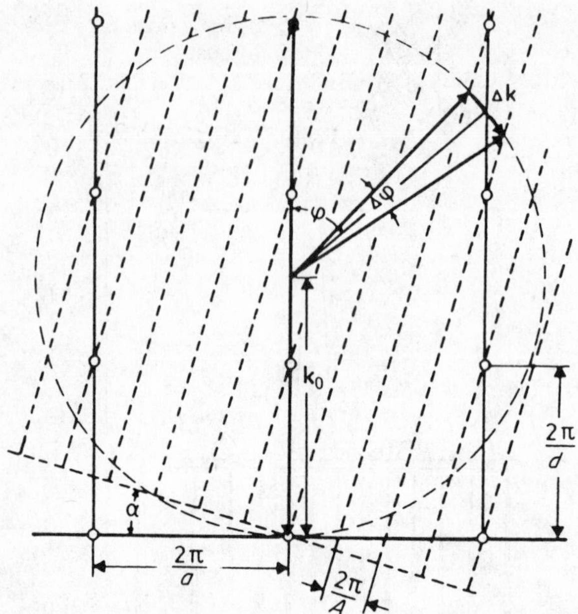

Figure 6.5. Ewald construction applied to stepped surfaces for normal incidence of the primary beam to the terrace plane. The reciprocal lattice of the step array is inclined by the angle α towards the reciprocal lattice of the terrace structure. The angular separation of the split (1, 0) beam is indicated by the wave-vector difference $\Delta\mathbf{k}$. The (0, 0) beam appears as a sharp single beam because the Ewald sphere intersects the reciprocal lattice rods in a reciprocal lattice point characterizing the third Laue condition. (Reproduced by permission of Springer-Verlag from Wagner[269])

diffraction pattern modulated by the terrace width structure factor. This is illustrated schematically in Figure 6.5. If the terrace widths are not too great, each diffraction beam may alternately take the form of a sharp single spot and a split pair of beams, as a function of incident beam energy. The energies at which the split beam appear are characteristic of the step structure. A beam will be unsplit whenever the scattering from successive terraces is in phase. This occurs near those energies at which the Ewald construction would indicate a point of the three-dimensional reciprocal lattice.

This approach is not sensitive to uniform displacements of the complete surface layer, since all terraces will be shifted equally. If the incident beam direction is inclined through an angle θ with respect to the terrace normal, and through azimuthal angles ϕ_a and ϕ_b with respect to the surface lattice vectors $\mathbf{a}_1$ and $\mathbf{a}_2$ respectively, then the characteristic energies for a single unsplit

(h, k) beam are given when S is an integer in the following equation:

$$e_{hk} + V_r = \frac{M}{4d^2}\left\{\frac{(S - hx - ky)^2 + (h\mathbf{a}_1^* + k\mathbf{a}_2^*)^2 d^2/4\pi^2}{(S - hx - ky)\cos\theta - (h|\mathbf{a}_1^*|\cos\phi_a + k|\mathbf{a}_2^*|\cos\phi_b)d\sin\theta/2\pi}\right\}^2$$

With

$$M = h^2/2m = 150 \text{ eV Å}^2 = 1.5 \times 10^{-18} \text{ eV m}^2$$

where $\mathbf{a}_1^*$ and $\mathbf{a}_2^*$ are lattice vectors of the reciprocal surface cell and x and y describe the lateral shift of subsequent layers as a fraction of the real lattice vectors $\mathbf{a}_1$ and $\mathbf{a}_2$. The characteristic voltages for roughly equal spot splitting are given when S is half-integral; this is the condition when wavefields from adjacent terraces are out of phase. This equation is generally valid for both reflected and transmitted beams, but since in LEED we only observe reflected beams we must choose values of S which are large enough to satisfy the following condition:

$$4\pi^2(s - hx - ky)^2 + 4\pi(S - hx - ky)\tan\theta(h|\mathbf{a}_1^*|\cos\phi_a + k|\mathbf{g}_2^*|\cos\phi_b) \geqslant (h\mathbf{a}_1^* + k\mathbf{a}_2^*)^2 d^2 \qquad (6.23)$$

For a specific crystalline structure, $\mathbf{a}$, $\mathbf{a}_2$, x, y, $\mathbf{a}_1^*$ and $\mathbf{a}_2^*$ may be directly related. For example, in the case of normal incidence upon a (100) face of a fcc structure crystal, equation (6.22) reduces to

$$E_{hk} + V_r = \frac{M}{4d^2}\left\{(S + \frac{h+k}{2})^2 + h^2 + k^2 + \frac{h^2 + k^2}{(2M + h + k^2)}\right\} \qquad (6.24)$$

Alternatively, one could observe characteristic energies of the specular beam ($h = 0$, $k = 0$) at various polar angles of incidence θ, for which equation (6.22) reduces, for any crystalline structure, to

$$E_{hk} + V_r = \frac{MS^2}{4d^2\cos^2\theta} \qquad (6.25)$$

Given a large number of observable beams and angles of incidence, the step height should be accurately determinable, although limited by the accuracy with which the inner potential V_r is known.

The terrace width is obtainable, in principle, from the spot splitting. Although this is often difficult to measure accurately, in cases where the terraces are relatively narrow yet regular it is only necessary to differentiate between members of a narrow range of integral values of atomic spacings. If the spot splittings occur parallel to one of the lattice axes (i.e. the step edges are non-kinked) it may be possible to deduce the splitting from the energies at which the individual spot components sweep through a Faraday-cup detector or in front of fixed-position telephotometer. The resolution of a LEED diffractor and its influence on the measurement of spot widths is discussed in section 2.2.

Alternatively, the mean terrace width may be deduced from the step height and knowledge of the inclination of the macroscopic plane with respect to the terrace planes. This inclination is probably best determined by a combination of X-ray crystallography to set the atomic planes and laser light which reflects the macroscopic plane; it could, however, be determined directly from LEED data if a series of photographs at successive energies is used to locate the inclination of the diffraction rods of Figure 6.5.

Step structures are not always regularly spaced or monotonically arranged. Examples of diffraction patterns characteristic of various types of step structure have been provided by Henzler.[270] These can be identified by the variation with energy of the different beam intensities and widths, as shown in Figure 6.6.

When the step distribution is not regular, the beams do not split simply, but become broadened due to the superposition of the different contributions. Although this may be recognized from the energy dependence of the sharpness of the various beams, quantitative estimates of this effect are not easy. Firstly there are the experimental difficulties of actually measuring the beam width, and secondly there is the problem that the measured width is not a direct measure of the broadening effect but rather some convolution of instrumental transfer widths (as treated in section 2.4). Once these difficulties have been overcome, the deduced beam widths may be compared with the results of theoretical calculations in order to estimate the mean terrace widths or step densities. A theoretical treatment of this effect has been made by Houston and Park.[271] If we consider steps in one dimension with terraces of width Γ (in multiples of the separation distance between adjacent atom rows), then, for a given surface, this width may occur with the normalized probability $P(\Gamma)$. The probability of any particular atom being found on a terrace of width Γ is thus given by the function

$$P'(\Gamma) = \Gamma \cdot P(\Gamma) / \sum_{\Gamma} \Gamma \cdot P(\Gamma) \tag{6.26}$$

In addition, there will be a probability distribution determining whether successive step heights are up or down. To start, we shall assume that only monoatomic steps are present. In the simplest analysis, as used to determine the distribution illustrated in Figure 6.6, intensity functions from regular step arrays are added incoherently with a weighting proportional to the number of atoms per terrace type $P'(\Gamma)$. This approach ignores the interference between neighbouring terraces of different widths, so cannot be expected to give quantitative results. A random distribution of neighbouring terraces with different widths must be treated with the use of an average autocorrelation or pair distribution function which is then Fourier transformed. Houston and Park[271] used such an approach, approximating the average autocorrelation function by the autocorrelation function of the average terrace (a function in real space) folded with the probability distribution of terrace types $P(\Gamma)$. This approximation is very good so long as the terrace width distribution is not too broad—if very narrow terraces and very wide terraces were to occur in close

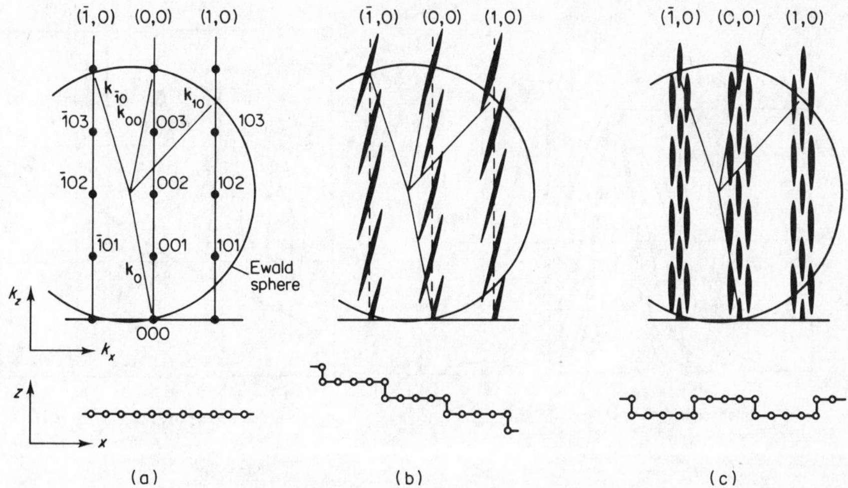

Figure 6.6. Reciprocal space with Ewald sphere for three forms of surfaces. Here, $\mathbf{k}_0$ denotes the wave vector of the incident beam, $\mathbf{k}_{00}$, $\mathbf{k}_{10}$ and $\mathbf{k}_{\bar{1}0}$ that of the diffracted beams with indices (0, 0), (1, 0) and ($\bar{1}$, 0) respectively. With the shown position of the Ewald sphere the (0, 0) beam reflects the in-phase condition, the (1, 0) beam out-of-phase condition between adjacent terraces. (a) Flat surface; (b) regular step array (monotonic); (c) regular step array (two layers). (Reproduced by permission of Springer-Verlag from Henzler[270])

proximity, the results would be less accurate. Using this approach, they found that spot splitting rather than just broadening occurs even in the presence of quite a broad distribution of terrace widths. Conversely, the observation of diffuse broadening without the appearance of distinct splitting indicates the presence of a wide variety of terrace widths. Henzler and Wulfert[272] have calculated the spot shapes for such a distribution in which widths vary from one to several hundred atom row separations, characterized by the mean percentage of step atoms, given by the average over the reciprocal terrace width

$$\left\langle \frac{1}{\Gamma} \right\rangle = \sum_{\Gamma} P'(\Gamma)/\Gamma \tag{6.27}$$

Results are given in Figure 6.7(a); assuming the same terrace width distribution in all directions on the surface, they used double the average value $(1/\Gamma)$ to give the percentage of step atoms.

These widths refer to the maximum values obtained at the out-of-phase condition (S = half-integer in equation 6.20). At the characteristic energies for which scattering from adjacent terraces is in-phase (S = integer), the width of the beam should be limited only by the instrumental transfer limit. As the energy of the incident beam rises, so the widths of the individual beams oscillate. The instrumental limit will increase with energy, but if this is monitored for a series of characteristic energies, a reasonable interpolation

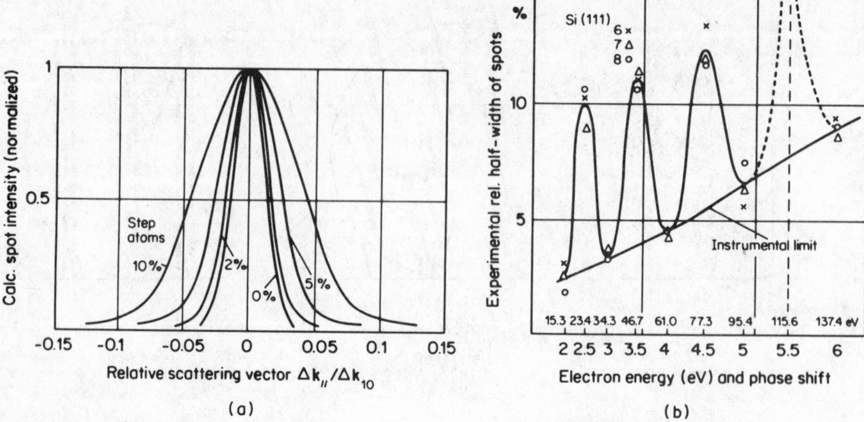

Figure 6.7. (a) Calculated spot shapes for terrace width distributions chosen to yield approximately a Gaussian shape. (b) Measured half-width of the 00 spot of a LEED pattern of an oxidized Si(111) face after removal of oxide versus electron energy. (Reproduced by permission of Springer-Verlag from Henzler and Wulfert[272])

may be made, permitting values to be set for the width-broadening effects which result from the presence of step distribution. This is clearly illustrated in Figure 6.7(b), taken from a study of Si(111) which was first oxidized and then etched. As a result of this treatment, Henzler and Wulfert[272] were able to show that the step edge density which resulted from this particular surface preparation was about 12 per cent.

Another way of determining step densities is to look for the characteristic modulation of the intensities of the beams with energy. This places the burden on to calculational accuracy rather than instrumental precision since such a modulation, if it exists, is superimposed on the already complex energy dependence of the beam intensities resulting from usual multiple-scattering diffraction processes. Laramore *et al.*[273] developed an algorithm to describe this modulation for the case in which there is a distribution of step heights based on a normalized Gaussian distribution:

$$P(N, d, z_0) = C \exp[-(Nd)^2/2z_0^2] \qquad (6.28)$$

where Z_0 characterizes the width of the distribution, d is the step height normal to the terrace planes, C a normalization factor and N an integer. The step modulation, $E\cos^2\theta = (n\pi/d)^2$, has its maxima V_r higher in energy than the kinematical Bragg peaks, which are given by $E\cos^2\theta = V_r + (n\pi/d)^2$. Although it is usual for multiple-scattering processes to introduce numerous additional features into the $I(E)$ profiles, the principal peaks are generally to be found in the vicinity of, or a few volts below, the energies predicted by Bragg theory. The predominant effect arising from a finite step density is thus a shifting of the major peaks towards higher energies, through the systematic diminution of their lower-energy parts and the augmentation of their higher-energy parts.

Since shifts upwards in energy may also be produced both by contraction of the surface layer and also by a reduction in inner potential as the incident beam energy increases, a very detailed comparison of $I(E)$ curves is necessary to permit a distinction to be made between each of these causes. Zanazzi *et al.*[274] examined an apparently stepped surface of Ag(110) with this method, but were unable to identify positive improvement in $I(E)$ curve comparisons. Their study was directed towards a crystallography of the flat Ag(110) surface, and they found it necessary to repolish the crystal, thereby preparing a new, smoother surface which could be satisfactorily analysed without recourse to any theory of step effects.

Recently, Gauthier *et al.*[37] examined a Ni(110) surface. The principal peaks observed from this surface appeared rather high in energy compared with the Bragg peak positions and this indicated the possible presence of steps. Quantitative *r*-factor comparisons (see section 7.3) were used to compare a large number of beams taken from a series of different incident beam angles. However, the introduction of even a few per cent of steps via the Laramore method worsened the agreement between the experimental and calculated intensity–voltage curves. An energy-dependent decrease in inner potential was used, which had also been successfully used in previous studies of the flat Ni(100), and with this contribution to the effect already included, the remaining shift could be attributed to a contraction of about 7 per cent of the surface interlayer spacing. It has been suggested[275] that a comparison of relative beam intensities at individual incident beam energies may also provide a sensitive indicator of the presence of steps (see section 7.3.5). Applying this $I(\mathbf{g})$ method to the same Ni(110) surface confirmed that, indeed, this method is more sensitive to the presence of steps than standard $I(E)$ methods. A comparison of the relative intensities of more than 100 beams concluded that the surface interlayer spacing was contracted by about 7 per cent, but that any regular step density would have to be less than a few per cent, in agreement with the associated $I(E)$ curve comparison.

In only one investigation has this step-dependent modulation of intensities been found to give a clear improvement of theory with experiment. This was in a study of Al(110) by Aberdam *et al.*,[276] where best agreement was achieved with a distribution characterized by $Z_0 = 0.6$, suggesting a distribution maximum around 20 per cent steps.

In this section we have been concerned primarily with step arrays which are characterized by relatively long terraces separated by monoatomic steps. Sometimes diatomic steps may be more stable, e.g. ZnO(0001),[277] and the analysis may be adapted fairly easily to treat such cases. Thermodynamically, a surface at finite temperatures can never be expected to be perfectly flat, and a few steps may occur amongst the assortment of possible thermally induced defects. However, if the macroscopic face is aligned parallel with the crystallographic planes and gentle annealing is used to remove those defects which were caused by the initial polishing and cleaning processes, the density of steps may be made negligibly small. Mica and cleaved GaAs are capable of producing almost perfectly step-free surfaces.

Steps may be introduced by cutting and preparing a macroscopic face slightly inclined with respect to a low-index crystal plane. On metals, a step–step repulsion, either due to electronic influences or strain-fields associated with the edge atom, will often cause the steps to assume a regular array with similar terrace widths, justifying the use of the Park–Houston model described earlier. However, irregular step arrays could be induced in GaAs(110) as a result of ion bombardment followed by gentle annealing.[278]

Steps may be formed in an otherwise flat surface by a number of processes. We have already mentioned ion bombardment. Henzler and Wulfert[272] induced a 12 per cent step density into Si(111) by etching off the surface oxide with fluoric acid, and stabilizing the structure with iodine. The adsorption of N on Fe(001) via repeated high-temperature treatments with NH_3 has been found to induce a step density of about 17 per cent (monoatomic step heights with terraces of about six atom rows).[279]

Often, annealing will reduce the density of defect structures in a surface. Surfaces of semiconductors which are disordered immediately following ion bombardment may first anneal into random step arrangements, then order progressively with rising temperature, yielding increasingly sharp LEED spots.

Even if a surface is initially formed with regular step arrays, by cleavage or deliberate surface orientation, these may disappear as a result of surface diffusion when the surface is heated. The macroscopic orientation is maintained by the formation of a low density of multi-atomic height steps. These large steps are really new terrace-like faces of some other low-energy plane. If the initially cut face is at a large angle with respect to the lowest-energy planes, the surface may 'facet' completely into a hill-and-valley-type structure of these other low-energy planes. The energetics of this process are described in section 1.5.8. Sometimes facet formation occurs on clean surfaces, in other cases faceting results from the presence of an impurity (e.g. O_2 causing Mo(111) to facet into {211} and {110} planes[66]). Occasionally, impurity atoms may reverse the process of faceting: on annealing, a stepped Fe(12 1 0) surface will facet into Fe(001) planes separated by occasional multi-atomic-height steps, or facets, but if N_2 is adsorbed, the surface reverts to a regular monoatomic-step array.[280]

An interpretation of LEED patterns from faceted surfaces can be readily made using the Ewald construction. We shall assume that the facets are large enough to produce sharp diffraction spots; if not broadening of the reciprocal rods will occur, but the basic form of analysis will remain the same. The observed pattern is then nothing more than the superposition of patterns from each facet. In the case of Mo(111), three {210} and three {110} facet directions are involved, producing a threefold symmetric pattern at normal incidence to the initial Mo(111) face.[66] The specular reflection or (0, 0) beam from each of the {210} orientated facets is visible on the hemispherical LEED grid used to make the observations, but the {110} facets are inclined at too great an angle for specular reflections from these facets to be observed at normal incidence. A typical sequence of patterns, formed as a function of energy, are illustrated in Figure 6.8.

As an illustration, we consider the Ewald construction in the plane defined by the normals to one pair of (210) and (110) facets (see Figure 6.9). Consider the vicinity of the $(0, 0)_{210}$ beam. At some particular low energy, the $(0, 0)_{210}$ reciprocal lattice rod intersects a reciprocal rod of the (110) lattice. A single spot is observed (see Figure 6.8(b)). As the incident energy is increased, the beam will split—the $(0, 0)_{210}$ beam remains fixed by the inclination of this face, but the (110) component will shift towards the centre of the diffraction pattern (corresponding to the former position of the (111) specular reflection, B and C respectively in Figure 6.9). Continuing to increase the energy will eventually bring the $(0, 0)_{210}$ beam to point D. As this point is approached, so will another (110) beam pass through the $(0, 0)_{210}$ beam.

In this particular case, the (111) face was converted totally into {210} and {110} facets, but the inclination of the $(0, 0)_{210}$ beam could be measured directly from its position on the screen. This could be achieved by comparing photographs of the unfaceted Mo(111) face and using standard diffraction theory to deduce the angle for beams of known energy and diffraction order which produced spots in the same place on the screen. From this, the identity of the (210) facets could be revealed, from which the rest of the analysis followed.

Using the concept of the Ewald sphere, it is easy to relate the inclination α and periodicity d_F of the facet faces to the energies at which the facet beams pass through the $(0, 0)_u$ beam of the original unfaceted face. This method may also be used to verify that facets associated with the observed {210} faces are indeed {110} planes.

A beam from the facet will pass the $(0, 0)_u$ beam whenever

$$n\lambda = 2d_F \sin \alpha \tag{6.29}$$

where n is the order of diffraction. The interference condition of the facet is given by

$$d_F(\sin \beta - \sin \alpha) = n\lambda \tag{6.30}$$

where β is the angle of diffraction. From this we see that

$$\frac{d\lambda}{d\beta} = \frac{d_F}{n} \cos \beta \tag{6.31}$$

This passes through the $(0, 0)_u$ beam when $\cos \beta = \cos \alpha$. Dividing by equation (6.29) at this condition gives

$$\frac{d\lambda}{d\beta} = \frac{\lambda}{2} \cot \alpha \tag{6.32}$$

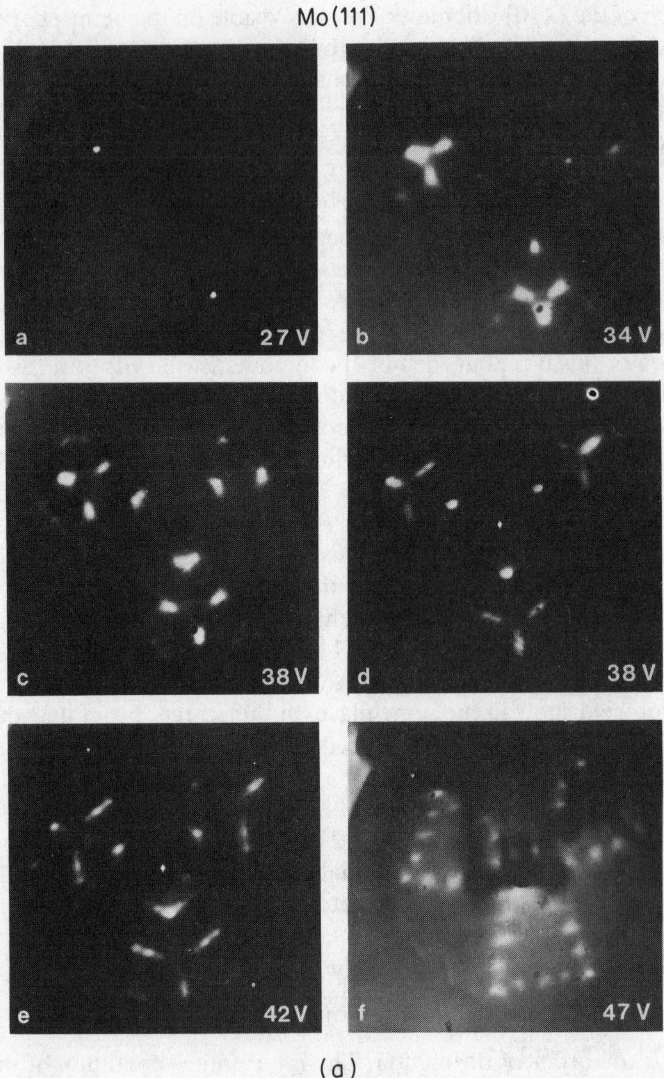

Figure 6.8. Diffraction patterns at a series of incident beam energies for a mo(111) surface faceted into (110) and (211) planes

from which α may be found, thence by substitution into equation (6.29) d_F may also be derived. Sometimes the order of diffraction, n, is not certain; in such cases it is possible to monitor the change in reciprocal wavelength $\Delta(1/\lambda)$, where $1/\lambda = (E(\text{eV})/150.4)^{1/2}$ Å^{-1}, as successive diffraction orders pass through $(0, 0)_u$. The following expression may then be used:

$$\frac{\Delta n}{\Delta(1/\lambda)} = 2d_F \sin \alpha \tag{6.33}$$

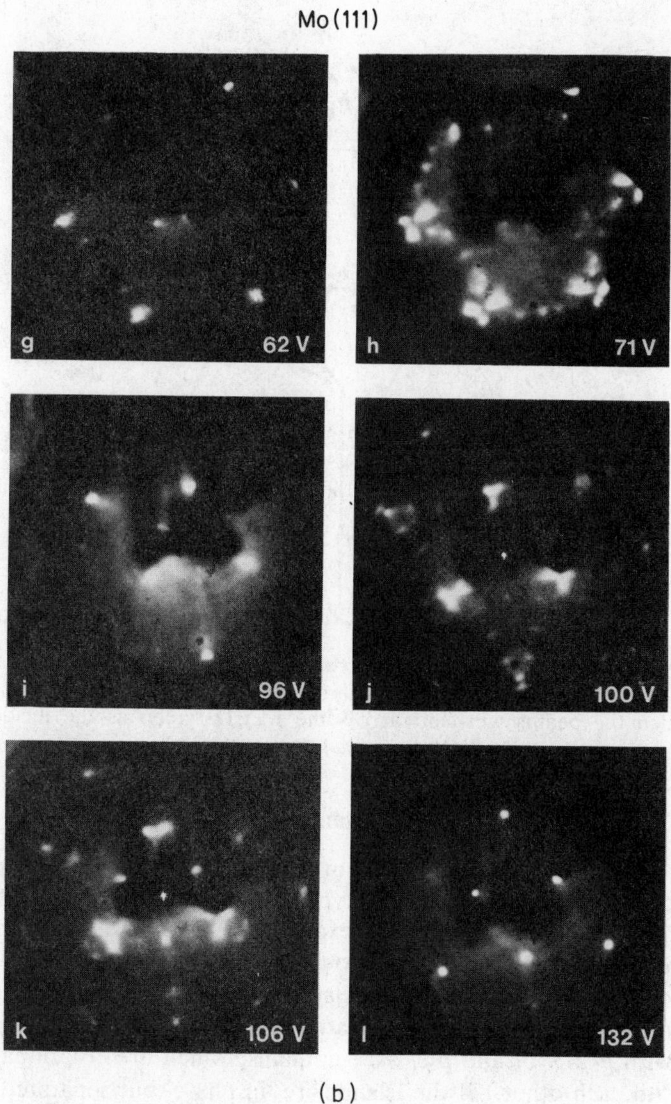

(b)

Although the principle is straightforward, in practice the movement of the spots may be difficult to follow due to the modulation of intensities with energy. Nevertheless, the assignment of the facet crystal structure is often helped by the fact that only a limited number of possible low-index planes have sufficiently low free energy to be eligible candidates for the formation of facets.

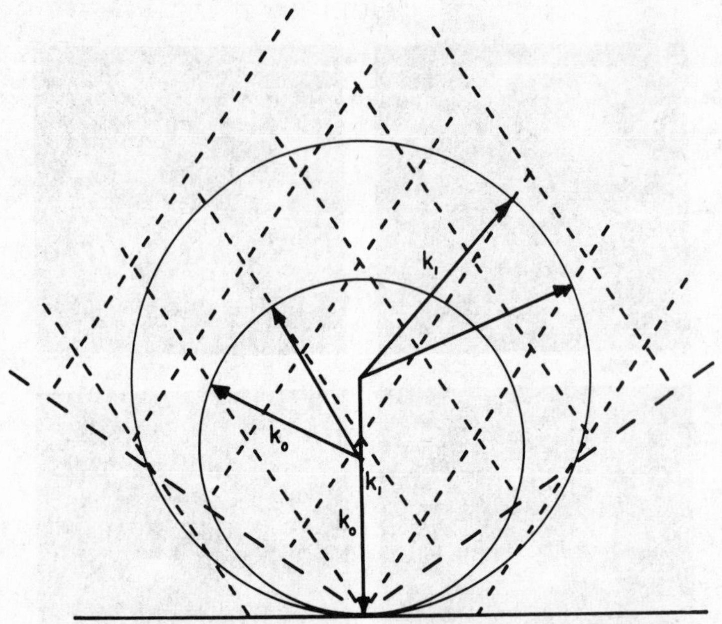

Figure 6.9. Ewald construction at two different energies, illustrating the form of diffraction patterns from a faceted surface, and the manner in which the beams will shift across the LEED screen as the incident electron beam energy varies

6.4 Antiphase domains

The presence of an adsorbed layer, or some form of surface reconstruction, introduces the possibility of further types of extended disorder. In particular, antiphase domains may occur. For example, if the adsorbate has a periodic structure with characteristic length double that of the substrate (e.g. $p(2 \times 2)$, or $c(2 \times 2)$) then it is possible for adjacent domains to grow which are not in phase with each other. This is illustrated in Figure 6.10. (Another possibility occurs with $p(2 \times 1)$ and $p(1 \times 2)$ domains, which are rotationally out of phase with each other.) If the islands are 'in phase' but separated, then the effect is that of a single domain with non-random vacancies; no beams are affected by destructive interference, but beams will broaden and the background intensity increase. A general approach for simple superstructures (in which the superstructure unit vectors $\mathbf{a}'$ and $\mathbf{b}'$ are given by a linear combination of the substrate unit vectors $\mathbf{a}$ and $\mathbf{b}$ with integer coefficients) has been given by Henzler.[270] If the in-phase component of the scattering vector $\mathbf{K}_{11}$ is given by $h\mathbf{a}^*_1 + k\mathbf{a}^*_2$, and the vector connecting the equivalent atoms in the two domains $\mathbf{I}$ is given by $l_1\mathbf{a}_1 + l_2\,\mathbf{a}_2$, then the phase shift between the domain is

$$\Delta\phi = \mathbf{K}\cdot\mathbf{I} = 2\pi(hl_1 + kl_2) \tag{6.34}$$

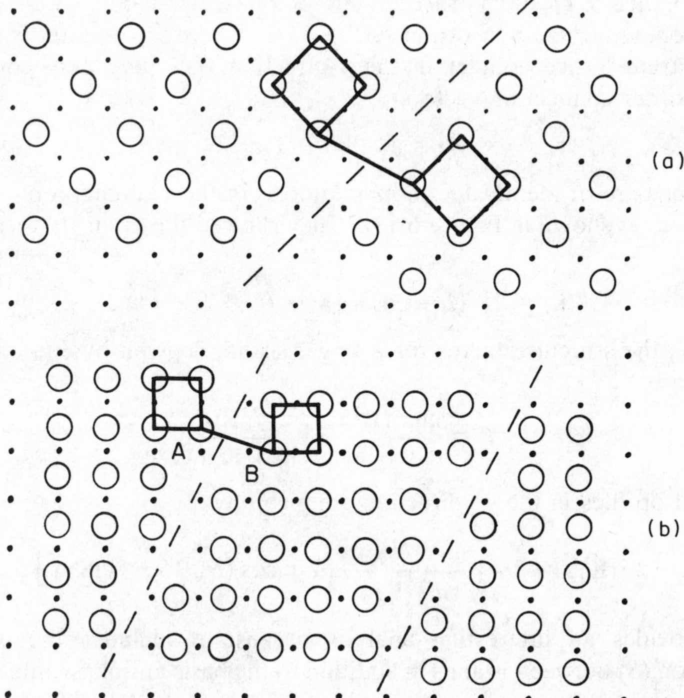

Figure 6.10. Examples of antiphase domains: (a)$c(2 \times 2)$ pattern in which the adatoms occupy fourfold symmetric 'hollow' sites; (b) in a $p(1 \times 1)$ pattern in which the adatoms occupy twofold 'bridge' sites

The domains scatter in phase if the sum in parentheses is integral, but out of phase if the sum is half-integral. If l, and l_2 are integers (**l** is a substrate vector) then the integral-order beams will be unaffected by the antiphase domain; certain half-integral beams, on the other hand, will be broadened or split. If l_1 and l_2 are non-integers (**l** not a substrate vector) then some of the integral-order spots may also be affected. This provides a valuable aid in determining the type of bonding sites which are being occupied by the adsorbate atom. For example, in Figure 6.10(a) $p(2 \times 2)$ adlayer domain A is out of phase with domain B. If the adatoms bond into fourfold sites then the connecting vector **l** can only compromise integral values for l_1 and l_2. By comparison, Figure 6.10(b) illustrates the two different types of antiphase domain relationship which could occur if the adatoms bond into bridge sites. If l_{AB} or similar exists then certain integral-order beams will be affected, and bridge-bonding must be occurring. McKee *et al.*[281] examined in detail the type of broadening or splitting that would occur in the presence of two $p(2 \times 1)$ antiphase domains. The general approach, and variety of types is similar to that summarized above for stepped surfaces. McKee *et al.* considered two islands of $p(2 \times 1)$ superstructure out of phase with one another in the x-direction, separated by a distance $(S - 1)\ \mathbf{a}_2$, where S is an integer and $\mathbf{a}_2$ is the substrate lattice contact in the y-direction (i.e. they only considered integral-order connecting vectors):

$$\mathbf{l} = \mathbf{a}_1 + (S - 1)\mathbf{a}_2 \qquad (6.35)$$

The islands are of identical size, M scatterers in the x-direction by N in the y-direction, as shown in Figure 6.11a. They showed the intensity variation to be

$$I(\mathbf{k}) = 2F^2\{1 + \cos \mathbf{k}\cdot[\mathbf{a}_1 + (N + S - 1)\mathbf{a}_2]\} \qquad (6.36)$$

where K, the structure factor for a single island, is given by equation (6.8) in this case

$$F = \frac{\sin M\mathbf{K}\cdot\mathbf{a}_1}{\sin \mathbf{K}\cdot\mathbf{a}_1} \cdot \frac{\sin \frac{1}{2}N\mathbf{K}\cdot\mathbf{a}_2}{\sin M\mathbf{K}\cdot\mathbf{a}_1} \qquad (6.37)$$

The spot profiles in the y^* direction are given by

$$I(\mathbf{K})_{y^*} = 2M_2 \frac{\sin^2 \frac{1}{2}N\mathbf{k}\cdot\mathbf{a}_2}{\sin^2 \frac{1}{2}\mathbf{K}\cdot\mathbf{a}_2}[1 \pm \cos (N + S - 1)\mathbf{K}\cdot\mathbf{a}_2]$$

This provides an interesting analysis because it includes the effect of separation (S) between islands in addition to the basic antiphase interference. The positive and negative signs refer to integral- and half-integral-order beams respectively. Figure 6.11 shows beam profiles in the y^*-direction for integral- and half-integral-order beams when narrow islands ($N = 3$) are, firstly, adjacent to one another ($S = 1$) then well separated ($S = 5$). Only the half-order features are split, but in the latter case, the integral-order beam gains satellite features which in practice, due to irregularities in island size and separation, would appear as broadening of the main peak. We see that, when $\mathbf{l}$ is integral, the integral-order beams are not split, but could appear broadened if the antiphase islands are small and well separated. As an example, if a gas is adsorbed on a surface and initial adsorption involves nucleation into islands, we would expect the integral order beams to be initially broad. As coverage increases, and if $\mathbf{l}$ is integral, the integral order beam may sharpen, but the half-order beams remain split. The extent of the splitting in the half-order beams may nevertheless decrease with coverage. A review of numerous examples of antiphase domain and other defect effects has been made by McKee *et al.*[282]

At this point, a word of caution is necessary. Although antiphase domains are capable of producing spot splitting, there have been cases where antiphase domains were assumed to be the cause of observed splitting where an incommensurate overlayer might more satisfactorily explain the observations. For example, iodine on W(110) forms a series of patterns in which spot splitting and continuous spot shifting are observed. Avery,[283] who observed these patterns, interpreted them in terms of the interference of antiphase

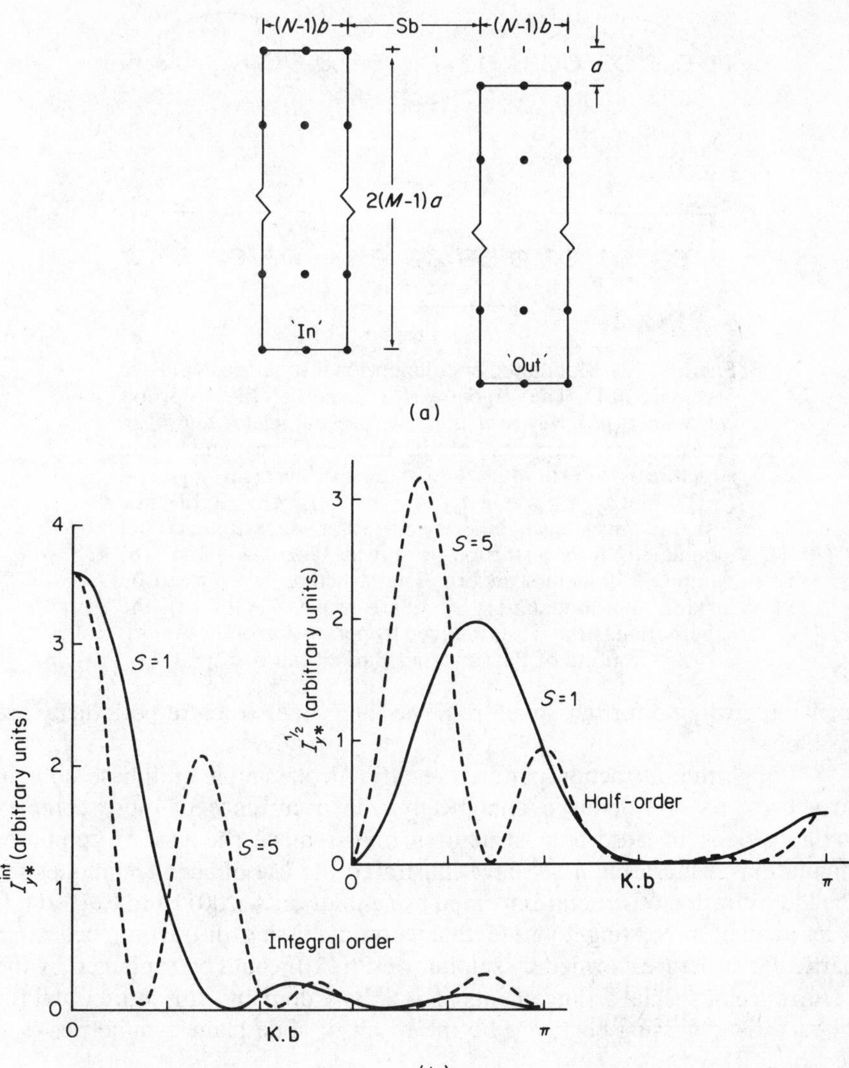

Figure 6.11. (a) Two out-of-phase separated islands of (2 × 1) structure; (b) intensity profiles for scattering by two islands with $N = 3$; (—) not separated ($S = 1$); (---) separated ($S = 5$). (Reproduced by permission of North-Holland Physics Publishing from McKee *et al.*[281])

domains. Dowben and Jones,[284] recognizing the similarity of these patterns with those they had previously observed in halogen adsorption on Fe showed that incommensurate 'rafts' of adatoms, continuously compressing as the coverage increases, could produce the spots as their first-order diffraction features, the weakness of the double-diffraction features (cf. section 1.4.2),

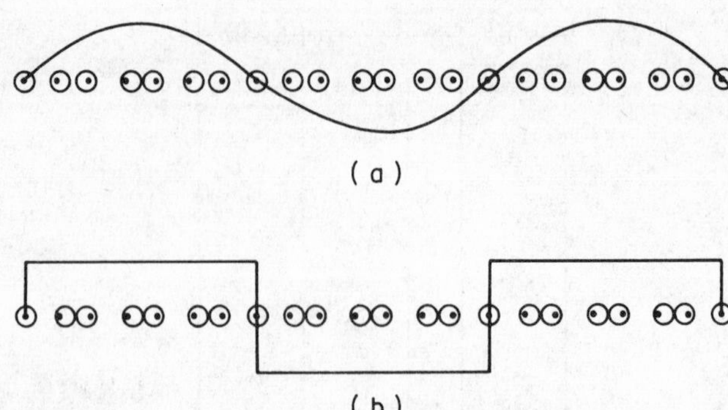

Figure 6.12. Modulated one-dimensional structures. (a) Pure harmonic PLD. The structure is produced by the distortion wave $\sin(q_1 na)$, where a is the lattice parameter and n an integer. For purposes of illustration a commensurate structure is shown with $q_1 = \geqq (2\pi/a)$. Since $\sin(q_1 na) = (-1)^{n+1}\sin(q_2 na)$, where $q_2 = \pi/a - q_1 = \quad (2\pi/a)$, the structure may also be viewed as a dimerized crystal modulated by a distortion with wavelength $\lambda = 14a$. (b) Antiphase domain structure. The structure is produced by making the modulation a square wave. As in (a) the superperiod is $14a$. (Reproduced by permission of the American Institute of Physics from Barker and Estrup[285])

explaining why no further spots from the long superstructure periodicity are visible.

Although the diffraction patterns resulting from simple antiphase domain structures may be deduced analytically, the occurrence of more complex juxtapositions of adsorbate structures may require the use of computer simulations. Maurice *et al.*[264] have illustrated the use of such techniques for the identification of structures formed by sulphur on Ag(001) and Cu(001). It is particular interesting to note that certain missing diffraction beams in particular structures formed by sulphur on Pt(110) could be explained by the occurrence of special arrangements of antiphase domains. It is more usual for characteristic missing beams to be the result of glide-plane symmetries (see section 1.7).

Antiphase domains are usually associated with discrete regions, each with the same periodicity, closely related to that of the substrate lattice. Barker and Estrup[285] have detected beam splitting in the W(001)–$c(2 \times 2)$H structure, and interpreted this as a result of periodic lattice distortions (PLD) which may or may not be caused by the presence of charge density waves. The similarities in periodic distortions caused by a harmonic PLD and the closest equivalent antiphase domain structure can be seen from Figure 6.12.

6.5 Order–disorder Phenomena

One of the most remarkable facts about adsorbate structures on single-crystal surfaces is the high degree of long-range order that is often observed. The relationship between the substrate periodic lattice and the periodicity adopted by the adlayer provides useful information concerning the types of interaction occurring either between adatoms or between substrate atoms and adatoms.

In some cases, the surface layer organizes itself into a close-packed hexagonal layer almost independently of the substrate, implying that the interactive forces within the surface layer ('lateral forces') are stronger than the adatom–substrate interactions. At the other extreme, the lateral interactions may be relatively weak, so that the adatoms assume no ordering, other than the fact that they sit randomly in the bonding sites provided by the substrate itself. In many cases, however, the adlayer will take on a periodicity that is different from, yet related to, the substrate periodicity. The $c(2 \times 2)$ overlayer structure is the commonest example of this. In this case, short-range lateral repulsions prevent adatoms occupying neighbouring sites, but are either not sufficient to prevent the occupation of next-nearest sites as coverage approaches of half-monolayer, or else the lateral interactions actually favour the occupation of such sites. Einstein and Schreiffer[286] have shown that perturbations of the substrate electron densities by adsorbed atoms may result in 'indirect' interactions (i.e. via the substrate) that produce alternating repulsive and attractive forces with distance from an adatom. If such an oscillatory interaction exists, a series of adlayer structures may be formed as a function of coverage, whilst other alternative structures are prevented. This could explain the special series of adlayer structures formed, for example, when sulphur is adsorbed on transition metal surfaces such as Mo[11] or W[57]. In general, such lateral interactions are most probable if the adsorbate chemisorbs with little charge transfer, as is the case with H, C or N. Alkali or halide adsorbates, on the other hand, chemisorb with a strongly ionic bond. The transfer of charge associated with such bonds results in surface dipoles that cause repulsive interactions. Consequently, such adsorbates will tend to spread across the surface uniformly, and will compress steadily with increasing coverage. There are numerous examples of this in the literature.[287]

Although much can be learned simply from observing the equilibrium adlayer structures formed with any particular substrate–adsorbate combination, for a more quantitative probe of the interactive forces present it is useful to monitor the manner in which as structure disorders as a function of increasing temperature.

In Chapter 5, the effect of increasing thermal vibrations on LEED beam intensities were considered. If an ordered overlayer is present, increasing temperatures will not only increase the vibrational amplitudes of the individual atoms, but also cause a certain proportion of the atoms to move to different bonding sites not included in the original ordered structure. At a sufficiently high temperature, complete disordering may result, although thermal desorption may also occur, complicating the interpretation.

To move from one site to another, an adatom must possess sufficient thermal energy to overcome the intervening potential barrier, E_p. The time τ

spent by an adatom in any one site is then related to the temperature T by

$$\tau = \tau_0 \exp(E_p/kT) \tag{6.39}$$

where k is Boltzmann's constant and τ_0 is a time constant approximately equal to the duration of one vibration (10^{-13} s).

The value of E_p depends on the nature of both the substrate and adsorbate species, but is usually small compared with the adsorption energy of the adsorbate upon the surface. If (E_p/kT) is greater than about 10, the adsorbate atom or molecule is highly unlikely to move from its initial bonding site during an experiment, whereas, if (E_p/kT) is less than unity, it will be free to move without restriction from substrate interactions. The temperature at which this occurs is called the critical temperature, T_c. The presence of other adsorbate atoms on the surface will, however, effectively modify E_p and thus cause certain adsorbate sites to be more favourable than others, and more complex theories of lateral interactions must be used to predict the temperature dependence of the resulting adlayer structures.

LEED is ideally suited to studying the order–disorder phenomena, due to its particular sensitivity to long-range order. Changing order affects both the intensities and widths of the corresponding diffraction beams.

The relationship between LEED beam intensities and disordering of the adlayer depends on the structure of the initial ordered structure. As an example we may consider the disordering of a $c(2 \times 2)$ structure. If p is the probability that any particular atom within the $c(2 \times 2)$ structure is in the 'correct' site then the probability of being in a 'wrong' site is simply $(1 - p)$. The diffraction beam intensities are the consequence of constructive interference between the scattered contributions from all 'correctly' sited atoms, but, in this case, all 'wrongly' sited atoms will lead to destructively interfering contributions. The net diffraction beam amplitude is thus directly proportional to $p - (1 - p) = 2p - 1$. The intensity I is therefore given by

$$I \propto (2p - 1)^2 \tag{6.40}$$

Of course, this only applies to the half-order beams characteristic of the adlayer; integral-order beams are formed by scattering from the substrate, even in the presence of a completely disordered adlayer.

From an analysis of the diffraction beam intensities as a function of temperature, the disorder may be quantified and compared with theoretical predictions. As an example, data obtained from Na on Ni(001)[288] are given in Figure 6.13, from which the complete extinction of the half-order beam at the critical temperature is quite clear, whereas the intensity of the integral-order beam passes through a small step, and then continues to drop slowly with temperature as a result of thermal vibrations, as predicted by the Debye–Waller factor. Andersson compared these observed results with predictions from two separate theoretical models. Predictions from the Ising model, given by the solid line, are seen to agree extremely well (once the T_c parameter is matched to the observed value), whereas the Bragg–Williams

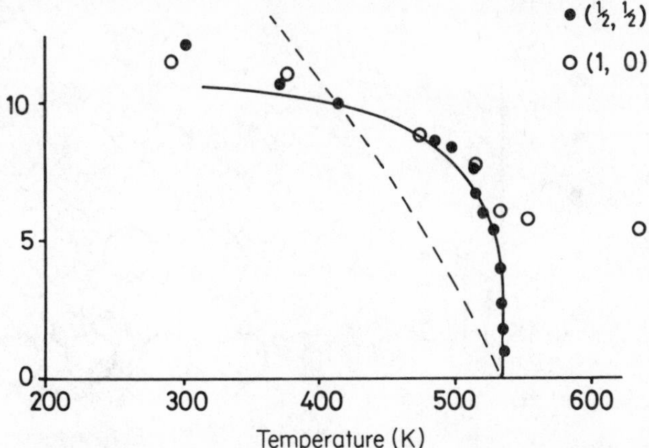

Figure 6.13. Variation of intensity with temperature for the (1, 0) and ($\frac{1}{2}$, $\frac{1}{2}$) beams from $c(2 \times 2)$Na–Ni(001). The solid line is that calculated using the Ising model, the broken line is from the Bragg–Williams approximation. (From Andersson[288])

model, given by the broken line, is clearly unrealistic. The adsorption of H on W(001) has been found to give very similar results,[289] and would therefore also correspond closely to the predictions of the Ising model. It is not appropriate to consider the many models of disorder here, but it is nevertheless clear that LEED used in this way provides a sensitive method for testing alternative theories.

Although most studies of order–disorder phenomena have made use of changes in the itensities of the diffraction beams, it is also possible to monitor the physical widths of the diffraction beams. As the disorder increases, so the beam may broaden. Roelofs *et al.*[290] have shown that these widths rise quite sharply above the critical temperature T_c, and the slope of this rise provides an additional probe of the lateral interactions. In a study of phase transitions in $p(2 \times 1)$O on W(110), Wang, Lu and Lagally[291] used both intensity and angular width variations of the diffraction beams to interpret the phase changes in terms of both order–disorder transitions and island dissolution.

Sometimes, particularly on non-square lattices, disorder may occur in one direction only. This is quite common during adsorption on deeply channelled surface such as bcc (211) or fcc (110) surfaces. This channelling produces an anisotropy in the lateral interactions, and in an appropriate temperature range, disorder may occur in one direction whilst being maintained in another. This effect has been clearly demonstrated in the case of oxygen on W(211)[292] and also with sodium on Ni(110).[281,293] Typical LEED patterns and their structural interpretations are given in Figure 6.14. Although the resulting LEED beam intensities can be predicted using analytical solutions in cases where the disorder is easily characterized, increasing complexity may

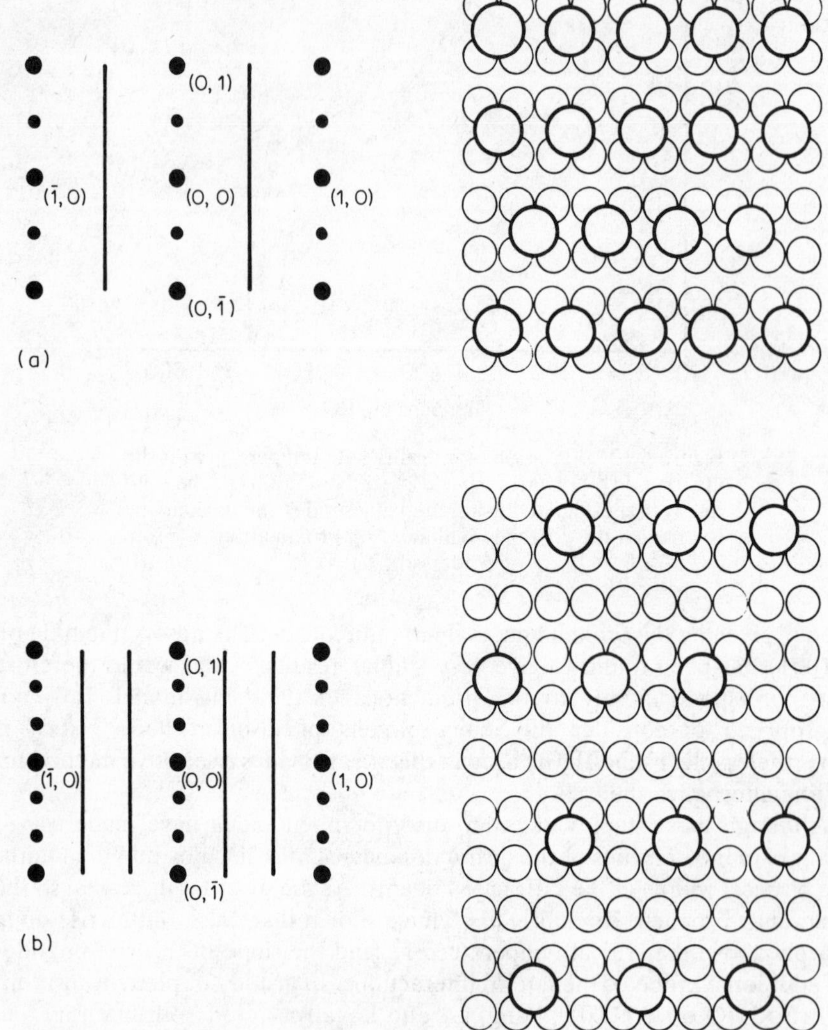

Figure 6.14. The effect of disorder in one dimension only on the diffraction patterns associated with an adlayer. (Reproduced by permission of North-Holland Physics Publishing from Gerlach and Rhodin[293])

favour the use of computer simulations using Monte-Carlo sampling techniques. The patterns observed when oxygen is adsorbed on W(211), for example, have been interpreted by Ertl and Plancher,[294] from which values of E_p appropriate to different crystallographic directions within the surface could be derived. An extensive review of work carried out on such anisotropic systems is contained in a recent paper by McKee *et al.*[295]

Of course, no surface structure is likely to be perfectly ordered: certainly, if the potential barrier energy is large compared to kT, the mobility may become negligible, but this is equally likely to freeze the adlayer into a partially disordered structure. In such a case, the temperature will need to be raised sufficiently high that the structure can order itself, but not so high that disorder is increased: clearly the resulting degree of order will be limited by statistical constraints. Nevertheless, Yang *et al.*[296] observed the effects of increasing disorder on intensity–voltage curves during the adsorption of oxygen on Cu(001), and concluded that, even when there is no apparent long-range order, the residual intensity–voltage variations may convey information about the adsorbate bonding geometry.

6.6 Data averaging

Because kinematic theory is so much simpler to handle than dynamic theory, it is clearly attractive to consider whether the kinematical features of the experimental spectrum could in some way be isolated from those contributions dominated by multiple scattering. To do this one must collect a large quantity of data whilst maintaining some fixed kinematical condition, and then average it all in the hope that most of the multiple-scattering contributions will be 'washed out'.

There are several different ways in which data suitable for averaging may be collected. For example, the kinematical wave function

$$\psi \propto \sum_j f_j(\mathbf{k}_0, \mathbf{k}) \exp[i(\mathbf{k} - \mathbf{k}_0)\cdot\mathbf{R}_j] \qquad (6.41)$$

is fixed for a given $(\mathbf{k} - \mathbf{k}_0) \equiv \mathbf{S}$, whereas multiple-scattering contributions in general depend on $\mathbf{k}$ and $\mathbf{k}_0$ independently. This restriction on the change of wave vector represents a restriction on the normal component of momentum of any particular beam (see Figure 6.15), so data averaging subject to this restriction is generally termed constant momentum transfer averaging or CMTA. The momentum transfer $\mathbf{S}$ is related to the energy E, incident angle θ and inner potential V_r by the following expression:

$$|\,\mathbf{S}\,| = 2(E \cos^2\theta - V_r)^{1/2} \qquad (6.42)$$

By recording $I(E)$ curves for a number of different beams and a set of different angles of incidence, both polar (θ) and azimuthal (ϕ), data may then be averaged by selecting all values of the diffraction intensity corresponding to all values of and E that together maintain a constant value of $|\,\mathbf{S}\,|$. Lagally *et al.*[297] first applied this approach using data collected in steps of 2° in polar angle and 10° steps in azimuthal angle ϕ. Since equation (6.42) is independent of ϕ, all data from different azimuthal angles that satisfy the appropriate restriction on E and θ may be averaged together directly. However, Aberdam *et al.*[137] have shown that the sensitivity of the intensity to azimuthal angle is very small when θ is near normal incidence, but when is increased to about 50° or more, high sensitivity to the azimuthal angle is obtained, causing marked variations in the intensity structure as it is varied.

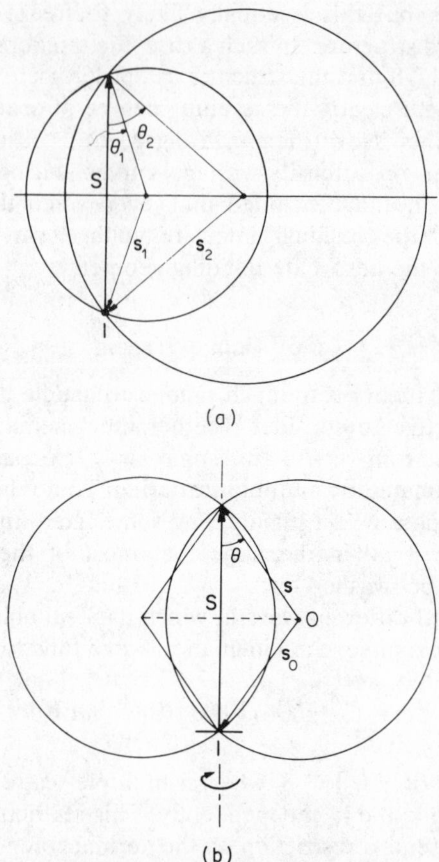

Figure 6.15. (a) The constant-momentum transfer averaging process according to Lagally and co-workers.[297] For each angle of incidence, an energy profile is recorded. Then, one selects in the various energy profiles the intensity at an energy such that the momentum transfer is constant. (b) The constant scattering factor averaging procedure. The basic data are rotation diagrams at constant incidence and energy. Averaging is made by continuous integration over the azimuthal angle. By this procedure, one avoids sampling as well as variations of the diffraction parameters with energy. (Reproduced by permission from Garland STPM Press from Aberdam *et al.*[137])

Clearly, as θ is increased, the sampling density of θ values should be increased correspondingly, ideally approaching a near-continuous set at large polar angles, and certainly more frequently than at the 10° intervals used originally.

In addition to this practical point, there are fundamental uncertainties associated with the CMTA method resulting from the fact that the data to be averaged are obtained from different values of E and θ. As a consequence, the intensities do not correspond to the same scattering factor $f(\mathbf{k}, \mathbf{k}_0)$ so that the average is taken from quantities that do not have equal weights. This difficulty can be remedied by maintaining both E and θ constant during the averaging process. Aberdam and Baudoing[298] proposed such a method, which may be termed a constant scattering factor averaging (CSFA) process, in which the averaged profile is built up from rotationally averaged points at successive polar angles of incidence. By keeping E constant, this method overcomes a second fundamental problem with CMTA, namely, that V_r plays an important role in evaluating the points that would maintain a constant $|\mathbf{S}|$, but V_r is not accurately known until after the analysis has been completed—this uncertainty may introduce systematic errors. A constant value of E also has the advantage that one can avoid normalization problems associated with a varying incident beam current.

Despite these shortcomings, the CMTA method has been used on a number of occasions for the analysis of surface structures. Early studies demonstrated that the averaged curves appeared kinematic in form but no quantitative estimates of the surface layer displacement were made[299–301] Quinto and Robertson,[300] studying Al surfaces, and McDonnel *et al.*[302] studying Cu, reported difficulties when using this approach, but Alff and Moritz[303] were more optimistic, when, in 1979, they used CMTA to estimate the surface contraction of Al(100) as 8 per cent, close to the value of 10 per cent obtained by Maglietta *et al.*[304] from a complete dynamical LEED analysis. In their investigation, Alff and Mortiz found that movements of the surface layer alone had very little effect on the energies of the major diffraction peaks, but had an appreciable effect of the peak shapes. In particular, a contraction of the surface interlayer spacing caused peaks to become rather asymmetric, with a shoulder developing on the high-energy side. This result appears to be quite general, and, for example, can be seen in the $I(E)$ curves for Mo(001) calculated using dynamical theory, as shown in Figure 6.16.

A fundamental limitation on the accuracy obtainable using any averaging process is the need for the volume of reciprocal space scanned by the diffraction sphere during the crystal rotation to be large relative to the three-dimensional reciprocal lattice unit cell. To satisfy this condition, the energy would need to be greater than about 500 eV, and θ greater than about 30°. All averaged LEED investigations fall short of this requirement, but these conditions can be fulfilled by shifting into the realm of MEED. Aberdam *et al.*,[137] using their CSFA method, studied Ni(001) at energies of 645 and 980 eV, and modelled the results with a pseudo-kinematical theory which neglected interlayer spacing but treated multiple interlayer scattering with up to third-order events. They found that reasonable accuracy could be obtained in an estimate of the surface normal relaxation, and also the real and imaginary components of the inner potential V_r and V_i. The neglect of intralayer scattering meant that no useful information could be gained concerning lateral shifts; it also led to poor correspondence between theory and experiment when the energy was so large that intralayer scattering became significant.

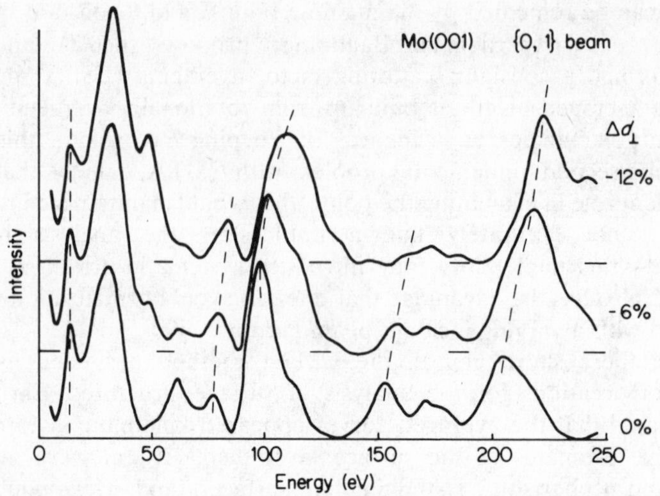

Figure 6.16. Effect of surface interlayer contraction $\Delta d_{\perp}$ on major peak shapes and energies, as exemplified by the (0, 1) beam from Mo(001) at normal incidence

If an average process is applied to the analysis of an adsorbate structure, a further difficulty is encountered. The different phases of the scattered waves from the different atomic types means that the averaging procedure is less effective in reducing multiple-scattering effects. A residue of multiple-scattering effects in the final intensity profile may introduce a systematic error in the value concluded for that adsorbate-substrate spacing.[305]

In the past, the main intention of data averaging has been to permit the relatively simple kinematic theory of LEED to be used for surface crystallography, thus avoiding the greater computational effort associated with full dynamical calculations. However, it is clear from the above discussion that serious limitations restrict its usefulness. Now that suitable computer program libraries are available, making dynamical analysis a viable option in many laboratories, the benefits of pursuing such an approach seem minimal. The most promising application appears to be the combination of MEED data with an 'enhanced' or pseudo-kinematical theory, using the CSFA approach.

An interesting application of data averaging which has provided information not readily available using current dynamical LEED theories is the use of energy-averaged diffraction patterns for the evaluation of Si(111)–(7 × 7) surface reconstruction models. This particular surface structure provides favourable conditions for data averaging, since Si is a relatively weak scatterer and the density of surface lattice points is especially high. Using kinematical theory to predict the diffraction beam intensities resulting from various vertical or lateral shifts in the surface atoms, Miller and Haneman[306] were able to show that certain reconstruction models gave very poor agreement, whereas others could give remarkably close agreement between theoretical and experimental relative beam intensities. In their study, Miller and Haneman averaged data over a range of 5 eV, in 0.5 eV steps starting at 105 eV. This method may be compared with the $I(\mathbf{g})$ method proposed for dynamical LEED analysis (see section 7.3.5). Certainly, if at least a simplified multiple-scattering method were used to provide the theoretical intensity data, this may well provide a valuable means of screening possible models of complex surface reconstructions.

CHAPTER 7

Comparison of theoretical and experimental results

7.1 Reproducibility of experimental data

The quality of experimental data is influenced both by the method of producing and recording the data, and also by the degree of perfection of the crystal surface structure itself. Nominally clean surface structures may be seriously affected by trace impurities, and in the case of an adsorbate with relatively weak interatomic interactions, the production of a perfectly periodic structure may be thermodynamically impossible at finite temperatures. Fundamental limits to the resolution of an experimental diffractometer, and basis techniques for aligning the crystal and preparing a clean, annealed surface have been discussed in section 2.2. Despite all reasonable care, some degree of uncertainty seems unavoidable. To date, only a small number of surfaces have been studied by more than a few different research groups so it is difficult to assess with any accuracy the variations which may stem purely from differences in experimental approach. The surface W(001) is one which has received considerable attention, and a recent comparison of experimental data published by six different sources[307] highlighted significant differences between data for certain beams. There are many possible reasons for the differences—incomplete cancellation of stray magnetic fields, poor surface cleanliness or imperfect topography, inaccurate characterization of the incident beam angle, the method and rate of data capture and so on. In the case of W(001) there is the added complication of a structural phase change which occurs slightly above room temperature. Apart from limitations in accuracy imposed by the physical nature of any particular piece of experimental apparatus, these differences emphasize the necessity of extreme care in performing LEED experiments if reliable data are to be obtained.

Accepting that some departures from ideality are inevitable, there are, nevertheless, certain precautions which may be used to minimize their influence on the final result. One precaution is *equivalent beam averaging.* In choosing the incident beam angle, it is advantageous to select highly symmetric directions so that the correspondence between symmetry-related beams can be used to help make fine adjustments. Once the symmetrically equivalent beams are as similar as practically possible, their intensity profiles

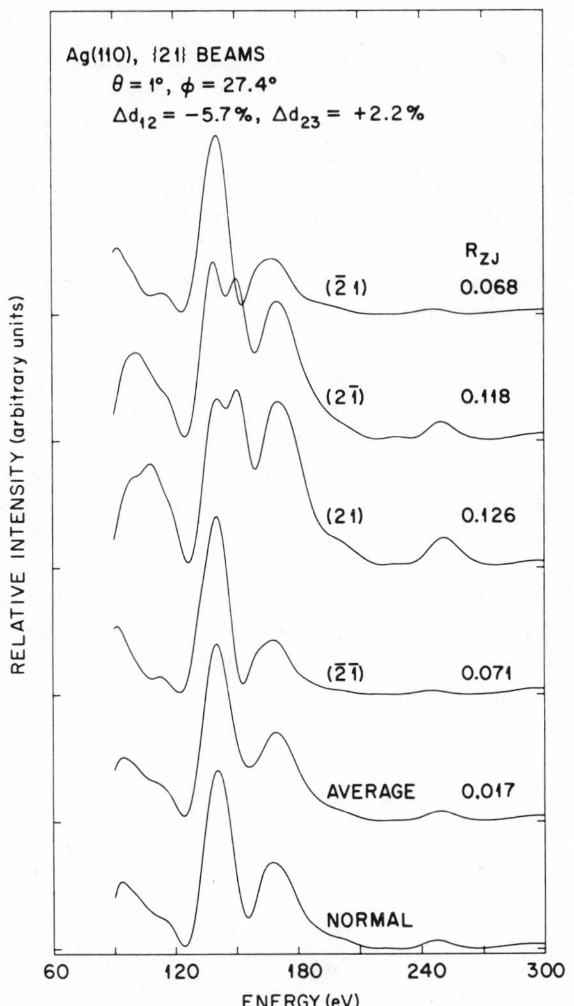

Figure 7.1. Equivalent beam averaging: the average intensity–energy curve taken from four equivalent beams produced when the incident beam is off-normal by 1% is very similar to the profile calculated for perfect normal incidence. The numbers written above each curve are the Zanazzi–Jona *r*-values obtained by comparing each curve with the normal incidence curve. (Reproduced by permission of North-Holland Physics Publishing from Davis and Noonan[308])

should be *averaged*. Davis and Noonan have shown that such averaging will not only reduce random errors associated with the data-collection process, but in addition may substantially reduce systematic errors which result from any incident beam misalignment. This is important because, no matter how carefully the diffraction beam symmetry has been matched at some incident beam energy, stray magnetic fields may affect this symmetry at other angles. Realignment at many different energies is not really very practical, because it would grossly increase the duration of the experiment, and the surface may become unacceptably contaminated. Moreover, if the crystal needs to be cleaned by heating, it is quite possible that the heat will affect the sample mount, causing further changes in alignment which will vary as the crystal and holder cool.

Equivalent beam averaging is particularly suitable when the data is to be collected at normal incidence, since three, four or even more equivalent beams may be averaged simultaneously, and any changes in one beam may be compensated by changes in another. Davis and Noonan[308] demonstrated quite clearly the benefits of this approach in reducing systematic errors, by comparing calculated beam profiles at normal incidence with individual and averaged beam profiles deliberately misaligned by up to 2°. Figure 7.1 illustrates quite strikingly the differences between various {2, 1} beams produced when the incident electron beam on a clean Ag(110) surface is aligned off-normal by 1°. The numbers on the right are a quantitative measure of the comparison of each beam with the beam profile appropriate to perfect normal incidence. The method used to calculate these numbers is that proposed by Zanazzi and Jona, and is described in detail in section 7.3.1. Clearly, the average beam profile agrees much better with the normal incidence data than any individual beam.

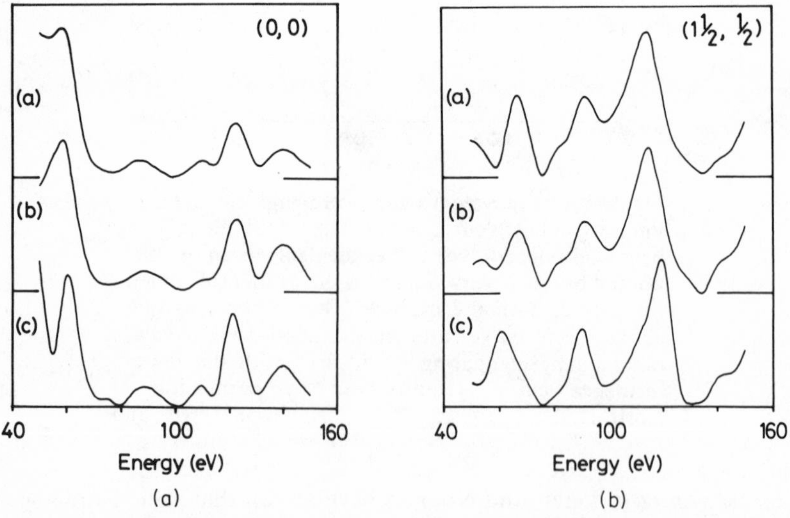

Figure 7.2. Two beams calculated for Mo(001)$c(2 \times 2)$Cu at normal incidence using (a) the combined space method of Van Hove and Tong, (b) the CHANGE method of Jepsen and (c) CAVLEED by Kinniburgh and Titterington. (Reproduced by permission of Plenum Press from Shih[209])

If the incident beam is off-normal, symmetry may be reduced to pairs of beams. In this case, averaging will be of limited benefit. Nevertheless, it is clear that, if beams from all over the screen are included in the comparison, systematic errors due to misalignment or stray magnetic fields will be reduced. Comparison of individual beams from different experiments as carried out by Stevans and Russell[307] is revealing but may provide an excessively pessimistic impression of the overall quality.

Conclusions drawn from the comparison of a limited quantity of data are prone to errors, and a number of controversies due to conflicting structural conclusions that have occurred in the past may be attributed to the use of insufficient data for reliable comparison. In section 7.3.5, it is shown that the total quantity of experimental data necessary for an accurate structural determination is large, and it is recommended that 20 or more beams be used, collected over a wide energy range. Quality and quantity are both necessary. In the case of W(001), extensive studies, subsequent to those considered by Stevans and Russell, have demonstrated good agreement amongst themselves, indicating a surface layer contraction of about 7 per cent.[32]

7.2 Comparison of theoretical methods

The diffraction of an electron beam by a crystal is an enormously complex process, and this is reflected in the considerable complexity of the associated mathematical theories and computer programs. Nevertheless, even the most detailed theories incorporate many assumptions, for example, to reduce the many-body scattering process to a spherical, single-electron model, to reduce the multiple-scattering process to manageable proportions or to model the real, imperfect crystal by a perfect single crystal. Different multiple-scattering theories were discussed in Chapter 4, and it was shown that the choice of approximation is usually a compromise between accuracy and economy (in terms of computer resources). However, the accuracy of a well-converged perturbation method is often sufficient to match the results of 'exact' methods and so need not be thought of as necessarily inferior.

In the case of certain assumptions, one has little option but to accept convention: the muffin-tin approximation, for example, is almost universally adopted—any attempts to improve on it would require major changes in the description of the electron scattering process.[309] With other assumptions, there is a degree of personal freedom—for example, in the choice of atomic wave functions from which to construct the ion-core potential, the energy dependence (if any) of the real and imaginary parts of the inner potential, the value of the surface Debye temperature and so on. However, even when all these choices have been made, the final result could depend on the precise way in which the computer programs have been encoded. Fortunately, a number of separate libraries of computer programs have been written, and it is possible to cross-check their results. Shih[209] has made a detailed comparison of results obtained from the CAVLEED, CSM and CHANGE program libraries, noting differences in language, storage requirements, flexibility and speed. In addition, he has made a number of direct comparisons of output for identical surface structures. As an example, curves for Mo(001)–$c(2 \times 2)$Cu generated from each of the three libraries are given in Figure 7.2. Differences are more pronounced in the fractional-order beams than in the integral-order beams. These differences partly reflect the consequences of using different approximation schemes, and partly result from the fact that the methods of data input did not permit exactly identical models to be generated. In practice, it is a useful exercise to calculate $I(E)$ curves from more than one program library, as an independent check on the input, which is complex and prone to errors.

7.3 Quantitative comparisons of theory and experiment

Initially, most surface structure analyses using LEED were determined visually, relying on the judgement of the experienced eye to weigh up the important features of experimental and calculated $I(E)$ curves. This may be adequate for preliminary comparisons of widely differing models, but lacks discernment when fine tuning is required. To improve the accuracy a large quantity of data must be considered, but as the data base is increased, it becomes difficult to retain the perspective necessary to make a meaningful assessment of all the different features. This problem is exacerbated by the fact that changing some parameter may appear to improve the agreement between certain features, whilst causing a detrioration in agreement between others. As we have seen, these may result from artefacts of the theoretical method or imperfections in the experimental technique in addition to the effects of changing the model itself. Consequently, a systematic and objective means of comparison is essential for detailed analysis.

The main problem is to decide which features of the data should be compared and what criteria should be used to quantify the agreement. An algorithm is used to provide a number and it is conventional to define this such that its numerical value decreases towards zero as the agreement improves towards perfection. The algorithm is called a *reliability factor* or *r-factor,* and the numerical result may be called the *r-value* of the comparison. The numerical result is sometimes loosely described as the *r*-factor of a comparison, but this is better reserved for the algorithm, not the result. To someone unacquainted with the method, it may seem confusing that a decrease in the *r*-value is equivalent to an increase in reliability. An alternative name, introduced for the comparison $I(\mathbf{g})$ data (see section 7.3.5) is a deviation- or discrepancy-factor (*D*-factor). This has the minor advantage that a decreasing discrepancy between the data is accompanied by a corresponding decrease in the numerical result. In either case, it should be noted that a decrease in the numerical result indicates that the data being compared are quantitatively more similar, but not that they are necessarily more reliable; indeed, for a meaningful assessment of reliability it is necessary to obtain additional statistical information based on the quantity of data used in the comparison.

Reliability factors had been developed for the comparison of X-ray diffraction data well before such methods were applied to LEED data, but unfortunately the nature and quantity of data available from typical LEED experiments is rather different from that attainable from X-ray diffraction. This is partly due to the complexity of the electron–solid interaction and partly due to the limited range of diffraction data accessible from LEED experiments.

An *r*-factor commonly used in X-ray crystallography is defined as follows:[310] the integrals of the intensities of both theoretical and experimental beams are normalized to a common value, and at regular energy intervals the square of the difference in intensity between the experimental and theoretical data is determined. The resulting *r*-factor is the mean-square difference in intensities between the two sets of data. Van Hove *et al.*[311] carried out one of the first quantitative comparisons of LEED data by using this *r*-factor from X-ray crystallography, supplementing it with four other *r*-factors, each comparing a separate feature of the data, such as the slope of the curves or differences in intensity, at each energy and so on. Provided the minimum value of each *r*-factor coincides with a common structural model, reasonable confidence could be placed on the result. They used data from a number of different surfaces (MoS_2, $NbSe_2$, W(110)–$p(2 \times 1)$O) to show that in each case a minimum *r*-value for each separate *r*-factor matched quite closely with a unique structure.

A limitation of such a method is the fact that the LEED curves are treated as though every energy point has the same significance as every other—the method could be equally well applied to match quite arbitrarily shaped curves. In fact, LEED curves are relatively highly ordered—they may be represented fairly accurately by a superposition of regular smooth peaks, each with a characteristic width of roughly 4 eV. (This width is a consequence of the inelastic scattering of electrons and is discussed in Chapter 3.) This means that, if the intensity is known at one point, the neighbouring point will be reasonably correlated in intensity and so should not really be treated as statistically independent.

For this reason, a number of *r*-factors specifically designed to cope with the actual form of LEED curves have been devised, and these are described and discussed in subsequent sections of this chapter.

7.3.1 Zanazzi–Jona *r*-factors

The first *r*-factor specifically formulated for the purposes of LEED is that devised by Zanazzi and Jona in 1977.[312] In developing their *r*-factor, they chose to emphasize the importance of matching peak positions (i.e. the energies of the peak maxima) as opposed to peak heights. The physical basis for this is that peak positions are highly structure-dependent, whereas peak heights are also strongly influenced by non-structural parameters such as the ion-core potential, vibrational amplitudes and so on. The purpose of devising a specific r-factor is to filter the more relevant features from those features that might be more strongly influenced by less relevant parameters. Of course, this division of the LEED curves into features influenced by structural and non-structural parameters cannot be especially rigorous, and ultimately the value of any particular r-factor can only be gauged by its success in practice.

Peak positions rather than heights are best compared by measuring the difference in slopes, i.e. the first derivatives of the intensities with respect to energy, at each energy point. As with all r-factors a meaningful comparison can only be made if the experimental and theoretical curves are both normalized to the same integrated intensity scale, and this can be accomplished by identifying the scaling factor c given by the ratio of the experimental intensity profile $I_e(E)$ integrated from the initial energy of the beam E_i to its final energy E_f, to the equivalent integral over theoretical intensities $I_t(E)$:

$$c = \frac{\int_{E_i}^{E_f} I_e(E)\,\mathrm{d}E}{\int_{E_i}^{E_f} I_t(E)\,\mathrm{d}E} \tag{7.1}$$

from which

$$\mathrm{r} = \frac{\int_{E_i}^{E_f} w(E)\,|cI_t'(E) - I_e'(E)|\,\mathrm{d}E}{\int I_e\,\mathrm{d}E} \tag{7.2}$$

where

$$I_t'(E) = \frac{\mathrm{d}I_t(E)}{\mathrm{d}E} \quad \text{and} \quad I_e'(E) = \frac{\mathrm{d}I_e(E)}{\mathrm{d}E}$$

In addition to the comparison of slopes at every energy point, equation (7.2) includes a weighting factor $w(E)$. This is designed to emphasize the extrema (peaks and hollows) of the curve. This is achieved using the following equation:

$$w(E) = \frac{|cI_t''(E) - I_e''(E)|}{|I_e'(E)| + \varepsilon} \tag{7.3}$$

where

$$I_t''(E) = \frac{\mathrm{d}^2 I_t(E)}{\mathrm{d}E^2} \quad \text{and} \quad I_e''(E) = \frac{\mathrm{d}^2 I_e(E)}{\mathrm{d}E^2}$$

As the slope tends to zero at each extreme, $w(E)$ increases accordingly. The factor ε is necessary to prevent excessively large values when $I'_e(E)\rightarrow 0$. From experience, a suitable value was found to be the maximum absolute value of the slope of the curve

$$\varepsilon = |I'_e(E)|_{max} \tag{7.4}$$

The second derivatives are introduced to emphasize differences between sharp features (the sharper the curve, the greater the value of the second derivative)—this term would, for example, reach a maximum where a hollow in one curve coincided in energy with a peak in the other.

In the form given by equation (7.2), the r-factor would be suitable for comparing one model with another provided the calculated intensities are compared with an experimental beam over a fixed energy range; but the absolute value of the r-factor would depend directly on the scale in which the intensities are expressed. This scaling effect can be avoided by dividing the integral by a function of intensity, and, by choosing this function to be the integral over the experimental intensities, uniformity with the scale fact c is also achieved. In addition, the constant of proportionality in equation (7.2) is chosen to be 1/0.027, so that a value of 1 would be obtained if random curves were compared. If a set of different models are compared, the most likely model is given by the one that produces the lowest r-value. But how reliable is that result? This is clearly affected by the total quantity of data used in the comparison. If an r-value of 0.2 is obtained from a comparison of a single beam, the agreement could be fortuitous, but if 10 or 20 beams each give good agreement, so that their mean r-value is 0.2, then one could feel more confident in the result. The mean r-value $\bar{r}$ from several different beams is found by using the energy range ΔE_g of each beam **g** as a weighting factor:

$$\bar{r} = \frac{\sum_g r_g \, \Delta E_g}{\sum_g \Delta E_g} \tag{7.5}$$

Although a value for the variance could be obtained from the scatter of individual r-values about their mean, Zanazzi and Jona chose to incorporate a function of the number of beams as a multiplier to mean r-value, so that the resulting 'overall r-value' (R) reduces with increasing numbers of beams in the comparison. This has the advantage of restricting the result to a single quantity, and also means that the result does indeed include a specific measure of reliability:

$$R = B\cdot\bar{r} \tag{7.6}$$

The factor they chose was

$$B = \frac{3}{2n} + \frac{2}{3} \tag{7.7}$$

where n is the number of beams. If n is less than 3, this factor will be greater than 1 and the r-value will be greater than the mean—this reflects the unrealiability of results taken from a limited data comparison. The second term provides a finite asymptotic limit so that the overall r-value does not decrease indefinitely as n is increased—otherwise an apparently good fit would be obtainable irrespective of the mean r-value if n were sufficiently large. Of course, the equation developed by Zanazzi and Jona is somewhat arbitrary, being based on experience of how many beams are necessary—in this case, 10 beams would appear to be sufficient, given the negligible change in R associated with greater numbers.

A disadvantage of the overall R-factor as defined above is that it only scales the results as a function of the number of beams, without regard to the energy range of each beam. Clearly, beams compared over, say, 250 eV would provide a more reliable result than beams compared over only 100 eV or less. Despite this disadvantage, the Zanazzi–Jona R-factor must be regarded as the most important means of quantitative comparison currently available, primarily because of its widespread adoption throughout the world and the large number of comparisons that have now been made using this factor.

7.3.2 Pendry *r*-factors

The Zanazzi–Jona r-factor represents a significant advantage over the original r-factors derived from X-ray crystallography in that it does not treat intensity–energy curves as a random set of data, but instead highlights the peaks and other major features that are most strongly affected by surface structure. A major drawback remains, however, in that no specific account is taken of internal correlations within the data. Slopes must be found explicitly by differentiating the data, and peaks are located by evaluating the second derivatives at each point. Although this does not necessarily reduce the suitability of the approach, it does make it rather cumbersome and slow.

As noted above, the shape of $I(E)$ curves can be considered to be a series of regular peaks, each having a characteristic width given by the magnitude of the imaginary potential $| V_i |$. If a typical peak shape can be given a specific mathematical form then it is only necessary to describe the energy of the centre of each peak and its intensity in order to describe a complete $I(E)$ curve. Overlapping of peaks may affect the resulting curve intensity, but this is simply found from a direct summation of neighbouring contributions. Once a specific mathematical form has been given, with peak positions given in terms of energies, the slopes are easily derived from the equation for the peak shapes.

The analysis of an $I(E)$ curve in terms of a series of Lorentzian peaks has been used by Pendry[313] as the basis for an alternative form of r-factor. A Lorentzian peak centred about an energy E_j with amplitude a_j and characteristic width V_i is defined as

$$I_j = \frac{a_j}{(E - E_j)^2 + V_i^2} \tag{7.8}$$

The $I(E)$ curve can be then described by a series of such peaks:

$$I(E) = \sum_{j} \frac{a_j}{(E - E_j)^2 + V_i^2} \tag{7.9}$$

The great value of deriving a precise mathematical form for the data to be compared is that an r-factor can be derived directly in terms of a suitable function of the original data—consequently, it is not actually necessary to perform the analysis of the curve into Lorentzians. Instead, the Pendry r-factor is derived as follows.

As with the Zanazzi–Jona r-factor, sensitivity to peak positions rather than intensities may be obtained by introducing the first derivative. This can be conveniently incorporated using the logarithmic derivative $L(E)$:

$$L(E) = I'(E)/I(E) \tag{7.10}$$

Using equations (7.8) and (7.9) this can be approximated by

$$L(E) \simeq \sum \frac{-2(E - E_j)}{(E - E_j)^2 + V_i^2} \tag{7.11}$$

If $I(E)$ reaches zero at any energy (which is indeed possible in the presence of multiple scattering) then $L(E)$ could become infinite. However, zeros are important features of $I(E)$ curves because the phase of the wave changes rapid in their vicinity, and this is particularly sensitive to structural features. On the other hand, intensity minima are often the least accurate part of experimental $I(E)$ curves due to the uncertainties of the thermal diffuse background (see Chapter 5) and the practical difficulty of tracking a beam that has essentially disappeared. It is therefore undesirable to give the zeros of the curves greater emphasis than the peaks themselves. Equality of emphasis can be achieved by means of a function $Y(E)$, defined as

$$Y(E) = \frac{L^{-1}}{L^{-2} + V_i^2} \tag{7.12}$$

This describes a function that can be readily derived from an $I(E)$ curve, but has a form that emphasizes certain important features such as the positions (energies) of peaks and hollows and suppresses other information such as absolute intensities.

An R-factor can be simply defined via the equation

$$R = \frac{\sum_{g} (Y_{gt} - Y_{ge})^2 \, dE}{\sum_{g} (Y_{gt} + Y_{ge})^2 \, dE} \tag{7.13}$$

This R-factor normalizes to unity in the case of random curve comparisons since Y is an oscillatory function (between $\pm V_{i}/2$) and the cross-product

$$\overline{Y_t \cdot Y_e} = \overline{Y_t} \cdot \overline{Y_e} \tag{7.14}$$

must therefore equal zero.

If there is a finite number of peaks, N, per given energy region, then elementary statistics shows that ratio of the variance (var R_N) to the mean R-value (R_N) is

$$\frac{\operatorname{var} R_N}{R_N} \propto \frac{1}{\sqrt{N}} \tag{7.15}$$

This implies that increasing the extent of the data base can only improve the accuracy of a comparison relatively slowly, assuming that the available information is not internally correlated. In practice, however, LEED data cannot be treated as statistically independent. Although the above relationship is undoubtedly a fair indication additional factors may also come into play. For example, in section 7.1 the effects of equivalent beam averaging were discussed, and the systematic consequences of slight misalignments noted. If the inclusion of more beams requires the collection of data from a wider area of the screen, then the implication is that certain systematic errors may be partially washed out. For such reasons, it is risky to rely too heavily on statistical analyses of reliability.

Nevertheless, in the absence of any better indicator, Pendry's approach can be used to estimate the accuracy of a comparison. For this he defines the 'double reliability factor' (RR) as

$$RR = \frac{\operatorname{var} R}{\bar{R}} = \sqrt{\frac{8|V_i|}{\Delta E}} \tag{7.16}$$

The 'statistical error' or reliability of the optimum r-value ($R_{\min}$) is thus

$$\operatorname{var} R_{\min} = RR \times R_{\min} \tag{7.17}$$

This can be used to estimate the error in the best model, by noting, for example, the range of interlayer spacings that yield r-values within the statistical error.

The advantage of the Pendry r-factor method is that it is sensitive to peak positions, yet only requires an evaluation of the first derivatives of the $I(E)$ curves. This uses considerably less computer resources than the Zanazzi–Jona r-factor. The analysis of the $I(E)$ curve into Lorentzians is a useful part of the argument, but it does not have to be carried out in practice. The r-factor is evaluated directly from the three equations (7.10), (7.12) and (7.13), with estimates for the error in the result obtainable from equations (7.16) and (7.17).

This r-factor is more recent than the Zanazzi–Jona r-factor, and has thus been used much less frequently. However, examples of use of the two methods are compared in section 7.3.5.

7.3.3 Metric distances

The r-factors of Zanazzi–Jona and Pendry were both written to emphasize specific features of $I(E)$ curves in the comparison of experimental and theoretical data. Recently, Philip and Rundgren[314] proposed an alternative method that does not really set out to highlight particular features, but does nevertheless constitute an interesting alternative approach to data comparison.

Although brought to bear on the problem of LEED data comparison quite recently, the method of Philip and Rundgren is a general approach that has existed for many years.[315,316] The principle is to compare the integral of the intensities over the whole energy range up to each point rather than any direct function of the intensity itself at that point.

The energy range from the initial energy E_i to the final energy E_f can be replaced by an integral 0 to 1 by the transformation

$$x = (E - E_i)/(E_f - E_i) \tag{7.18}$$

so that normalization of the total intensity range can be written as

$$\int_0^1 I_t(x)\, dx = \int_0^1 I_e(x)\, dx = 1 \tag{7.19}$$

The integrals from 0 to x can then be defined by the functions

$$F(x) = \int_0^x I_e(t)\, dt$$

and

$$G(x) = \int_0^x I_t(t)\, dt \tag{7.20}$$

A simple illustration is given in Figure 7.3. Comparison of $F(x)$ and $G(x)$ can be effected in one of many possible ways. Consider two idealized functions $F(x)$ and $G(x)$ as illustrated in Figure 7.4. Their features are exaggerated for clarity. One possible way to compare such curves is to sum the difference directly as a function of x

$$R_2 = \int |F(x) - G(x)|\, dx \tag{7.21}$$

Another way is to compare the distance between the two functions parallel to a diagonal, that is, along directions with slope (s) – 1. This is called the Levy distance.[315] This r-factor may be referred to as R_s with $s = -1$. Alternatively, the integral of the shortest distance between the two functions—if this is determined by considering the distance from the upper of the two curves, as illustrated in Figure 7.5, it is known as the Hausdorff distance.[316]

Because these r-factors are defined with respect to the integral function of the intensity data, any differences or errors are accumulated in the final result. The nature of $F(x)$ and $G(x)$, and the fact that derivatives are not used means that the result is quite stable with respect to noise in the data and curve smoothing is not required.

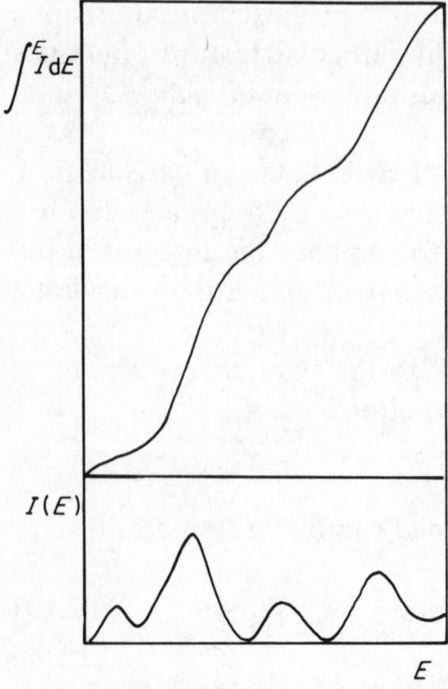

Figure 7.3. Schematic illustration of the relationship between a typical $I(E)$ curve and the integral function $^E\!\int I \, \mathrm{d}E$ used in the metric distances approach to curve comparisons

Although it is not possible to emphasize individual peaks or features as in the Zanazzi–Jona or Pendry r-factors, it is nevertheless possible to impose some control over features that vary in a more general sense with energy. Inelastic damping, for example, causes a gradual decrease in mean intensity with energy, but for a structural determination peak positions may be equally significant throughout the whole range of energies. To maintain a suitable balance, a weighting function $\mu(x)$ may be included into the r-factor:

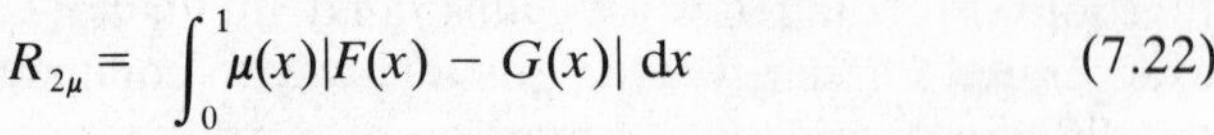

$$R_{2\mu} = \int_0^1 \mu(x)|F(x) - G(x)| \, \mathrm{d}x \tag{7.22}$$

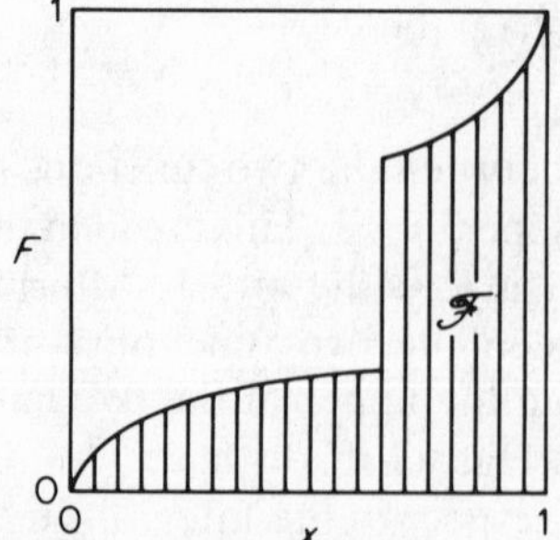

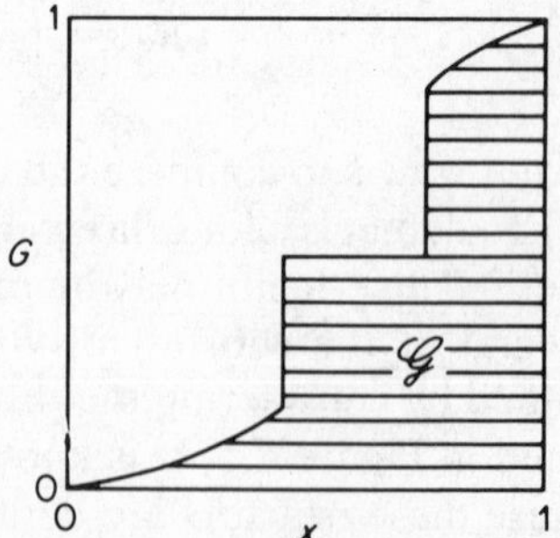

Figure 7.4. Highly schematic illustration of two sets of integrated intensity data prior to comparison using the method of metric distances. (Reproduced with permission from Y. Gauthier, 'These d'Etat', l'Université Scientifique et Medicale de Grenoble, 1981)

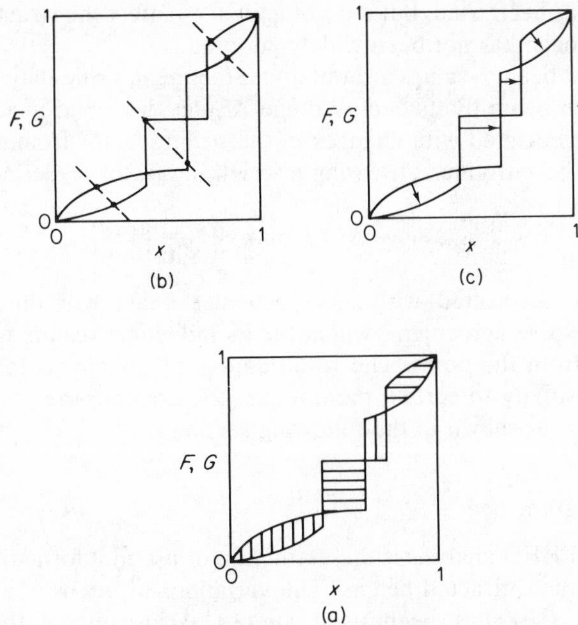

Figure 7.5. Various schemes for comparing two curves given in Figure 7.4, using metric distances: (a) R_2; (b) R_3 (Levy); (c) R_4 (Hausdorff). (Reproduced with permission from Y. Gauthier, 'These d'Etat', l'Université Scientifique et Medicale de Grenoble, 1981)

Further variants can be developed by varying the slope s to a value other than -1 in the R_s-factor.

Normalization of all beam intensities or intensity functions to unity is necessary for meaningful comparison, but clearly suppresses a potentially significant source of information. Indeed, the relative intensities of beams may be used exclusively as the basis for an alternative method of structural determination, as described in the following section. It may seem reasonable that the scaling factor relating the integral over theoretical intensities to the integral over experimental intensities for any beam j,

$$c_j = \frac{\int \Delta E_j I_e^j \mathrm{d}E}{\int \Delta E_j I_t^j \mathrm{d}E} \tag{7.23}$$

(where ΔE_j is the total energy range) should be the same for all beams. In practice, it is usual to determine a scaling factor for each beam separately, and similarly generate on r-value for each individual beam. The alternative is to consider all the data simultaneously and calculate a single, common scaling factor. This has been tried, but did not appear to affect the structural results, and this approach has not been widely adopted.

If individual beam r-values are obtained, the mean value may be found by weighting each beam by its energy range ΔE_j, as described in section 7.3.1. Information associated with changes in the scaling factor from one beam to the next may be introduced by using a weighting factor w_j defined as

$$w_j = \Delta E_j + \max\left(c_j, \frac{1}{c_j}\right) \tag{7.24}$$

The r-value associated with any particular beam will thus be larger, representing worse agreement, whenever its individual scaling factor differs substantially from the norm. The resulting overall r-factor is found to have enhanced sensitivity to certain parameters, for example the position of the surface barrier, as shown in the following section.

7.3.4 The I(g) method

The aim of a LEED analysis is the extraction of useful information from the intensities of the diffracted beams. The variations of intensity with incident beam energy *I(E)* has been used almost exclusively as the basis for comparisons between theory and experiment, partly due to the ease with which the electron energy may be varied but also because early interpretations were based on Bragg theory or simple kinematical theories, which were specifically appropriate to *I(E)* profiles. The ability of multiple-scattering theories to reproduce faithfully most details of experimental *I(E)* curves has demonstrated in recent years, and this has opened up the possibility of using intensity data in forms that cannot be related to simple theories. For example, instead of tracing the variation of intensity of each beam with energy, the variation of intensity with azimuthal angle could be used, the energy being held constant.

Data in this form was not generally considered to be structurally sensitive, until applied to the analysis of the unreconstructed W(001) surface in 1978 by Feder and Kirschner.[317,318] One reason for the recent success in using $I(\phi)$ data (or 'rotation diagrams') may be the fact that structural sensitivity is minimal at small polar angles, and only appreciable when the polar angle is greater than about 40°. Feder and Kirschner carried out their analysis using data collected at a polar angle of 47.5°. A more practical reason is that few types of LEED apparatus are designed to facilitate the collection of $I(\phi)$ data. Previous applications of such data were as input for LEED data reduction, using averaging over azimuthal angle in preparation for analysis with kinematic theory calculations (see section 6.6), or for MEED analysis.[319] An example of the comparison between experimental and theoretical rotation diagrams is given in Figure 7.6. It is found that peak shapes and positions are not sensitive to the surface structure, but the relative peak height ratios are particularly sensitive.

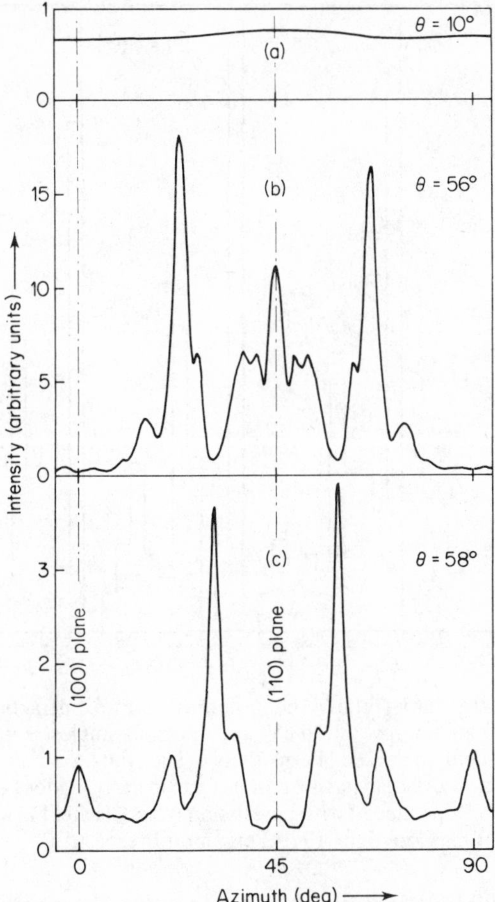

Figure 7.6. Sensitivity of LEED intensity to co-latitude and azimuthal angles of incidence. (a) $\theta = 10°$. Sensitivity to azimuth is small, since incidence is not far from normal. (b) $\theta = 56°$. Sensitivity to azimuth is very large. Sampling in azimuth every 10% is inadequate. (c) $\theta = 58°$. The situation is the same. In addition, the sensitivity to θ is also very large, since within 2°, both the shape and the scaling factor are drastically modified. (Reproduced by permission from Garland STPM Press from Aberdam *et al.*[137])

An alternative is to measure the relative intensities of all the beams simultaneously at some fixed incident beam energy and angle:[320,321] this may be called an $I(\mathbf{g})$ method, since $\mathbf{g}$ is generally used to designate reciprocal lattice vectors and hence diffraction beams. The number of beams that can be compared at any one energy is generally not large, but the quantity of data

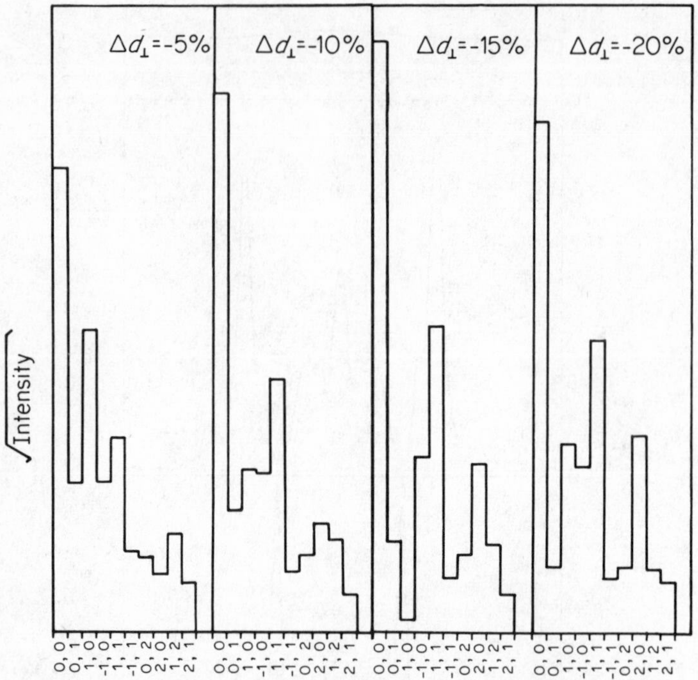

Figure 7.7. $I(\mathbf{g})$ data (relative beam intensities) for diffraction from Mo(001) at an energy of 75 eV and incident angles $\theta = 8°$ and $\phi = 270°$. The high degree of sensitivity of such data to small changes in the surface interlayer spacing ($\Delta d_{\perp}$) is particularly evident from the (1, 0) beam. (Reproduced with permission from Clarke, *Vacuum*, **29**, 405. Copyright 1979 Pergamon Press Ltd.)

may be increased indefinitely by repeating the exercise at a variety of different energies or angles.

As an example of $I(\mathbf{g})$ data, Figure 7.7 shows the intensities of the diffraction beams produced by diffraction of a 105 eV electron beam upon a Mo(001) surface. Sensitivity to the surface interlayer spacing is clearly demonstrated by the relative intensity of the (1, 0) beam, which changes from being the strongest beam to one of the weakest as the interlayer spacing is changed by 10 per cent, equivalent to only about 0.16 Å.

Quantitative comparison is easily achieved, since the data is already in the form of discrete intensities. As with the formulation of *r*-factors, some care must be taken to weight the contributions so that errors in a few beams do not dominate the result. For example, a direct comparison of the beam intensities will naturally emphasize differences between high-intensity beams and be relatively insensitive to differences between low-intensity beams. A variety of different comparison schemes have been suggested from which the comparison of the square roots of the intensities has been found to be very suitable:[32] this method reduces the dominance of the strongest beams without giving undue emphasis to the weakest beams (which can be adversely affected by uncertainties in the subtraction of the background level). In this case, the numerical difference of discrepancy between the beams is described by a D-factor defined by the equation

$$D = \frac{1}{\sqrt{N}} \sqrt{\left[\sum_{i}^{N} (\sqrt{I_t} - \sqrt{I_e})^2 \right]}$$

where N is the total number of beams. This method has been used in a number of surface studies, the most extensive being an investigation of the clean Ni(110) surface.[275] It was found that the structural result obtained was very close to that found from standard $I(E)$ methods using r-factors for quantitative comparison. A discussion of these results is given in section 7.3.5. The number of beams necessary for an accurate description was found to be about 100, which could be acquired by the measurement of about 10 beams at 10 different incident beam angles or energies.

The $I(\mathbf{g})$ method has a number of advantages, both experimental and theoretical.

1. It uses experimental data specifically chosen to suit the output of the computational method. Calculation of diffraction beam intensities necessarily involves the manipulation of layer-scattering matrices and so, even though only a few beams are used (as commonly occurs in $I(E)$ studies) it is necessary to calculate the intensities of every propagating beam. Unless every one of these calculated beams is included in the $I(E)$ study, this may result in a considerable quantity of unused computed information. The $I(\mathbf{g})$ method was devised to make full use of all the information produced for any specific incident beam angle and energy. Efficient use of the data reduces the computational effort necessary to obtain a reliable result.
2. The inclusion of a sufficient number of diffraction beams necessitates the use of relatively high incident beam energies ($\gtrsim$100 eV) when dealing with clean surfaces, but, conversely, the method has the capability of efficiently analysing complex surface structures for which many diffraction beams are produced, at fairly low energies. As an extreme example, consider the (7 × 7) reconstruction of the Si(111) clean surface. In this case there are 49 beams produced to every one of the (1 × 1) unreconstructed surface. It is currently beyond the capabilities of most computers to perform multiple-scattering calculations for this surface at anything above the very lowest energies, and so $I(E)$ curves are out of the question. At low energies, normal incidence calculations are just about feasible with modern large-capacity machines, and the problem lends itself to treatment using the $I(\mathbf{g})$ method. Similar arguments favour the use of the method for most very complex surface structures.
3. Uncertainties in the inner potential make it necessary to compare intensities not at an individual energy but over a narrow range of energies.

This, however, can be accomplished by comparing theoretical data for a particular energy with experimental data from a range of neighbouring energies. This avoids having to increase the quantity of calculation necessary, but does not greatly increase the burden of the experimentalist.

4. Experimental data can be acquired conveniently by photographic means. A problem with photographic methods when applied to *I(E)* analysis is that it is a fairly lengthy process to extract data for any individual beam as a function of energy, since each increment of energy involves a different photograph. Errors may be introduced if the exposure or the external factors differ from one frame to the next. By contrast, the $I(\mathbf{g})$ method uses data from individual frames separately. This also means that data may be acquired by photographic or computerized methods more rapidly than for *I(E)* analysis.

7.3.5 Comparison of *r*-factor methods

An *I(E)* curve may be reduced to a series of peaks with characteristic energies, height and widths. To a good approximation, the characteristic widths are determined by the absorptive potential, which does not vary much from material to material, so the structure can be determined from the set of peak heights and energies. This, we have already seen, is the starting-point for Pendry's development of an *r*-factor. If a non-structural parameter (e.g. scattering potential), is altered, this must cause changes to occur in the peak heights or energies, and in some way affect the structural result. If the influence of 'reasonable' changes in a parameter (i.e. within the bounds of possible error in the original estimated value) is relatively weak, then its influence on the structural result is also likely to be small. To a good approximation, changing the real part of the inner potential results in a uniform shift in the energies of every peak, but does not affect their heights. If another variable were found that affected peak heights but not their energies, then these two variables could be considered to be independent— if one is varied until best agreement with the experimental data is found (i.e. it is optimized), subsequent variation of the other variable would not alter the original conclusion.

Variation of the interlayer spacing changes both energies and heights of the peaks. These changes are complex but certainly not random—if the interlayer spacing is decreased, the major peaks will tend to shift to higher energies. This underlying systematic shift may be approximated by a corresponding shift in inner potential; consequently, there is a strong interdependence between inner potential and interlayer spacing. It has become common practice to calculate *r*-values for a range of different interlayer spacings and inner potential values simultaneously, and to plot the results in the form of contour diagrams, as illustrated in Figure 7.8. Clearly, uncertainty in the energy scale will cause a systematic shift in all peak energies irrespective of the source of the uncertainty; although we may describe the energy variation

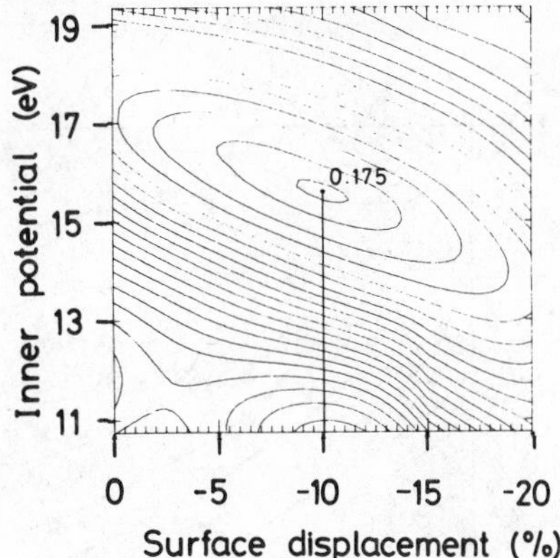

Figure 7.8. Counter plot of the overall r-value (R) for Mo(001), as the inner potential and surface contraction are varied, in 0.02 steps, apart from the first step, which is 0.01, starting 0.003 above the minimum. The figure near the centre is the minimum r-value

in terms of inner potential, this could equally well account for any errors that arise from the data collection or generation process, such as any uncertainty in the incident beam energy, gun filament or sample work functions. Other non-structural parameters, or structural parameters such as lateral shifts may have weaker or less systematic influences on the *I(E)* data, and the extent of the interdependence between each parameter can only really be found by trial and error.

An interesting consequence is that different r-factors or comparison methods are sensitive to different features of the intensity data and will therefore exhibit differing degrees of sensitivity to the various structural and non-structural parameters.

Recently, an extensive study of the sensitivity of different comparison methods was carried out by Gauthier *et al.* using a clean Ni(110) surface.[322, 37, 275] To provide statistically reliable results, 35 beams were included in the *I(E)* curve comparison, and more than 100 beams in an $I(\mathbf{g})$ comparison.

Results of the comparison using both Zanazzi–Jona (Z–J) and Pendry r-factors are illustrated in Figure 7.9 in the form of contour diagrams. The first thing to note is that both r-factors lead to a minimum for the same value of the interlayer spacing, that is, for a 7 per cent contraction from the typical bulk value. Thus although these r-factors have been constructed using rather different approaches, giving different emphasis and treatment of the various features of the *I(E)* curves, they both favour the same surface structural model. However, the relative sensitivity to structure and inner potential is different. For example, if the interlayer spacing is varied by 1 per cent, this

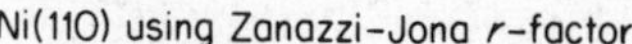

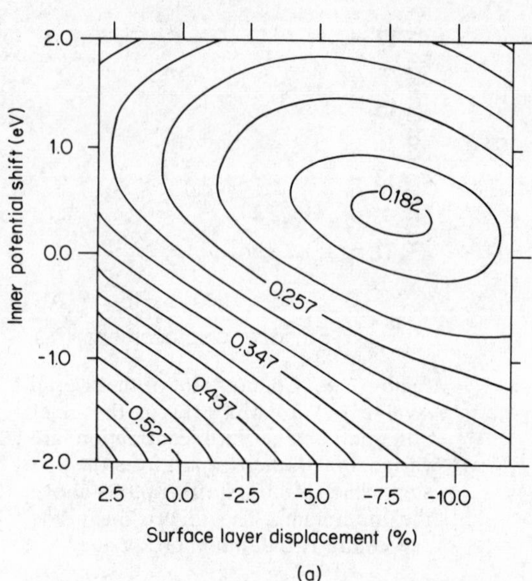

(a)

Ni(110) using Pendry r-factor

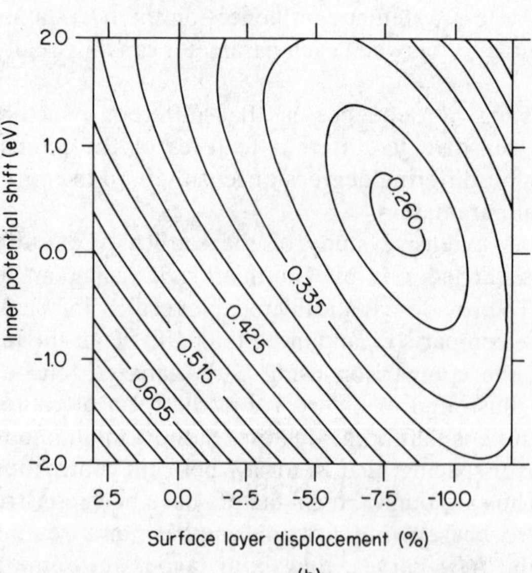

(b)

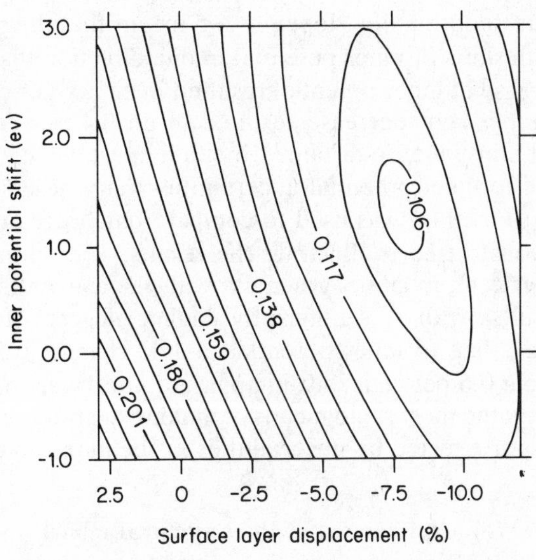

(c)

Figure 7.9. Intensity data analyses for clean Ni(110) using: (a) the Zanazzi–Jona r-factors for $I(E)$ curve comparisons; (b) the Pendry r-factor for $I(E)$ curve comparisons; (c) $I(g)$ data. The lower the contour value, the better the agreement. The relative sensitivity of each method to shifts in the inner potential and surface layer displacement is different, but the optimum value of the surface layer displacement is about 7.5 per cent contraction in each case. (Reproduced by permission of the Institute of Physics from Clarke *et al.*[275])

results in a 1.6 per cent change in the Z–J r-value, but 4.8 per cent in the Pendry r-value. On the other hand, a 1 per cent change in the inner potential will alter the Z–J r-value by 1 per cent, but will alter the Pendry r-value by only 0.5 per cent. These reflect the shape and density of the contours about the minimum in the two diagrams.

The Z–J r-factor clearly emphasizes features affected by inner potential at the expense of features influenced by the surface interlayer spacing, whereas the Pendry r-factor is about three times more sensitive to interlayer changes and only half as sensitive to the inner potential in this particular example. Results from the 'metric distance' approach of Philip and Rudgren depend somewhat on the particular definition used in the comparison, but if the R_s method described previously is used, in which the slope (s) of the line along which differences are summed is set to -5, then the minimum r-value was found to correspond to an interlayer spacing contracted by 8 per cent.[323] The sensitivity of this method to changes in interlayer spacing was very similar to that found for the Pendry r-factor: 4.4 per cent change in r-value for a 1 per cent change in interlayer spacing. The asymmetry of the comparison process introduced by the adoption of a slope $s = -5$ means that the r-value changes more rapidly with shifts of inner potential in one direction than in the other: a 1 per cent increase in inner potential resulted in an 5.9 per cent increase in r-value, but a 1 per cent decrease resulted in an 0.4 per cent increase in r-value. Overall, this 'metric-distance' method must be considered to be rather less stable to inner-potential uncertainties than the Pendry r-factor.

Finally, the $I(\mathbf{g})$ method was used to compare data once again, a contour diagram being constructed to illustrate the results. The optimum interlayer spacing was 8 per cent, in close agreement with the conventional $I(E)$ curve comparison results, with a significantly higher degree of sensitivity to interlayer spacing than to inner potential: a 1 V change in inner potential corresponding to a 0.6 per cent shift in apparent interlayer spacing. The $I(\mathbf{g})$ method is not only the most rapid means of making quantitative comparisons, but is also the least affected by uncertainties in the inner potential.

7.4 Refinement of the structural model

Diffraction beam intensities are affected by both geometric (or structural) features of the surface and also by non-structural parameters such as the scattering potential. On the one hand, this provides an opportunity for extracting additional information about the surface (for example, the surface Debye temperature) but, on the other hand, this introduces the possibility of systematic errors affecting the structural result. If the elucidation of surface crystallographic structures is the main aim of the analysis, then these additional inflences are undesirable and their effects must be minimized. From this point of view it is fortunate that most non-structural parameters have a relatively weak influence on the diffraction beam intensity profiles. The real part of the inner potential stands out as the most influential, and it is for this reason that it has become common practice to compare curves for a range of inner potentials to find the one that gives the best overall agreement. In section 7.3.5 we saw that this is readily achieved by plotting r-values in a contour diagram with surface layer displacement and inner potential as axes, and then locating the minimum. Watson *et al.* were amongst the first to employ this technique, in their analysis of the structures of Rh(100) and Cu(111).[324]

Although the quantitative comparison scheme may act as a partial filter of intensity data, the effect of varying inner potential is too similar to that of changing surface layer displacement to be significantly reduced. This fact alone accounts for many of the discrepancies that have arisen in past structural determinations using LEED. If we return to the results of section 7.3.5 to assess the importance of the inner potential we see that an error of 1 V is sufficient to change the apparent surface layer displacement by several per cent. If the Z–J r-factor is used (which is currently the most popular scheme for quantitative analysis) then errors of +1 V and −1 V would lead to the conclusion that the surface layer is displaced by −5.5 per cent and approximately −11 per cent respectively for the surface of Ni(110). This is clearly not an insubstantial range. Using Pendry's *r*-factor the layer displacements would range from −6.3 to −9.5 per cent for the same uncertainty in inner potential, whilst with the *I*(**g**) method the resulting displacements would be reduced even further, to between −7.1 and −8.3 per cent. In all these examples we use the common convention of describing surface layer displacements in terms of the typical bulk interlayer spacing, the minus sign indicating that the interlayer spacing at the surface has reduced or contracted from the bulk value.

Over the past few years, a large number of independent investigations have been made of the W(001) high-temperature phase surface structure. Results ranging from −4.4 to −11 per cent were found, from which Stevans and Russell[307] concluded that LEED analysis is incapable of resolving the surface contraction within practical limits without extensive examination of the influence of the scattering potentials on the comparison. In fact, there are numerous minor influences that must be carefully taken into account if one is to arrive at an accurate result, but the starting-point has to be the optimization of the inner potential, via the quantitative comparison of an extensive set of intensity profiles. In 1979, Heilmann *et al.*[325] used the Z–J *r*-factor to analyse a fairly extensive set of profiles, but assumed that the inner potential was fixed at 13 eV. Using this, they concluded that the surface layer was contracted by 10 per cent from the typical bulk layer spacing. At about the same time, in an independent study, Clarke and Morales[32] also analysed the W(001) surface, but using contour diagrams found the optimum inner potential to be rather smaller, about 11.5 eV. The resulting surface contraction was correspondingly smaller, about 7 per cent, in close agreement with other groups that had taken a fixed value of 11 eV for the inner potential.

It is not necessary to recalculate *I(E)* curves for each possible value of the inner potential—to a good approximation, the effect can be simulated by simply shifting the curves directly in energy, and re-evaluating the *r*-values for the new energy scale. The contour diagrams illustrated in Figure 7.8 were produced by calculating *r*-values at 0.5 eV intervals, and interpolating on a 100 × 100 mesh—this results in quite smooth contours without an excessive increase in computer usage.

In the preceding discussion it has been assumed that the inner potential has a single, fixed value. In fact, it tends to vary somewhat for different incident beam energies, being greatest for low incident beam energies. The greater the inner potential, the deeper the ion-core potential well, and hence more attractive. This energy dependence occurs because the inner potential compensates for any errors or variations in the ion-core potential. For high-energy incident electrons, the effects of correlation are weakened, reducing screening and thus increasing the net repulsion experienced from the valence electron. This reduction in correlation, and associated exchange effects enters the calculations in the form of a reduced inner potential for high-energy electrons.

More detailed theoretical considerations indicate that this variation is not a linear function of energy, but is instead of a form similar to that given in Figure 3.6, the inner potential changing by as much as 3 or 4 eV as the energy increases from 20 to 120 eV. In a series of studies of Mo(001),[15] W(001)[32] and Mo(110),[19] this variation has been found to improve r-values by about 10 per cent compared with a single-valued inner potential, and at the same time the apparent surface layer contraction was lessened: in the case of Mo(001), the best surface layer displacement with a single-valued inner potential was about -11.7 per cent, whereas with the energy-dependent inner potential, a value of -10.7 per cent was found. Although the structural conclusion is not highly sensitive to the precise form of the inner-potential variation, a form similar to that in Figure 3.6 is a necessary ingredient, particularly in the case of the heavier transition metals.

Once the inner potential has been suitably accommodated within the comparison process, attention may be focused on parameters that have a weaker influence. Concentrating initially on structural features, we should consider whether LEED analysis is capable of determining atomic shifts laterally within the surface layer, or of movements of layers beneath the surface. In practice, little work has been done to clarify these points, primarily because their influences are relatively weak and corresponding displacements are likely to be small (compared with typical surface normal displacements), so the relevant information is easily obscured in the *I(E)* curves. As an example, consider the second-layer displacement of clean Mo(001). This was found[15] by using contour diagrams to find the minimum r-values for each of a series of different second-layer displacements. Two features became apparent: first, the surface layer displacement was found to vary as the second layer was shifted; secondly, the overall minimum r-value was found to pass through a minimum for a given combination of surface and second-layer displacements. The result is given in Figure 7.10. The displacement of the second layer is small, -1.5 per cent, i.e. contracted by about 0.025 Å, but its incorporation affects the surface layer conclusion.

As with the inner potential, these two parameters—the surface and second interlayer spacings—are interdependent. Fortunately, deviations from bulk spacing are very small for most materials—however, this does illustrate the point that LEED is sensitive to this parameter and in cases where the surface interlayer spacing is significantly different from the bulk, shifts of the second layer should be investigated.

Although second-layer displacements may be very small and rather difficult to detect for clean metal surfaces, significant displacement of several layers may occur at the surface of semiconductors. Displacements of up to five atomic layers have been detected at the surface of Si(001),[326] but it is clear from the above discussion that values of the displacement for each separate layer would be difficult to determine very precisely. On the other hand, a model in which only the surface layer was assumed to shift from the typical bulk spacing was found to be quite inadequate, as would also be expected in the presence of significant subsurface reconstruction.

Second-layer displacements have been detected at the surfaces of several different clean metals,[323] but in contrast with the above Mo example, the second layer has usually been found to shift outwards with respect to the bulk, causing the outermost two layers to move close together.

If a gaseous species is adsorbed on the surface of a metal, sensitivity to the bond lengths between the adatoms and substrate atoms depends on the relative scattering strengths of the two atomic types. If the adatom is a weak scatterer, the intensity is dominated by scattering from the substrate—the extreme example being hydrogen, which is almost totally transparent to the electron beams. In certain cases, hydrogen induces a reconstruction of the substrate surface; its presence is deduced from the changed substrate scattering rather than from its own scattering effect. W(001), for example reconstructs to produce a $c(2 \times 2)$ pattern—and for a while some controversy existed concerning the relationship between this hydrogen-induced reconstruction and the spontaneous surface phase change at low temperatures that also results in a $c(2 \times 2)$ pattern.[327]

In most cases, determination of the adsorbate bond length is straightforward. Because the substrate is frequently a stronger scatterer than

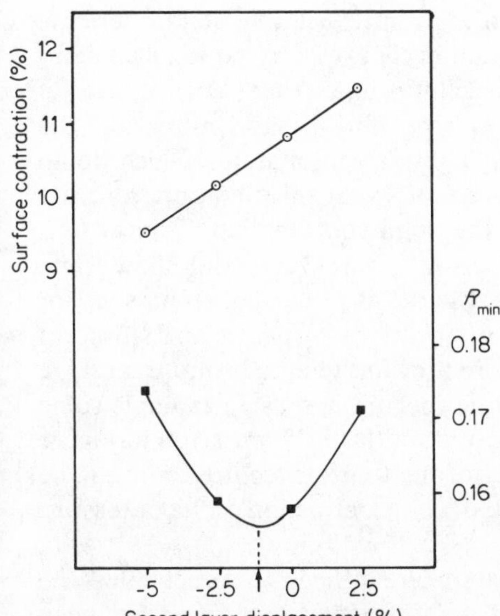

Figure 7.10. Variation in R_{min} with second layer contraction of Mo(001), showing the corresponding surface layer contraction at each value. (Reproduced by permission of North-Holland Physics Publishing from Clarke[15])

the adsorbate, the substrate surface interlayer spacing also has a strong influence on the diffraction beam intensities. These two interlayer spacings tend to be less interdependent than the surface and second interlayer spacings of a clean metal surface because the scattering factors of the two atomic species are different and, more particularly, because the adlayer usually has a periodicity that differs from the bulk, and thus introduces fractional-order beams that may have a greater sensitivity to the adlayer than the integral-order beams. From the point of view of analysis, the substrate interlayer spacing is not particularly sensitive to the adlayer separation. Physically, however, the substrate may be strongly affected by the presence of the adatoms. The reconstruction of W(001) by hydrogen, as described above, is a clear example of this effect. More often, the influence is less dramatic. In a number of cases the adsorbate species causes the outermost substrate layer to move away slightly from the bulk. For example, with sulphur adsorbed into a $c(2 \times 2)$ structure on the Mo(001) surface, the outermost Mo layer shifts from the 10 per cent contraction from the typical bulk interlayer spacing to only about 5 per cent.[11] This is not really surprising—if the surface layer of the metal shifts towards its bulk as a consequence of the absence of atoms on the outer side of the layer, the adsorption of a layer of adatoms will at least partially restore the bulk-like condition and hence result in an outward shift. The adsorption of oxygen on Fe(001) does produce a more interesting result: in this case the outermost Fe layer is actually pulled away beyond the typical bulk interlayer spacing,[328] reflecting a greater affinity for the oxide layer than the base metal. The influence of the adsorbate on the substrate surface is clearly affected by the extent of the interaction between the species.

Lateral shifts (shifts parallel to the surface plane) are generally more difficult to detect or quantify than displacements normal to the surface, because the diffraction beams are less strongly influenced by the former type of shift. Nevertheless, large shifts that result in changes in bond-site symmetry are usually easy to distinguish. In Chpter 1 it was noted that (110) surfaces of bcc materials could occupy either of two possible bond symmetries; the surface layer could be sited symmetrically above a pair of atoms, which would occur if the bulk structure were preserved, or could shift laterally to a site above three atoms, thereby increasing the bond coordination. Typical $I(E)$ curves obtained from Mo(110) are given in Figure 7.11, and show quite clearly that the bulk-like twofold symmetric site is preferred. Bond sites for adatoms can also be distinguished, often by a visual examination of appropriate $I(E)$ curves. The fractional-order beams tend to be more sensitive to adatom geometry than the integral-order beams, and as an example some fractional-order beams obtained for sulphur on Mo(001) are given in Figure 7.12. In these figures, a clear preference for the fourfold 'centred' site can be seen. From these and other curves the 5 per cent contraction of the outermost Mo layer may also be deduced.

Small lateral shifts about a symmetrical site are difficult to detect unless the shifts are systematic and thereby affect the diffraction pattern. It has been suggested that the W(001) surface at high energies comprises atom shifted slightly away from the central symmetrical site, but with the shifts oriented randomly between each of four symmetric directions.[329] The author has investigated this hypothesis, but a quantitative comparison of *I(E)* curves using the Z–J *r*-factors revealed changes in *r*-factor from 0.175 to 0.170 in going from the symmetric to the proposed structure, an insufficiently large change to warrant any firm conclusion.[330] Perhaps a future investigation using a much wider experimental data base (eight beams were used in the analysis mentioned here) might succeed in clarifying this ambiguity. By comparison, the low-temperature phase of W(001) contains systematic lateral shifts that can be deduced from the resulting glide-plane symmetry and consequent diffraction beam asymmetry (see Chapter 1). A LEED analysis of this structure has been successfully undertaken, and a lateral shift of about 0.15 Å determined.[331] There are two main problems with the investigation of lateral shifts. Firstly, they break the symmetry of the two-dimensional lattice, and thus any potential savings associated with the incorporation of beam symmetry cannot be obtained. Secondly, the effect is relatively weak, and thus requires the comparison of a larger data base than normal, to improve

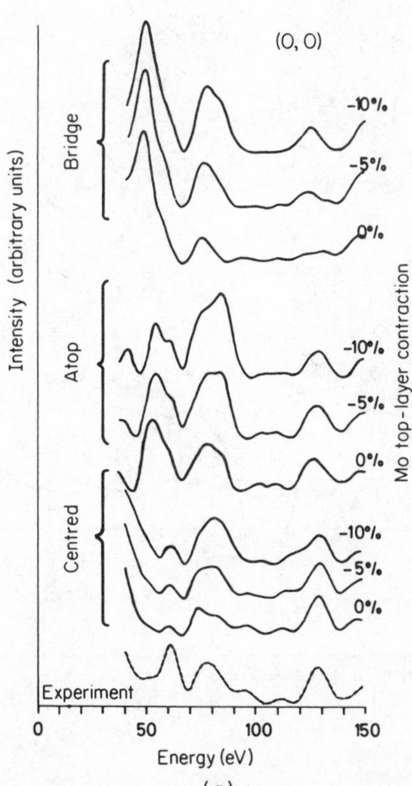

(a)

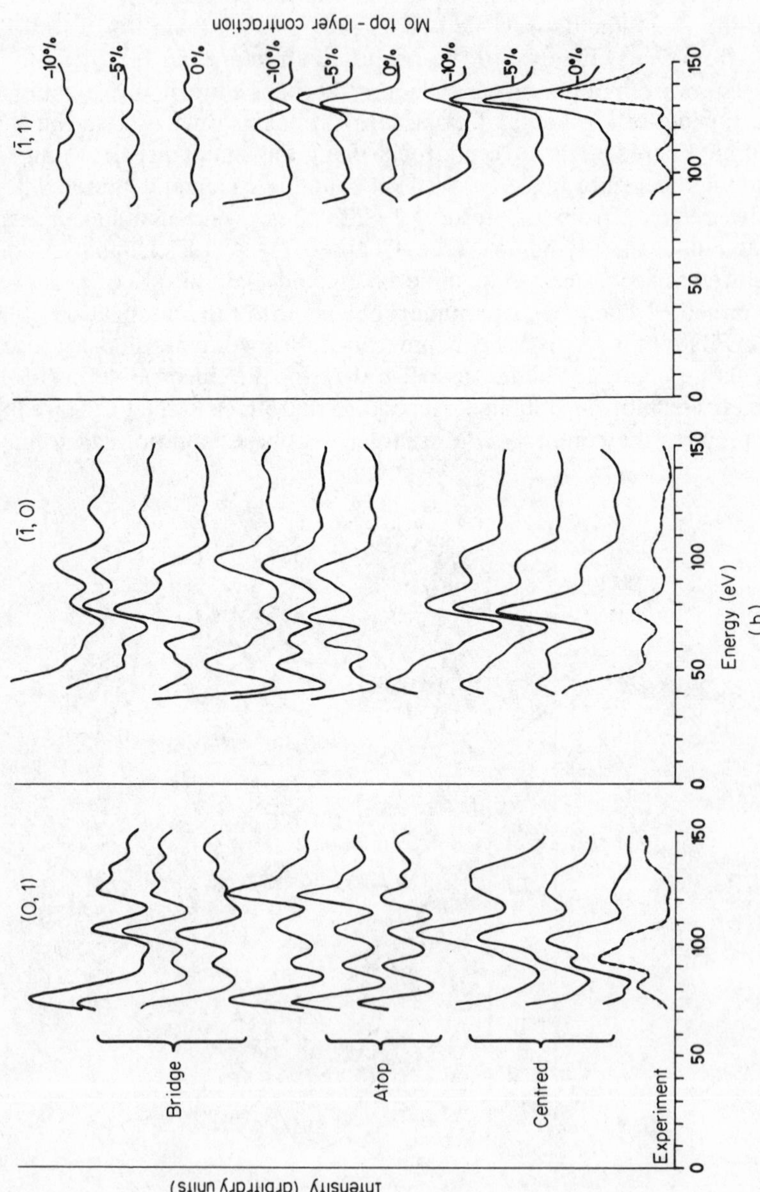
Intensity (arbitrary units)
Bridge
Atop
Centred
Experiment
(0, 1)
(1̄, 0)
(1̄, 1)
Energy (eV)
(b)
Mo top - layer contraction
-10%
-5%
0%
0
50
100
150

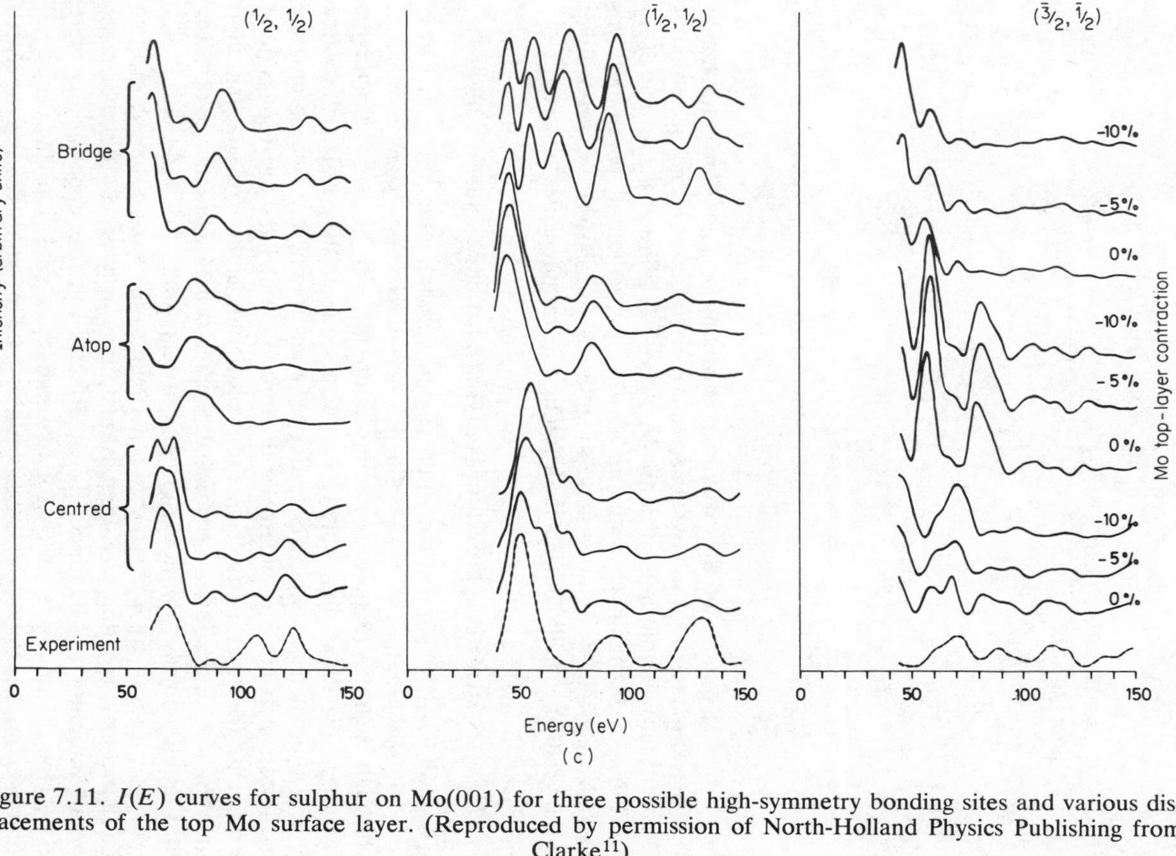

Figure 7.11. $I(E)$ curves for sulphur on Mo(001) for three possible high-symmetry bonding sites and various displacements of the top Mo surface layer. (Reproduced by permission of North-Holland Physics Publishing from Clarke[11])

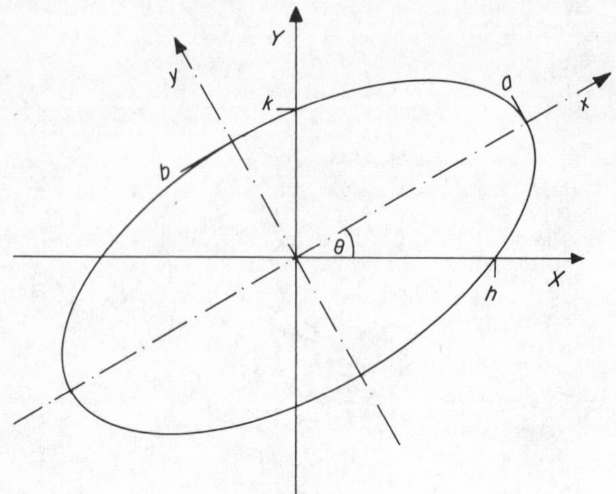

Figure 7.12. Schematic representation of an ellipsoidal r-value contour

the statistical quality of the available information. These two drawbacks combine to make this a relatively expensive operation in terms of computer resources and without any certainty of success. A prerequisite is that every effort has been made to optimize all the non-structural parameters, to maximize the quality of agreement possible and thus improve the sensitivity to the relatively small changes in diffraction beam intensities that result.

7.5 A systematic approach to the refinement of structural analyses

In section 7.4 it was noted that there are many parameters, both structural and non-structural, that may influence the diffraction beam intensities. Although each parameter may affect the intensities in its own characteristic way, it is inevitable that changing one parameter may affect the value determined when subsequently testing another parameter. Two parameters that are strongly interdependent are the inner potential and the displacement of the surface layer away from the typical bulk interlayer spacing; their interdependence is so strong that it has become common practice to optimize both simultaneously via the use of contour diagrams.

Ideally, if there are n parameters, then one would like to vary all of them simultaneously, creating an n-dimensional array of r-values from which the overall minimum may be found. If five different values of each variable were tested (the minimum necessary for reasonable interpolation) then the total number of models to be calculated would be 5^n. This would require the use of substantial computer resources for values of n greater than, say, 4.

Standard mathematical methods exist for locating minima in multidimensional arrays, but most do not seem to be appropriate to this problem. It is better to make use of prior knowledge—that some variables are more influential than others, most are not particularly interdependent and some are much easier to vary than others—to construct an *ad hoc* scheme specifically to aid the process of LEED analysis. The aim of such a scheme would be to take an initial model and, by varying one parameter at a time, converge towards the overall minimum by some form of linear progression. If thus linear optimization process were iterated m times, then the time to reach the minimum would scale as $5nm$ which, if m were small, would be considerably more rapid than the $4n$ of the global approach. The problem is to determine the best order in which to vary the parameters.

When a single parameter is varied, its interactive effect with any other parameter may be displayed as a section through a corresponding two-dimensional contour. It is therefore useful to consider the relationship between the influence of a parameter and the shape of the corresponding contours. Typical contour diagrams are given in Figure 7.9. Provided a large quantity of data has been used in the comparison, the curves may be smooth and quasi-elliptical, particularly in the vicinity of the minimum. For the present, let us assume that the contours are indeed ellipses centred on the r-value minimum. The shape of these elliptical contours is characterized by their eccentricity and orientation. The orientation is a measure of the interdependence of the two parameters: for example, if the two parameters are totally independent then the ellipse axes will be parallel to the axes of the diagram. In this case the eccentricity would be a direct measure of the relative strengths or influence of the two parameters. Of course, the two parameters may be characterizing two quite different sorts of variable—Debye temperature and inner potential, for example. The shape of the contours would therefore depend primarily on the scales chosen for each axis. Consequently, it is necessary to decide on the plausible limits for each parameter, or maximum range of uncertainty that must be considered. If these limits determine the length of the axes, and the axes are given physical identical lengths, so that any two-dimensional contour is square, then further general conclusions may be drawn from the visual appearance of the contours. For example, Figure 7.13 shows contours that would result from the simultaneous comparison of two interdependent parameters. For convenience, the origin is chosen to coincide with the central minimum. If we were to fix the variable $X = 0$ and vary Y we would obtain an r-factor variation given by the curve $X = 0$ in Figure 7.14. Similarly, the curve $Y = 0$ would be obtained if we scanned through values of X at this fixed Y value. The two lines define parabolas $x^2 = h^2r$ and $y^2 = k^2r$ where we have chosen $x = \pm h$ and $y = \pm k$ for the unit contour value r. The relative strengths of each parameter may be given by the values h and k. Of course, any of a whole family of ellipses could be defined which would pass through these points, each characterized by a different orientation angle θ with respect to one of the primary axes (see Figure 7.15).

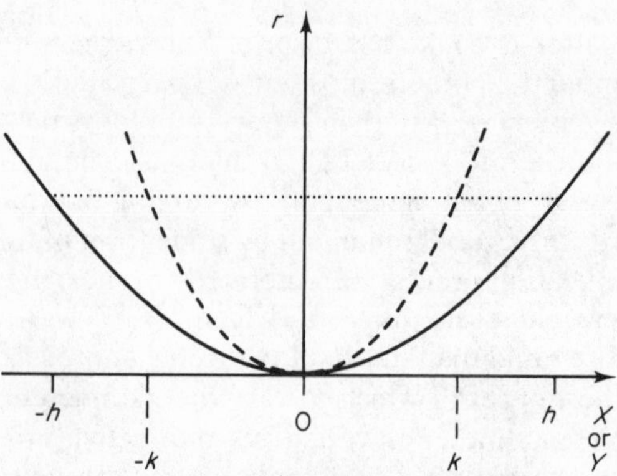

Figure 7.13. Parabolic variation in r-value, passing through the minimum of the idealized contour plot, along either the X- or Y-axis

The standard equation for an ellipse is

$$\frac{x^2}{a^2} - \frac{y^2}{b^2} = 1 \tag{7.26}$$

where a and b are the semi-major and semi-minor axes. Rotation through θ with respect to the axes X and Y can be described by the rotation matrix

$$\begin{pmatrix} X \\ Y \end{pmatrix} = \begin{pmatrix} \cos\theta & -\sin\theta \\ \sin\theta & \cos\theta \end{pmatrix} \begin{pmatrix} x \\ y \end{pmatrix} \tag{7.27}$$

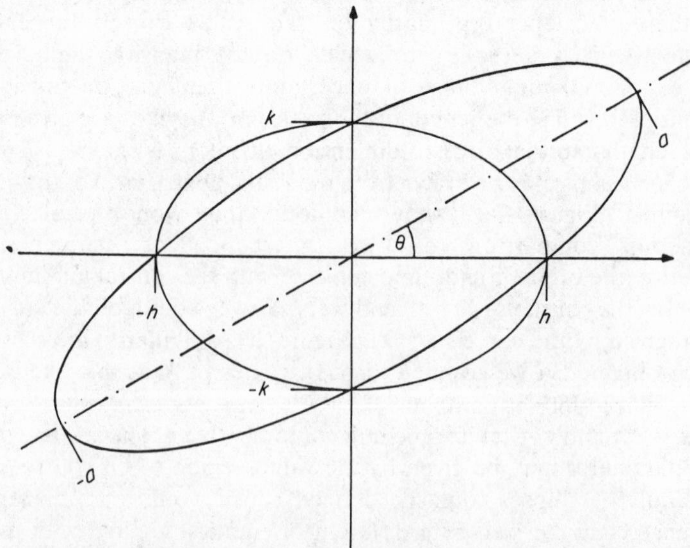

Figure 7.14. Illustration of different ellipses with different eccentricities, that nevertheless pass through identical intercepts along the X- and Y-axes

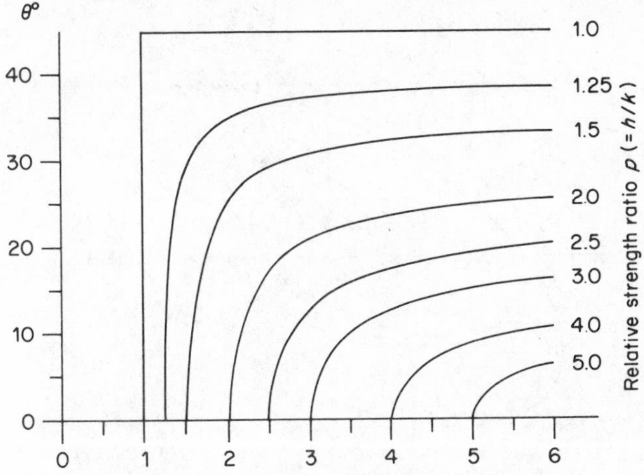

Figure 7.15. Relationship between the ratio of the semi-major to semi-minor axes, $q(=a/b)$ and the inclination of the major axis θ, to the 'relative strength ratio' $p(=h/k)$

from which the reader may satisfy himself that

$$\frac{h^2}{k^2} = \frac{a^2 + b^2\tan^2\theta}{a^2\tan^2\theta + b^2} \tag{7.28}$$

If the relative strength ratio $(h/k) = p$ and the ellipse parameters ratio $(a/b) = q$, then it can be shown that

$$\tan^2\theta = \frac{(q^2 - p^2)}{(p^2q^2 - 1)} \tag{7.29}$$

Consequently, the angle θ is intimately connected with both the degree of interdependence and the relative strengths of the parameters. If the ratios are defined such that a is greater than b, and h is greater than k, then their relationship with θ may be plotted as in Figure 7.16. From this, it is clear that θ can only equal 45° if the two parameters have identical influence ($p = 1$). In any other case one parameter will be more influential; even if the two are highly interdependent, θ can never exceed its asymptotic limit. For example, if $p = 2$, θ cannot exceed 26°. The major axis of the quasi-elliptical contours will therefore lie closer to the more influential parameter.

Let us assume that we wish to find the combination of values of X and Y that give the minimum r-value, in other words, that optimize the model parameters X and Y. In general, it would be too laborious to calculate a whole array of r-values for all permutations of these two. Indeed, the accuracy of the comparison process for a weak parameter such as the Debye temperature would generally be insufficient to justify such an extensive survey, and it is common practice to vary the parameter in question until the r-value passes

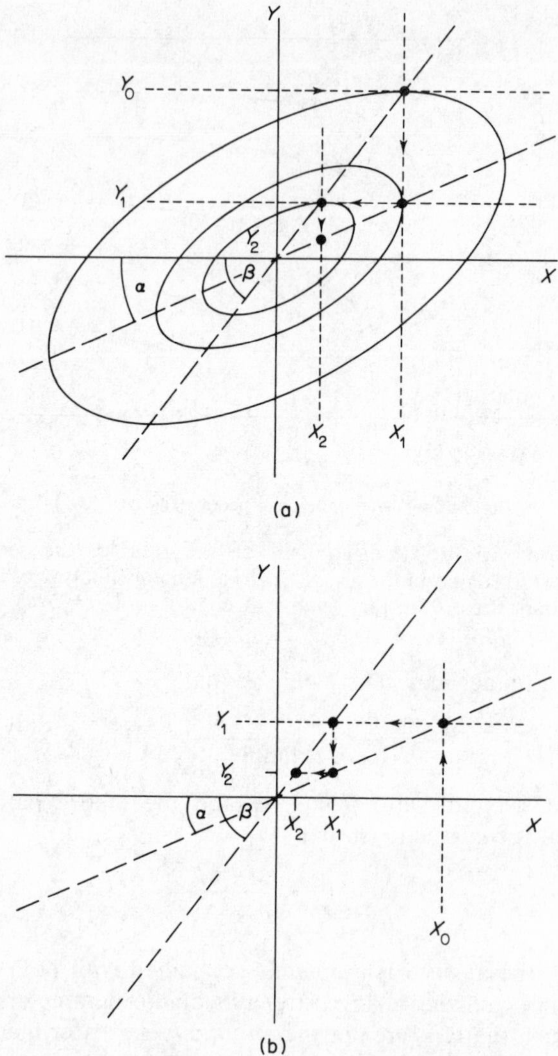

Figure 7.16. Progress towards the contour minimum if one starts with: (a) the original guess (Y_0, X_1) and then optimizes Y first; (b) the original guess (Y_1, X_0), and then optimizes X first

through a minimum, and leave it at that. The mathematical analogue described above enables us to check the validity of such an approach, and to assess the best course of action to be taken.

Considering the simultaneous optimization of two parameters, it is first necessary to decide which should be compared first. Consider what happens if we start by guessing a value for Y and vary X until an r-value minimum is

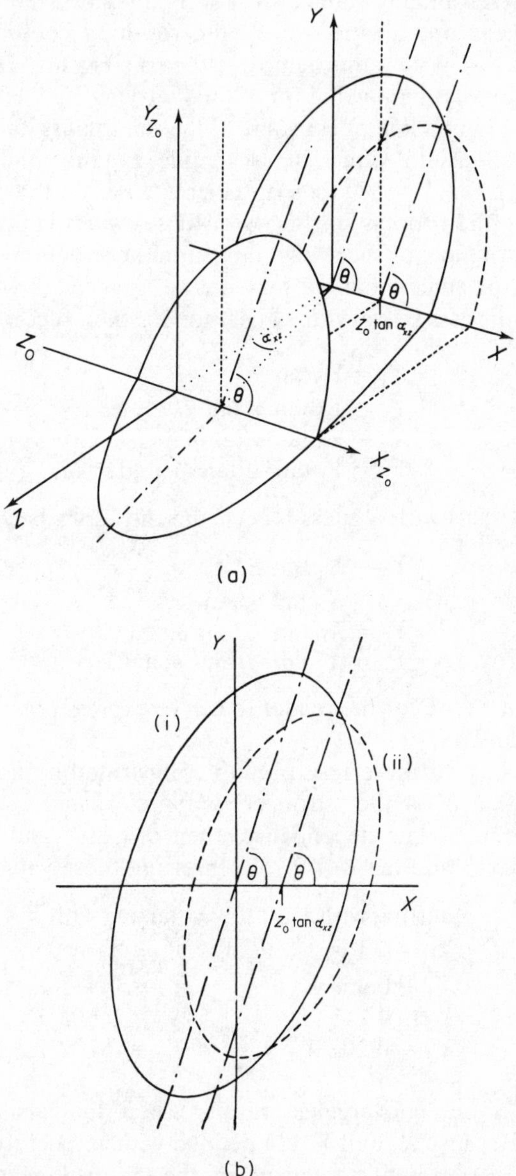

Figure 7.17. If a third variable, z, is present, but is not optimized, the effect of varying x and y can be pictured as an optimization process in a plane parallel to the 'best' plane. In (a) the relationship between these planes is illustrated in three dimensions; in (b) the net result is shown to be a systematic error between the X–Y plane optimum and the 'true' minimum

found. The corresponding value of X, say X_1, may be used as a starting-point for a subsequent optimization of Y, the resulting combination giving the r-value $R(X_1, Y_1)$. Would this lie nearer the true r-value minimum than if we had started by guessing a value for X, and varying Y?

As we scan along a line of constant Y, the minimum r-factor will belong to that contour-ellipse to which the scan line is tangential. The process of optimizing X, given $Y = Y_0$, then optimizing Y, given the new value X_1, and then repeating the process starting with these values is illustrated in Figure 7.17. Clearly, the loci of successive tangential intersections lie on one of two lines, inclined at angles that we may define as α and β with respect to the X-axis. If the process starts with a guess for Y_0, then successive points will be

$$\begin{aligned} X_1 &= Y_0/\tan\beta \\ Y_1 &= Y_0 \tan\alpha/\tan\alpha/\tan\beta \\ X_2 &= Y_0 \tan\alpha/\tan^2\beta \\ Y_2 &= Y_0 \tan^2\alpha/\tan^2\beta, \text{ and so on} \end{aligned}$$

If the process starts with a guess for X_0, the the series becomes

$$\begin{aligned} Y_1 &= X_0 \tan\alpha \\ X_0 &= X_0 \tan\alpha/\tan\beta \\ Y_2 &= X_0 \tan^2\alpha/\tan\beta \\ X_2 &= X_0 \tan^2\alpha/\tan^2\beta, \text{ and so on} \end{aligned}$$

It is left as an exercise for the reader to derive expressions for $\tan\alpha$ and $\tan\beta$ in terms of q and p.

The significance of these results can be illustrated by a specific example. If $p = 2$, so that Y is the 'stronger' parameter, and $q = 3$, indicating a moderate degree of interdependence, then $\theta \simeq 20°$, and it can be shown that $\tan\alpha = 0.315$ and $\tan\beta = 0.752$. Inserting these values one obtains

Starting with X_0	Starting with Y_0
$Y_1 = 0.315\ X_0$	$X_1 = 1.33\ Y_0$
$X_1 = 0.455\ X_0$	$Y_1 = 0.455\ Y_0$
$Y_2 = 0.143\ X_0$	$X_2 = 0.605\ Y_0$
$X_2 = 0.202\ X_0$	$Y_2 = 0.202\ Y_0$

Thus more rapid convergence results when the weaker parameter is guessed first. Because X_0 and Y_0 are defined with respect to the true contour minimum, they represent the error in the original guess; in the above example, if Y is guessed first, the value obtained for X would be proportionately more in error than the original guess of Y. If X were guessed first, both X and Y would be relatively nearer the true minimum. Repetition of the process further reduces the error by more than half.

Although the above is a single, specific example based on the assumption of perfectly elliptical contours, it is clear that certain broad assumptions may be drawn: (i) it is best to start the optimization process by varying the strongest parameters first; and (ii) once each parameter has been considered, variation of the strongest parameters should be repeated to check that the initial conclusions have not been affected. As a corollary, it may be stated that the correct value for a certain parameter cannot be obtained if its effect on the intensity data is related to that of another parameter which has not itself been optimized. The extent of the resulting error, however, is difficult to predict and can only be estimated from experience.

Nevertheless, provided an extensive set of experimental data is gathered and care is taken during the comparison process to optimize most of the more influential parameters, then structural results may usually be obtainable with an accuracy ±0.03 Å or better. In many cases, particularly when there is extensive surface reconstruction or a molecular species is adsorbed, the likely range of each variable may be narrowed by using an approximate theoretical technique to carry out a preliminary survey of suitable values. The suitability of various theoretical methods has been discussed in Chapter 4. It is clearly advantageous to begin the detailed comparison process with the best possible choice of parameter values. Only when the model is relatively close to the correct structure can the analysis of quasi-elliptical contours be assumed valid. On the other hand, once in this region it would be possible to develop suitable scanning methods to permit the computer to automatically search for, and locate, the correct structure in a relatively efficient manner.

CHAPTER 8

Comparison of LEED with related surface analysis techniques

8.1 Introduction

The importance being attached to improving our understanding of all aspects of surface behaviour is reflected in the enormous growth in surface analytical techniques that has occurred in recent years. The interface between solid and gaseous phases provides the site for catalytic reactions, corrosion and many other interactions. As the interest in micro-electronics shifts towards ever more complex and miniaturized circuitry, surfaces will assume an increasingly significant role in the characteristics of circuit operation. The development of suitable vacuum technology and the successful implementation of certain surface sensitive techniques such as LEED has resulted in the widespread adoption of such techniques in laboratories throughout the world and, at the same time, the development of many novel techniques capable of providing complementary or confirmatory information.

In the summer of 1973 a review of all known surface-sensitive techniques listed 56 separate methods.[332] Now, one decade later, a comparable list would certainly extend beyond 100. To attempt to describe them all would require a series of books, and even then would probably leave the reader uncertain as to the relative virtues of each possible technique as a means of investigating a particular surface system or phenomenon. In this chapter I have therefore attempted to describe only those techniques that are well established, or can be related closely to LEED either in the information they provide or in the experimental or theoretical methods that they use.

A problem with brief reviews is that they rarely do justice to the scope and potential of the methods discussed. This is particularly apparent when attempting to categorize techniques and when assessing the relative merits of each technique. Within this book I have outlined many of the possible applications of LEED, showing that it can be used to provide much information beyond the elucidation of two-dimensional periodic structures. Primarily, LEED is a surface crystallographic technique, able to detect the geometrical structure of the topmost atomic layer at a surface and the structure of relatively simple adsorbate species upon that surface. In addition, it is sensitive to certain topographical structures such as periodic steps and kinks, facets and adsorbate domains. It is also sensitive to thermal vibrations in both the bulk and surface atomic species. Through extensions to the technique (SPLEED) spin-polarization and surface magnetic features may also be studied. A knowledge of the chemical species present at the surface is necessary for LEED theory and the analysis of beam intensities, but it is not feasible to work back from the LEED data to determine the chemical composition. Auger electron spectroscopy (AES) can be performed using precisely the same experimental hardware (provided the electronics are suitably modified) and this, fortunately, yields chemical compositions and thus provides this essential complementary information. Because commercially available LEED systems almost always include the means for carrying out rudimentary AES, this capability may be reasonably included in the list of LEED attributes.

This should not detract from the fact that dedicated AES systems, using sensitive detection devices such as cylindrical mirror analysers (CMA) and glancing incidence, high-current guns (for surface sensitivity) are capable of providing a much more detailed chemical 'fingerprinting' of bonding types, and the quantitative measurement of surface coverages. When exploited fully, AES can then be considered as an independent technique able to provide quite separate information, complementary to that available from a standard LEED/AES system.

8.2 Electron and positron diffraction methods

8.2.1 Low energy positron diffraction (LEPD)

The diffraction of low-energy positrons (LEPD) was first demonstrated as recently as 1980, by Rosenberg *et al.*[333] This was made possible by the development of suitable low-energy positron sources. The basis of these sources is the β^+ decay of ^{58}Co, and details of such a source are provided in the original publication (see Figure 8.1).

LEPD is very closely related to LEED. It has the advantage that the scattering theory necessary to interpret the data is simplified by the non-identicality of the particles: there is no exchange term for positrons, so the problem discussed in section 3.2.3 can be ignored (since $\alpha = 0$). Also the correlation of positrons with electrons is relatively small. These differences affect the overall scattering behaviour, so the data obtainable from LEPD and LEED is slightly different. The mean free path for positrons is rather less than for electrons with comparable kinetic energy (about 5 per cent less at 100 eV, and 30 per cent at 50 eV, and this should provide the method with a slightly greater surface sensitivity. At present, however, LEPD has poorer resolution than LEED, because of the greater complexity of the experimental apparatus, but technical developments currently in progress could perhaps succeed in improving the resolution to a comparable standard.[334]

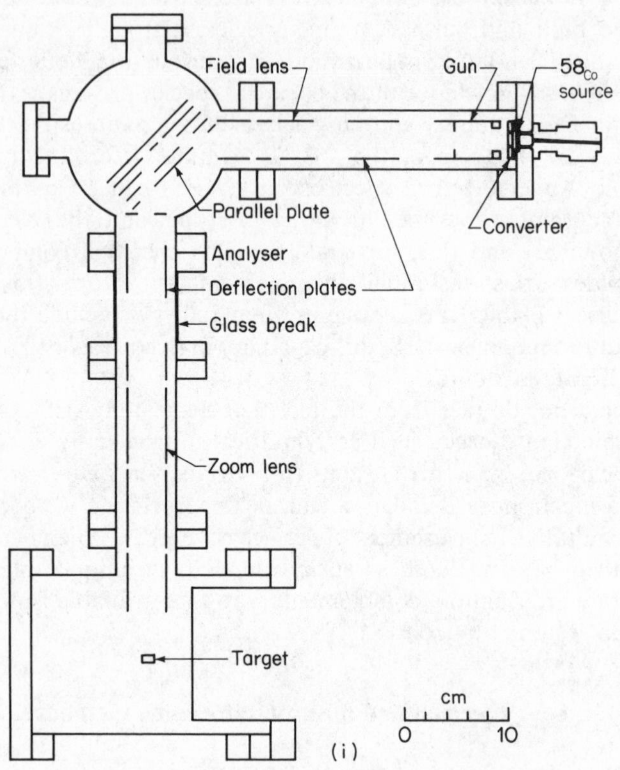

Figure 8.1. (i) Low-energy positron source, converter and beam-transport system. (Reproduced by permission of the American Physical Society from Rosenberg *et al.*[333])

8.2.2 Medium energy electron diffraction (MEED)

When LEED analysis is used to determine surface structures, it is desirable to compare intensity data collected over a wide range of energies. This increases the quantity of data available for comparison, both in energy range and in the number of different beams observable. It also reduces the influence of imperfect modelling of the scattering potential, which is greatest at low primary beam energies.

A practical restriction on the upper energy limit is imposed by the available computer resources. Using a typical current generation computer, not more than about 10 phase shifts may be used in a LEED intensity calculation for a simple clean surface structure, and, for a more complex surface structure, the number of phase shifts may be restricted to 5 or 6. Depending on the

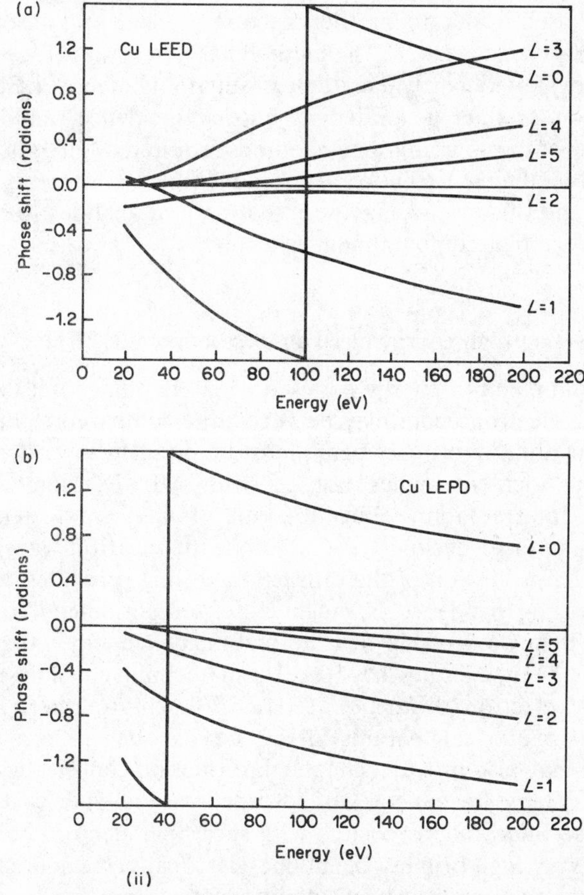

(ii) Comparison of phase shifts for Cu: (a) for LEED applications including the Kohn–Sham exchange-correlation term; (b) for LEPD applications, having no exchange-correlation term. (Reproduced by permission of the American Physical Society from Weiss *et al.*[334])

computational demands of the structural model, the maximum primary energy may be at best about 300 eV and at worst less than 100 eV, using the usual multiple-scattering theories and comparison techniques.

If the primary beam is directed towards the surface, near grazing incidence, it may be possible to include further approximation within the multiple-scattering theory that may permit 20 or 30 phase shifts to be incorporated. This theory, known as the chain method (as outlined in Chapter 4), provides the possibility of extending the energy range to 1 or 2 keV, subject to certain limitations on the incident beam direction.[188] Known as medium energy electron diffraction (MEED), this method has been successfully used to determine the structure of clean and adsorbate-covered Al surfaces by Masud *et al.*[335] The method has not, however, been extensively adopted or exploited, partly due to the unsuitability of most LEED systems to the collection of data generated at grazing incidence, and partly to the unfamiliarity of the computational methods required. MEED is, nevertheless, more suitable than LEED for data reduction methods, as described in Chapter 5, and this in turn may permit structural analyses to be carried out using rather simpler computational schemes.

8.2.3 Reflection high energy electron diffraction (RHEED)

If the incident electron energy is raised to 10 keV or more, the penetration depth of the electron beam may be very large compared to the interatomic layer spacing and the primary beam must be directed towards the surface at near grazing incident angles not, as with MEED, simply to ease the theoretical interpretation, but for the fundamental requirement of maintaining surface sensitivity. In addition, the electron wavelength is very short, and the projection of the diffraction grating presented to the incident beam at grazing incidence is much closer in periodicity to the electron wavelength than would occur near normal incidence. Operating in this mode with primary beam energies 10–50 keV, the technique is known as reflection high energy electron diffraction (RHEED). This method is particularly sensitive to topological features in the surface such as facets or epitaxial growth, but may also provide valuable information concerning surface steps. It is not so suitable for the study of adsorbed gaseous species due to the high electron desorption rate associated with such high electron beam energies.

At these very high primary beam energies, scattering is dominated by the strong Coulombic repulsion of the ion core, resulting in a far simpler scattering approximation term than possible in LEED. Nevertheless, only occasionally have attempts been made to extract precise crystallographic information from RHEED. An interesting example of the application of this method is the study of Si(111) (6×1)–Ag by Ichikawa and Ino.[336] With the incident beam aligned along a certain azimuth, certain characteristic diffraction beams were absent, indicating the existence of twofold glide planes (see Chapter 1 for the origin of this effect). A general analysis of the diffraction beam intensities was used to identify two possible structural models.

Although RHEED intensity analysis has not been refined to an extent comparable with LEED intensity analysis, the simpler scattering model permits the analysis of complex surfaces such as the reconstructed clean Si planes or epitaxial systems which are currently beyond the capabilities of full multiple-scattering LEED analysis. LEED and RHEED together provide complementary information covering a wide range of crystallographic features.

8.3 Surface crystallographic methods using electrons or radiation

In this section, we shall consider techniques that can provide crystallographic information using either electrons or radiation as the probe. Unlike the methods described in section 8.2, which are closely related to the LEED process, energy loss or transfer plays a crucial role in each of the methods described here.

Several of the techniques we shall be considering are called spectroscopies, because they are based on the analysis of the energy distribution (spectrum) of the emitted particles. These were developed primarily to provide electronic, vibrational or chemical information. However, crystallographic information may be derived either by the indirect relationship between vibrational modes and bond lengths or symmetries or else by using the intrinsic emission processes as sources of electrons which subsequently diffract as they attempt to leave the crystal. Diffraction is an interference phenomenon associated with the elastic (zero energy loss) scattering of waves from a periodic structure. Structural sensitivity is realized when the wavelength of the propagating waves is similar to the dimensions of the periodic structure—differences in phase between waves scattered from successive scattering centres cause strong interference effects that are highly sensitive to the interatomic distance. In LEED, the interference from successive scattering centres in the periodic crystalline structure produce a diffraction pattern. With surface EXAFS (see section 8.3.2), structural sensitivity is obtained directly from the interference effects occurring between neighbouring atomic centres. If electron emission processes are used, subsequent diffraction provides the crystallographic information, but because the emitter provides a radial distribution of electrons rather than a uniquely oriented beam, this information is contained not in a set of discrete diffraction beams but in an angular distribution of intensities. To extract this information, the electron detector must be able to resolve the angular distribution as well as the energy distribution of the emitted electron flux.

8.3.1 Angle-resolved ultraviolet photoemission spectroscopy (ARUPS), X-ray photoemission spectroscopy (ARXPS) and Auger emission spectroscopy (ARAES)

Electron emission from the core levels of surface atoms can be stimulated by irradiation by X-rays. Electron emission from less tightly bound energy levels can be stimulated by ultraviolet light. Electrons photoemitted using ultraviolet light (UPS) have energies characteristic of the atomic species being examined, but are generally not more than a few electron-volts or tens of electron-volts, and are thus comparable to the primary beam energies used in LEED experiments. The angular distribution of the photoemitted current is affected both by the radial distribution of electrons emitted from each atom and also by diffraction effects caused by their subsequent scattering within the crystal. Because the electron sources are located within the crystal surface itself, a significant proportion of the photoemitted current may result from forward scattering—this contrasts with LEED, in which only back-scattered contributions can be seen by the observer. However, at very low energies, the back-scattered contribution may be comparable to the forward-scattered contribution (see Chapter 3) and the resulting angular distribution of photoemitted current may be very sensitive to the geometry of the neighbourhood of the emitter, particularly its symmetry.

The diffraction of the emitted electrons can be calculated by the suitable adaptation of the multiple-scattering theories developed previously for LEED.[337]

In LEED, it is usual to treat the scattering of an incident plane wave by measuring the intensities of the discretely diffracted beams as a function of the incident beam energy. In angle-resolved electron emission studies, the emergent beam has a fixed energy, characteristic of the emission process, but its intensity is measured over a continuous range of angles. This can be readily accommodated within LEED theories because an integral step in calculating the effect of ion-core scattering is to decompose the incident plane wave into spherical (partial) waves—in this case, the partial wave distribution is determined directly from the photoemission process itself. The angular distribution of this atomic photoemission current has been the subject of considerable theoretical and experimental work[338,339] and a suitable multiple-scattering computer program has been developed by Pendry and co-workers.[340]

A particular advantage of ARUPS over LEED is that, by observing the current emitted at a specific energy, emission from the substrate and adsorbate atoms may be separated, selecting out specific surface contributions. In addition, disordered adatom systems may be studied, because each atom acts as an independent emitter (assuming that scattering from the surface layer is relatively weak).

Disadvantages of the method, compared with LEED, include uncertainties in the final state photoemission angular distribution and the additional experimental complexities associated with photon source (ideally, a synchrotron should be used) and the angular resolved electron detection system.

Electrons emitted from certain Auger electron emission processes have relatively low energies and in principle may be used to provide crystallographic information. The Auger electron emission process, however, involves not one but two electron transitions, and thus introduces additional complexities and uncertainties in the angular distribution of electrons emitted from each atomic source. For this reason, few detailed calculations of angle resolved AES current densities have been carried out. Recently, however, a detailed investigation[341] has yielded close agreement between theory and experiment, determining the anisotropy of the emission processes and confirming the adatom geometry of sulphur on Ni(110), as previously determined by LEED and MEIS (see section 8.4). Currently, ARAES is not well established, and lacks the accuracy of LEED, but provides a valuable complementary technique that deserves further research.

The energies of photoemitted electrons produced by X-ray irradiation are substantially higher than produced in UPS experiments—typically about 1 keV. The scattering processes that result are dominated by forward scattering out of the crystal and so are not so closely related to the LEED multiple-scattering situation. Angle-resolved XPS is particularly suited to the study of the orientation of adsorbed molecules. For example, it has been found that adsorbed CO can be adequately modelled assuming a single-scattering process,[342] from which the bonding site symmetry and molecular orientation may be determined.

8.3.2 Surface extended X-ray adsorption fine structure (SEXAFS) and X-ray adsorption near-edge structure (XANES)

The process of X-ray photoabsorption involves the extraction of a localized core-level electron and, by raising its energy, converting it into an outgoing spherical electron wave. If a beam of X-rays passes through a material, this photoabsorption process will extract energy, and thus attenuate the emergent beam. If the energy of the X-rays is just less than that required to extract an electron from a core level (the core-level binding energy), attenuation may be relatively low, but, as the energy of the X-rays is gradually increased until it exceeds the binding energy, electron emission becomes possible, a new channel for energy transfer is opened up and the X-ray attentuation abruptly increases. As the X-ray beam energy increases still further the photoabsorption process becomes less efficient, and the attenuation decreases steadily, until a new absorption edge is encountered. For an isolated atom, the absorption coefficient (μ) above an edge decreases monotonically with increasing X-ray energy, but, if two or more atoms are in close proximity, the absorption coefficient above the edge will have a complex fine structure superimposed (see Figure 8.2). It is this fine structure that is called EXAFS (extended X-ray absorption fine structure).[343]

The origin of this fine structure lies in the final state (outgoing spherical wave) of the photoemitted electron. If the outgoing photoelectron encounters another atom, it may be partially back-scattered. This will result in an incoming wave at the origin which may interfere with the initial outgoing spherical wave. As the energy of the X-ray is varied, so the energy, and hence wavelength, of the photoelectron will also vary, and the alternating constructive and destructive interference will produce a sinusoidal modulation in the absorption coefficient.

If an atom is located at the surface of a crystal, scattering from all its neighbours will result in a complex EXAFS, but the extraction of the relevant crystallographic data is in principle relatively easily accomplished using Fourier analysis, provided the phase shifts associated with the emission and

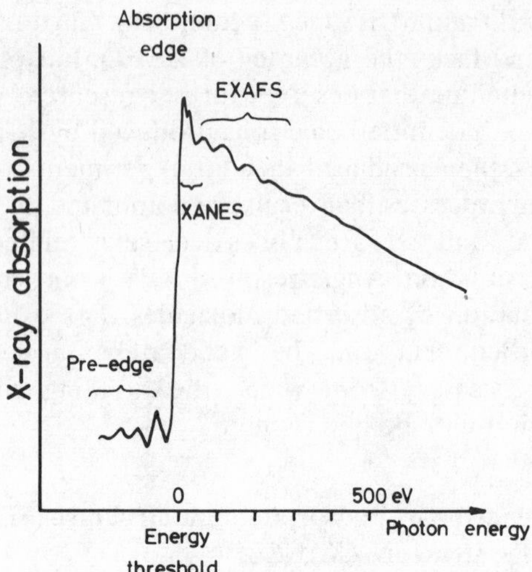

Figure 8.2. Schematic X-ray absorption edge, and associated fine structure

back-scattering processes can be determined. To obtain surface sensitivity, the usual method of measuring the attenuation of the transmitted X-ray beam is not suitable, since the number of the atoms within the bulk greatly exceeds the number located at the surface. The Auger and secondary electron yields produced during the photoadsorption process are, however, also subject to the same final state interference effects, and the short mean free path of the emitted electrons effectively filters out signals from deeper than a few atomic layers of the surface. The unfortunate consequence of this restriction in the number of emitters is that the resulting signal is very weak, so data-collection times of several hours are necessary to obtain adequate signal/noise ratios. Such lengthy collection times may be reduced in the future as more powerful X-ray synchrotron sources become available.

The first application of SEXAFS to an adsorbate structure was made in 1978 by Citrin *et al*.[344] who monitored the intensity of the adsorbate core Auger electrons as a function of incident X-ray beam energy to elucidate the bond lengths and adsorption site geometry for iodine on Ag(111). To determine the phase shifts necessary for the interpretation of the data, they used an empirical derivation by comparing EXAFS data from a bulk AgI compound of known crystallographic structure.

Recently, Bunker and Stern[345] have examined the influence of using both theoretically and empirically derived phase shifts on the accuracy of the EXAFS method, and concluded that interatomic distances may be determined within ±0.03 Å if theoretical phase shifts are used, and accuracies better than ±0.01 Å could be obtained if suitable compounds can be found to provide an empirical parameter fit.

In EXAFS, the mean free paths of the scattering electrons is short, so that the observed fine structure is dominated by the pairwise interaction between neighbouring atoms. Consequently, the method provides very accurate interatomic distances but does not directly provide information on bond angles. This must be deduced independently, aided by the information available from the superposition of signals caused by scattering from next-nearest neighbours.

Despite the impressive accuracy obtainable from SEXAFS, exploitation of the method has been limited, partly because it requires the extensive use of synchrotron facilities, and partly because the method is best suited to a rather limited number of adsorbate–substrate systems. The criterion is that the relevant adsorption edges should be well separated in energy from each other, so that the fine structure is not obscured by overlapping data. SEXAFS is not well suited to the study of clean surface structures, since bulk effects interfere with the information specific to the surface. Nevertheless, one can envisage many applications of this method for the generation of surface crystallographic data, particularly in situations where an accurate determination of bond lengths within a known bonding geometry is required.

The fine structure used in EXAFS occurs beyond about 50 eV above the adsorption edge, but there are additional intensity modulations within about 50 eV of the absorption edge which also contain structural information (see Figure 8.2). This near-edge structure results from strong scattering with long mean free paths. The use of this data comprises a separate technique, called XANES, and its use has been demonstrated for an oxygen on Ni(001) system by Durham *et al.*[346] Multiple scattering is a feature of XANES, and this provides the method with a sensitivity to three-body correlations, and hence bond angles, in addition to atom–atom separations. The theoretical interpretation of this data is much more complex than for EXAFS, but a method related to the cluster model described in Chapter 4 has been shown to be suitable for this purpose.[347] It remains to be seen whether the capabilities of the method can be exploited to provide valuable crystaglographic information. It has the potential of providing information on short-range structures within liquids and amorphous metals.

8.3.3 High-resolution electron energy loss spectroscopy (HREELS)

Characteristic electron energy loss processes can be observed using a standard LEED/AES equipment, as noted in section 8.1. In that case, the high energy of the incident beam stimulates surface plasmons which in turn absorb energy from the primary beam and re-emit electrons with energies characteristic of the plasmon frequencies. In HREELS, the incident energy is dropped below the plasmon threshold (to about 5 eV above the Fermi energy). By employing high-energy resolution (roughly 10 meV or less) the technique is sensitive to, and able to distinguish between, surface phonon modes.

The long-range interaction between an incident electron and the surface oscillating dipole is very similar to that between a photon and the surface dipole, so HREELS and infra-red (IR) spectroscopy (see section 8.3.4) are closely related in the vibrational modes that they can detect. However, the dipole selection rule that normally restricts excitations to those vibrational modes containing an oscillating dipole moment with a component perpendicular to the surface, can be broken by short-range 'impact' electron scattering. Under suitable conditions, therefore, all possible normal modes are observable using HREELS. The information is contained in the energy distribution of the specularly reflected beam. To obtain the required high resolution, band-pass energy analysers must be used to both monochromate the primary beam and to analyse the emergent beam. For this purpose 180° hemispherical analysers, 127° cylindrical sector analysers and cylindrical mirror analyzers have all been used successfully.

Identification of the vibrational modes provides a simple means of distinguishing between competing models of surface structures. It may permit certain models to be ruled out, because the associated vibrational modes do not exist or would be incompatible with the observed modes. However, it may not necessarily be able to provide a unique, unambiguous model. In an investigation of oxygen adsorption on Si(111), Ibach *et al.*[348] observed three distinct loss peaks, indicating that oxygen bonded in an associated or quasi-molecular form rather than in a disassociated, atomic form. Dubois and Somorjai[349] have recently reviewed the subject, providing a number of examples of molecular adsorption on rhodium, whilst in the same volume Thiel and Weinberg[350] have described the results of an investigation of molecular adsorption on ruthenium. In these examples, a clear distinction could be made between molecular adsorption into linear 'atop', 'bridge' or 'hollows' sites. This method could be linked with LEED by using a two-level vacuum chamber and an extended travel precision manipulator. In this way standard LEED and AES could be used to monitor the preparation of the sample prior to lowering it down to the electron loss spectrometer.

HREELS is a valuable and powerful technique that is sensitive to a wide range of vibrational energies without the need to change analysers or recalibrate (in contrast with IR spectroscopy—see section 8.3.4). It is quite sensitive, being able to detect components present in concentrations less than 0.1 per cent of a monolayer if they are strong scatterers and it is well suited to the study of clean metal surfaces. Unlike some spectroscopies (such as AES) it is able to detect hydrogen, and because the primary electron beam has a low energy, it is a relatively 'soft' probe, suitable for studying weakly adsorbed molecular species. Like most spectroscopies, it is capable of studying disordered surfaces, but can only be used in the UHV regime due to electron–gas collisions.

LEED and HREELS can be used together to great advantage. Andersson,[351] for example, has used both methods within a single apparatus to study the adsorption of CO on Ni(001).

8.3.4 Infra-red (IR) spectroscopy

Infra-red (IR) spectroscopy gives very similar information to HREELS: it, too, provides a means for measuring vibrational spectra of adsorbed molecules and from these, valuable structural information may be deduced. IR radiation is one of the gentlest probes of a surface, causing little more than a slight increase in the vibrational energy of the adsorbed molecules. On the other hand, because it uses photon beams, it can be used to study surfaces in high-pressure environments (unlike most surface techniques, which need a high vacuum).

IR radiation may be used in a number of different ways to provide surface information. In some cases, the standard method of observing adsorption lines arising from transmission through the surface may be used. This approach was used as early as 1954 by Eischens *et al.*[352] in a study of small metal particles supported in a matrix of silica spheres. In this example, transmission–absorption could be used because silica is transparent to the radiation. However, this method is not suitable for the study of single crystals. For this purpose, it is necessary to examine the adsorption spectrum of the reflected radiation. Used in this way, the method is generally known as reflection–absorption (RA) spectroscopy, or simply IR absorption spectroscopy (IRS). Important developmental work in this field was carried out by Greenler *et al.*[353] and has since been applied to many surface studies. Tomkins[354] has written a useful introductory review, in which the basis of this method and its application to surface problems is outlined.

More recently, an alternative method of using IR radiation for the study of surfaces has been described by Fedyk and Dignam.[355] They studied the

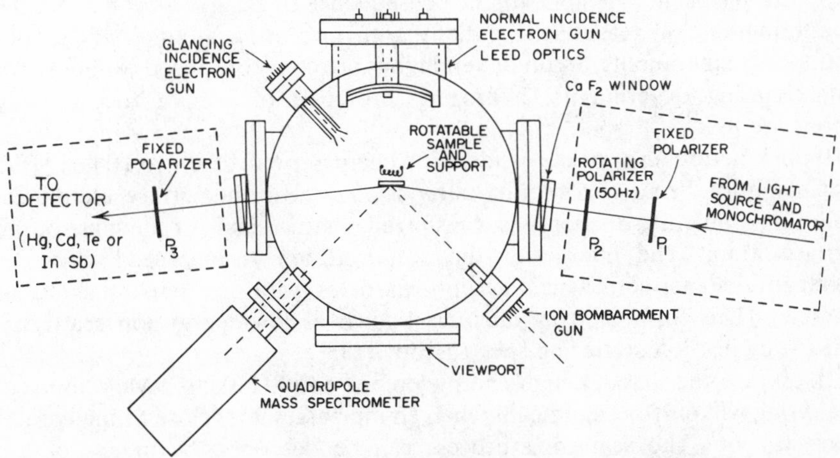

Figure 8.3. Illustration of a UHV system, incorporating LEED, IR and other surface-sensitive techniques. (Reproduced by permission of North-Holland Physics Publishing from Fedyk and Dignam[355])

polarization modulation of the specularly reflected beam, using a method that may be described as IR ellipsometric spectroscopy of IRES.

IRES has the advantage of better resolution than HREELS (up to 10 times better), but is less sensitive. From an experimentalist's point of view it is much less convenient to use, because a single monochromator is insufficient to enable the whole spectral range to be covered, and several different gratings, and possibly even detectors, are required to scan from say 4000 cm^{-1} to 250 cm^{-1}. By comparison, a wide spectral range can be covered using HREELS without modifications to the apparatus. An example of an experimental system combining IRES with LEED for surface studies is given in Figure 8.3. The advantage of the high resolution obtainable with IRES is that interactions between adsorbed species may be observed, and the distinction made between bonding sites and substrate identity. IR spectroscopy also has the advantage that IR radiation can pass through windows and can thus be mounted externally to any UHV system. If required, an IR spectrometer could be moved from one system to another.

8.4 Surface crystallographic methods using ion- and atom-scattering methods

8.4.1 Low-, medium- and high-energy ion scattering

There are numerous ways in which ion beams may be used to probe the structure of surfaces. A simple distinction may be made between those that use the incident ion beam to knock, or sputter, atoms from the surface for analysis (as outlined in section 8.5) and those methods in which the ion beam itself is detected following its interaction with the surface.

Sputtering is an almost inevitable consequence of using ion beams, but the sputtering rate increases substantially with increasing atomic weight. For sputtering experiments, argon or xenon ions are commonly used, whereas, for ion-scattering experiments, hydrogen (protons!) or helium (occasionally neon) are used.

If the kinetic energy of the incident ion beam is low (100 eV to 10 keV), the impinging ion beam will interact fairly strongly with the surface atoms. On collision, a quantity of energy is transferred, characteristic of the mass of the surface atom. The energy of the scattered ion is measured using an electrostatic analyser, which accepts particles within a narrow band of energies. This form of experiment is known as low-energy ion scattering (LEIS) or just ion-scattering spectroscopy (ISS).

If, on the other hand, a high-energy ion beam (500 keV to 2 MeV) is used, scattering will only be significant when an ion passes very close to the core of a crystal ion. The scattering process can be described by means of the Rutherford back-scattering model, involving a Coulumb scattering potential. If the angle of incidence is parallel to a crystal plane, the ions may pass almost unhindered along the channels between the layers, a phenomenon known as

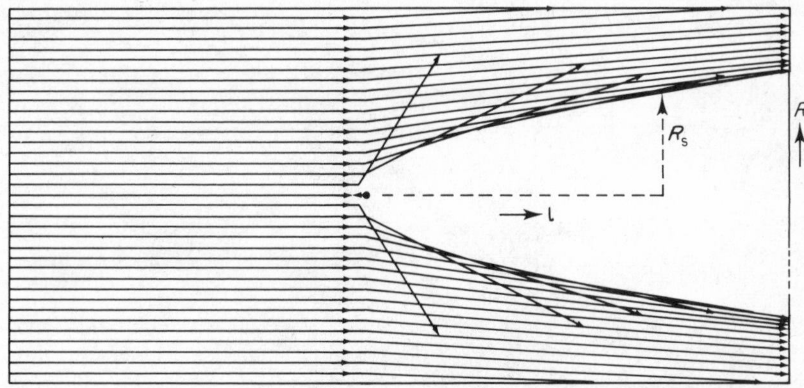

Figure 8.4. An example of a shadow cone behind a target atom bombarded by energetic ions; R_S is the radius of the cone at a distance l from the atom. (Reproduced by permission of North-Holland Physics Publishing from Turkenburg *et al.*[358])

ion channelling. By comparison, the back-scattered contribution may be much greater when the ion beam is incident at an oblique angle to the major crystal planes. With high-energy ion scattering (HEIS), which is also known as Rutherford back-scattering (RBS) or ion channelling, the predominant source of surface information is contained, not in the energy of the scattered ions, but in the angular distribution of the scattered intensity.

When a beam of energetic ions bombards an atom the repulsive potential has a strong effect on ions passing near the atom centre, but a much weaker effect on those passing further away. A typical distribution of trajectories in any two-dimensional plane through the scattering centre is given in Figure 8.4. A shadow cone is produced, the shape of which will depend on the energy and mass of the scattering species. If the interaction between the beam and target can be described by a potential of the form

$$V = A/e^n \tag{8.1}$$

where A is a constant, r the distance from the atom centre and n characterizes the power of the scattering potential, the radius of the shadow cone, R at a distance l behind the atom centre is related to the incident beam energy E by the following equation:

$$R = f(n, A)\left(\frac{l}{E}\right)^{\frac{1}{n+1}}$$

For example, $n = 1$ for a Coulomb potential, and in this case

$$R \propto \left(\frac{l}{E}\right)^{1/2}$$

If the atomic potential can be accurately described in terms of n and A, the simple geometrical shape of the shadow cone will provide a relatively

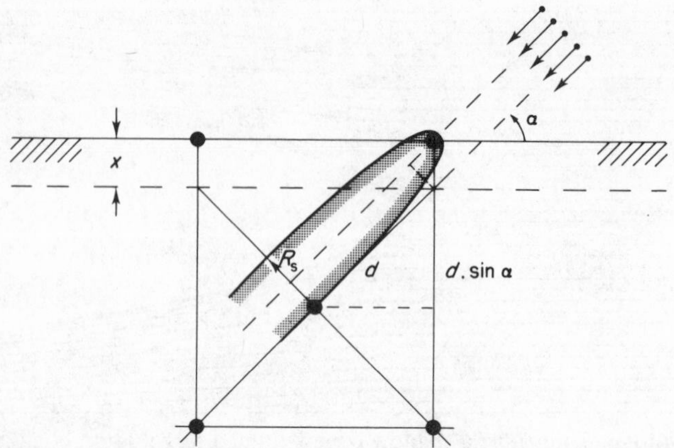

Figure 8.5. A two-dimensional model illustrating the relationship between the displacement x and the radius R_s of the shadow cone belonging to the primary energy E_{crit}. The incident beam enters along a crystallographic direction that makes an angle α with the surface. The bulk spacing is d. (Reproduced by permission of North-Holland Physics Publishing from Turkenburg *et al.*[358])

straightforward means of determining surface layer displacements. The basis of this method is illustrated in Figure 8.5.

If the incident beam is parallel to a major crystallographic plane (inclined at some angle α to the surface) and the surface layer is not displaced from the bulk truncation position (d), then near-specular back-scattering can only occur from this one layer near the surface. If the surface is displaced from the bulk truncation position (in this example, relaxed through a distance x), there will be a critical beam energy E_c at which strong specular scattering from the second layer will be 'switched on', as it emerges from the shadow cone.

The width of the shadow cone decreases with increasing energy so, to resolve very small surface layer displacements it is advantageous to utilize high (MeV) ion energies.[356,357] An alternative is to fix the ion beam energy and vary the angle.

Experiments have been carried out using medium-energy ion beams by Turkenburg *et al.*[358] The angular variation of the emergent ions will be affected not only by the above-mentioned back-scattering process, but also by the scattering and blocking processes that affect the outgoing, scattered ions. Once details of the shadow cone dimensions have been determined, it is relatively straightforward to compute the relationship between the angular variation of the scattered ions and surface geometry.[359]

Unlike LEED, which is predominantly sensitive to surface layer displacements normal to the surface plane, and less sensitive to lateral displacements, ion scattering is sensitive to displacements in either direction, and may therefore provide valuable information for the study of reconstructed clean surfaces or epitaxially grown layers.

8.4.2 Atom scattering

In common with LEED, the earliest demonstration of atom scattering was carried out in 1929,[360] but the development of the technique for quantitative surface studies only occurred in the 1970s, following necessary improvements in vacuum technology. Atom beams with kinetic energies as low as a few milli-electron-volts (meV) may be used, and at such low energies the atoms are sensitive to the weak van der Waals forces that form an attractive potential of less than 10 meV at the surface. As the impinging atom approaches within the range of the van der Waals force and its electron cloud overlaps the surface electronic density, it experiences a strong, but short-range, repulsion, sufficient to turn the beam away whilst it is still roughly 2–3 Å from the surface ion cores.

Atom scattering is therefore particularly sensitive to the geometry of the electron densities at a crystal surface. The ion cores at a surface provide a discrete set of scattering centres, but the valence electrons at the surface are smoothed out by forces analogous to surface tension at the surface of a liquid. This smoothing process is not, however, total, and the surface 'seen' by the atoms takes the form of a continuous, but corrugated, scattering medium, rather like a mattress.

The electron density corrugations are particularly small for clean metal surfaces, and until recently He diffraction had only been detected from the relatively open-structures (112) face of tungsten.[361] Adsorbates may cause more pronounced corrugations of the electron density distribution. The resulting atom or molecule scattering may be sensitive to the bonding sites and symmetries of the adsorbed species, from which the surface structure may be deduced. The simplest method of interpretation uses the assumption that the weak, attractive force is negligible and the repulsive potential has the effect of a hard, corrugated wall. Typical results, obtained from clean Ni(110) and hydrogen adlayers on the same surface,[362] are illustrated in Figure 8.6.

Used in this way, atom scattering provides a valuable complementary method to LEED, but does not provide information concerning interlayer spacings. In theory, it could be possible to learn something about surface interlayer spacings by varying the kinetic energy of the atom beam. With increasing beam energy, the atoms will probe deeper, and the corrugations of the effective hard-wall boundary will become more pronounced. The change in corrugation amplitude will reflect the contours of constant electron density which are in turn linked to the relative positions of the adatoms and upper layers of the substrate. Unfortunately, on increasing the beam energy to a few hundred meV, the ion cones themselves will be affected, as the scattering becomes increasingly inelastic. This will interfere with the simple diffraction processes characteristic of lower beam energies, and thus make interpretation of the surface structure more difficult. At this stage, it would appear that LEED and atom scattering provide an important pair of quite complementary techniques—one predominantly sensitive to interatomic spacings normal to

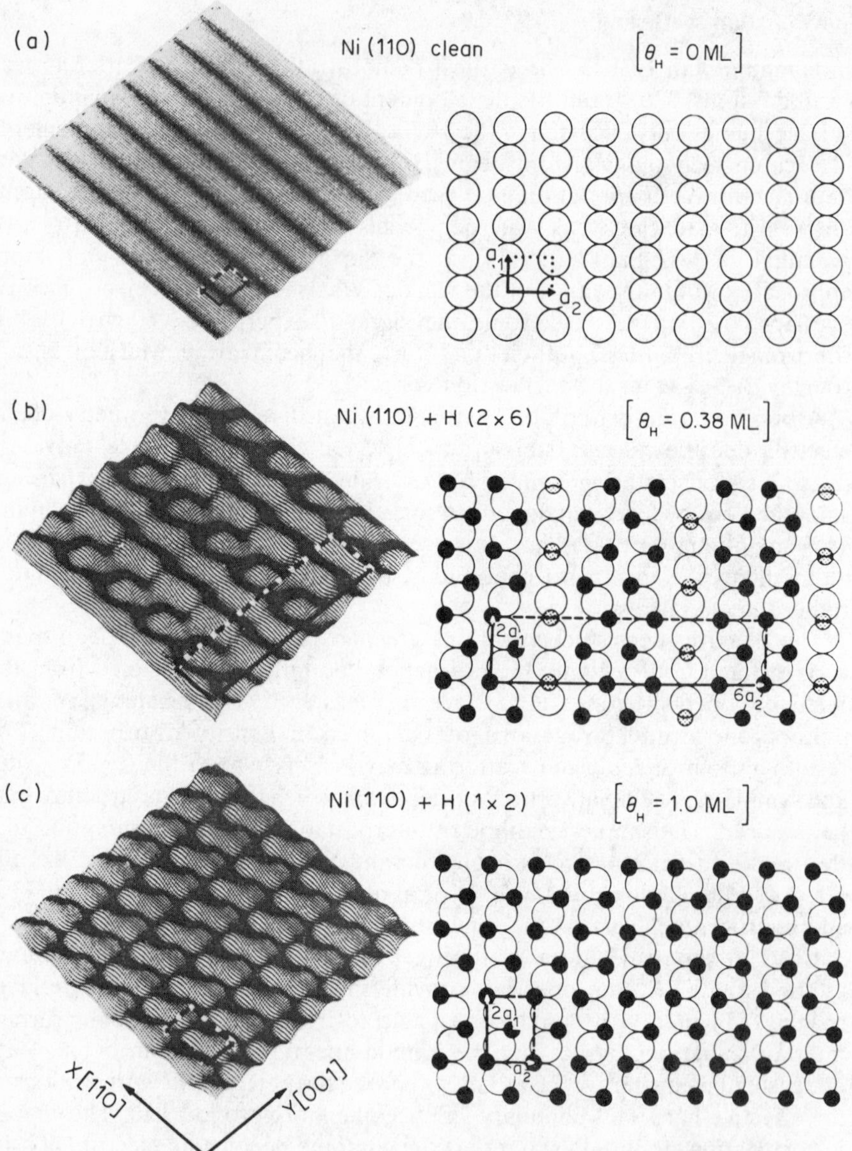

Figure 8.6. Atom-scattering corrugation functions and corresponding hard-sphere models for (a) the clean Ni(110) surface, (b) the (2 × 6) and (c) the (2 × 1) adsorbate phases of hydrogen on Ni(110). The surface unit cells are indicated. The corrugations are expanded by a factor of six in the vertical direction. The figures refer to scattering data obtained with low He energies (~60 meV). (Reproduced by permission of the American Physical Society from Rieder and Engel[361].)

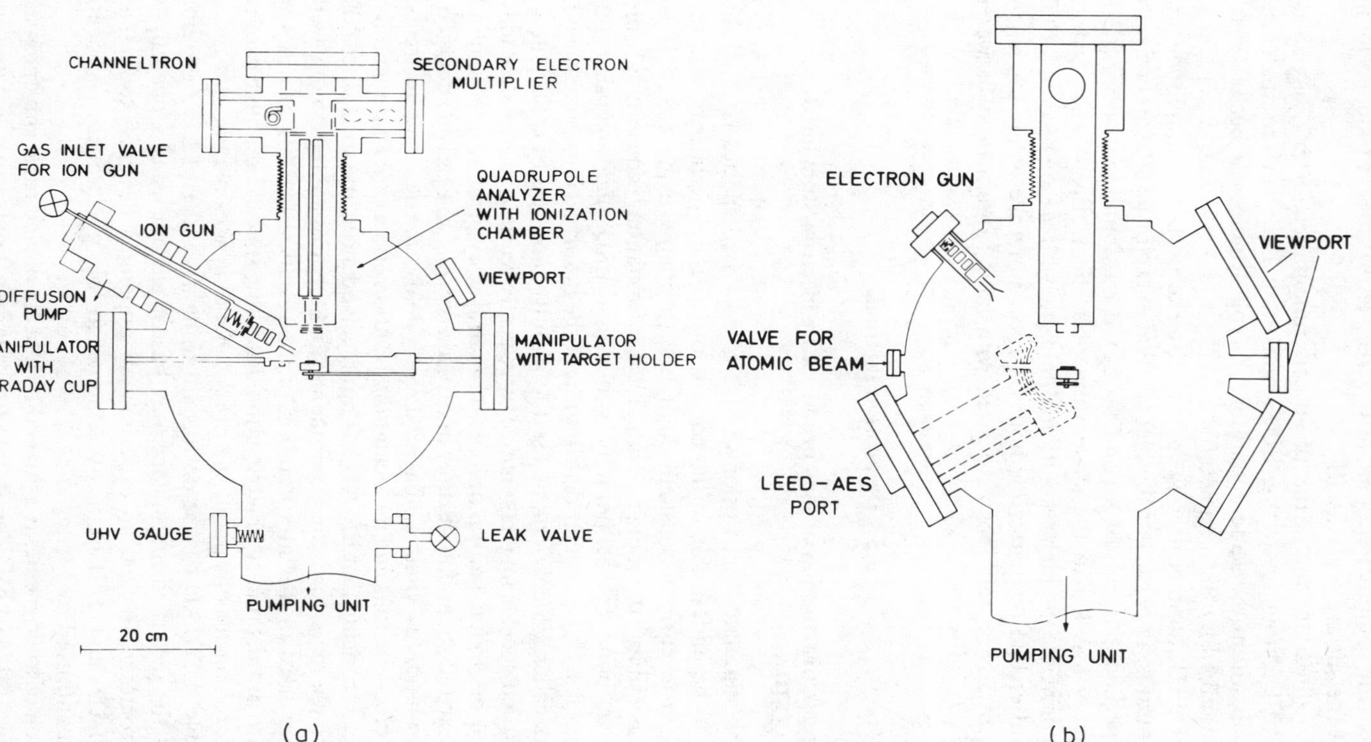

Figure 8.7. Schematic diagram of a UHV apparatus permitting the LEED, atom scattering and other surface-sensitive techniques to be used in conjunction. (Reproduced by permission of North-Holland Physics Publishing from Estel *et al.*[362])

the surface, the other predominantly sensitive to the two-dimensional structure of the surface layer itself.

Figure 8.7 illustrates the layout of an experimental chamber designed to combine LEED, atom scattering, SIMS and other basic surface analytical techniques all within the same apparatus. A typical molecular beam source has been described by Bush and Raff.[363]

As with LEED, atom scattering may be used to probe the amplitudes of surface vibrations, in particular the Debye–Waller factor (see Chapter 5). In a recent experimental study of Xe on Cu(001), it was found that both LEED and He energy loss led to the same values for the Debye–Waller factor, although in principle the high resolution obtainable with He energy loss (~0.5 meV) should enable this latter method to probe more accurately the small changes in dispersion relations associated with lateral interactions between adatoms.[364]

8.5 Desorption methods

8.5.1 Angular distribution of ions produced by electron-stimulated desorption (ESDIAD)

The electron-stimulated desorption (ESD) of adsorbed species is generally considered to be undesirable in the context of LEED experiments, and the primary electron current is usually kept low to minimize the effect. It is, however, possible to obtain valuable information concerning the chemisorptive bond by deliberately generating ESD and analysing the desorbed species. If the angular distribution of the desorbed ions is measured, this method, called ESIAD, can be used to identify specific adsorption sites.

The desorbed species may take any energetically feasible form—they may be molecular or atomic, neutral or ionic. If neutral, they may be in the ground state or an excited state; if ionic, they may be positively or negatively charged. Much information is thus available if ion yields, masses and energies, desorption cross-sections and molecular dissociation effects are all measured.

It is rather surprising that ESIAD has not received more attention since it is relatively simple to perform the experiments, and theories are reasonably well developed to assist in the interpretation of the results. ESDIAD provides a direct picture of the bond symmetry of the desorbing species, although care is needed before conclusions about the precise bonding sites may be drawn.

In principle, ESDIAD can be performed using a standard LEED system. Analysis of the desorbed ion masses may be carried out using a quadruple mass spectrometer. In fact, an experimental apparatus designed to enable both ESDIAD and LEED measurements to be performed has been constructed and used by Madey *et al.*[365]

Desorption of ionic species is achieved by a beam of electrons with energies in the region 100–440 eV. The simple form of detection system, devised by Madey (see Figure 8.8) involved the use of a hemispherical LEED grid to

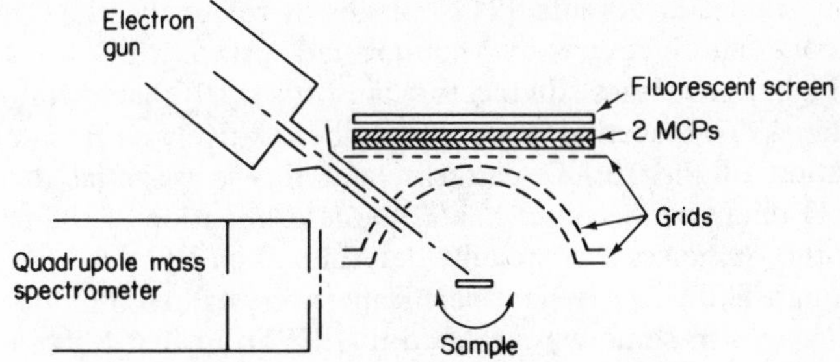

Figure 8.8. A simple experimental configration, using LEED grids and multichannel electron multipliers, to permit ESDIAD experiments to be carried out. (Reproduced by permission of North-Holland Physics Publishing from Madey *et al.*[365])

provide a field-free region into which the ions emerge, followed by a second, flat grid in front of two channel plates. The amplified signal is then displayed on a phosphor screen. It is possible, by simply changing the potentials on the grids, to switch between ESIAD and LEED to correlate observations from the two techniques. This configuration of grids and channel plates is not, however, particularly good for LEED, since distortions are produced by having only one hemispherical grid, and the fixed angle of the electron gun plus the small area of the channel plates results in a very restricted field of view. The relative simplicity of the system and direct visibility of both ESIAD and LEED patterns are definite advantages for qualitative experiments.

By contrast, the dedicated ESDIAD system of Niehus[366] separates the ESDIAD and LEED mechanisms within the same UHV chamber, using a movable Faraday cup and channeltron to detect the ions, with a computer display of the angular profiles. Time-of-flight techniques enable the energies of the emergent ions to be determined, and combine with the angular information to give very precise quantitative ESDIAD data. A standard LEED system may be used since it remains separate from the ion-detection system.

Because ESIAD uses the desorption behaviour of individual atoms from specific surface sites to produce its patterns, it is possible to obtain information about disordered or dilute adsorbates where LEED data may be blurred or uninformative. A most interesting example of this was given by studies of oxygen on W(111).[366] As the W(111) surface is exposed gradually to oxygen at room temperature a blurring of the clean surface LEED beams is observed, but no new features appear. Simultaneous ESDIAD experiments display sharp features at all stages, which moreover change their character substantially between low and high oxygen coverages. From this one may conclude that at low coverages the oxygen atoms sit in fourfold hollows, whilst at higher coverages they form a mixture of atop sites and pairs of atoms on either side of an atop site. When the surface is annealed, well-ordered structures may be observed with the LEED—the W(111) surface, which is rather unstable, facets into {211} planes at relatively low O coverages; at higher coverages it reverts to a non-faceted surface with (4 × 4) periodicity, and at higher coverages still facets again with {110} planes forming on {211} facet planes. The faceting processes can be relatively easily interpreted from observation of the LEED patterns, and this is essential to interpret the ESDIAD data; on the other hand, the determination of the actual bonding sites of the oxygen is most readily determined by ESDIAD.[367]

Although ESDIAD is not specifically a crystallographic technique, it is clear that it is in some ways related to LEED, in that it can be performed using the same electron-gun energy range, and can be used to provide very valuable complementary information concerning chemisorption phenomena.

8.5.2 Secondary ion mass spectrometry (SIMS)

If surface chemical and compositional information is of particularly important concern to the experimenter, then SIMS may be a more appropriate technique to use. SIMS is one step further removed from LEED than ESIAD in that an ion beam is used to stimulate the desorption processes. SIMS is a well-established method and various forms of experimental apparatus are commercially available, often referred to as ion microprobes. Although incident ion energies may be raised to 20 keV or more, surface sensitivity is enhanced by reducing the incident energy to a minimum—at 3 keV about three or four atomic layers may be sampled, at 1 keV only one or two layers. An advantage of SIMS is its ability to detect very low concentrations of surface species—less than 10^{-6} monolayer may be detected under favourable conditions.

In its standard form, SIMS provides a sensitive measure of coverages and surface chemical species, but cannot be regarded as a crystallographic tool. Recently, however, it has been shown that angular-resolved SIMS can be used to identify surface atom site symmetries, and may even be capable of determining bond lengths.[368] If this extension of SIMS is developed, this may provide yet another means of deriving crystallographic data.

8.6 Other methods providing surface crystallographic data

X-ray diffraction is the established basis for bulk crystallography, and is generally considered unsuitable for surface crystallography, due to the deep penetration depth associated with X-rays. Surface sensitivity may, however, be achieved using glancing incidence. This method is capable of providing quantitative measurements of lateral displacements of atoms in surface layers, and so would have specific value in aiding the examination of clean surface reconstructions. As an example, this method has been recently used to characterize the reconstructed (2×1) surface of Au (110).[369]

Field ion microscopy (FIM) is another well-established method for the study of surfaces, but is primarily used as a means of determining the short-range two-dimensional ordering of surface atoms, rather than as a method capable of providing quantitative three-dimensional surface structural data. It has the unique distinction of being able to resolve individual atom sites, and has been used to provide valuable information on the mutual interaction between atoms adsorbed on surfaces. Severe constraints upon this method include the fact that the sample must take the form of a very sharp tip, prepared by etching the end of a fine filament, and that the observation is made in the presence of a very strong electric field. Attempts to correlate FIM observations of the low-temperature $c(2\times2)$ reconstruction of W(001) with LEED and ion-scattering results were inconclusive, and the electric field effect could not be ignored as a possible cause of the discrepancies found.[370] Estimates of the third dimensional atomic spacing could, in principle, be made from the observation of field desorption from consecutive layers,[371] but this cannot be regarded as an accurate crystallographic method comparable to others discussed earlier in this chapter, not least because of the effects of intense electric fields on surface geometries.[372]

Another method capable of providing a high-resolution, real-space image of surfaces is scanning tunnelling microscopy (STM). This method was first introduced as recently as 1982 by Binnig *et al.*,[373] and has been subsequently used to provide a clear picture of the (7×7) reconstruction of Si(111).[374] The method involves scanning a metal tip maintained at a constant tunnel current over the sample surface. The tunnel current is maintained constant by adjusting the separation between the surface and the tip, and, as the tip scans the surface, the variation in tip height will reflect the topography of the surface. This method is particularly sensitive to the position of atoms possessing dangling bonds, although at this stage a precise microscopic tunnelling theory does not yet exist. Nevertheless, the Si(11) example cited illustrated clearly how such a novel method may provide very valuable information complementary to that obtained from LEED.

Low-energy current image diffraction (CID) is a new method[375] in which the specimen current is measured whilst a low-energy electron beam scans the surface, changing angle as it does so (see Figure 8.9). By displaying the absorbed current synchronously on a cathode ray tube, a pattern is produced which can be related to the total back-scattered LEED intensity as a function of energy and incident angle. The interpretation of such data can therefore be made directly using standard LEED computer programs, provided non-linear effects due to secondary emission processes are ignored. Whether such a method can provide crystallographic information comparable to that obtainable using more conventional methods remains to be seen. Even if such data could be obtained, it is not clear that this method would demonstrate any obvious advantages over standard LEED techniques. Closely related to this approach is the observation of diffraction effects in the angular distribution of secondary electrons. This latter method has been used in the study of Ag(111).[376]

As noted at the beginning of this chapter, the number of different surface sensitive techniques currently available is well over 100. The results of many

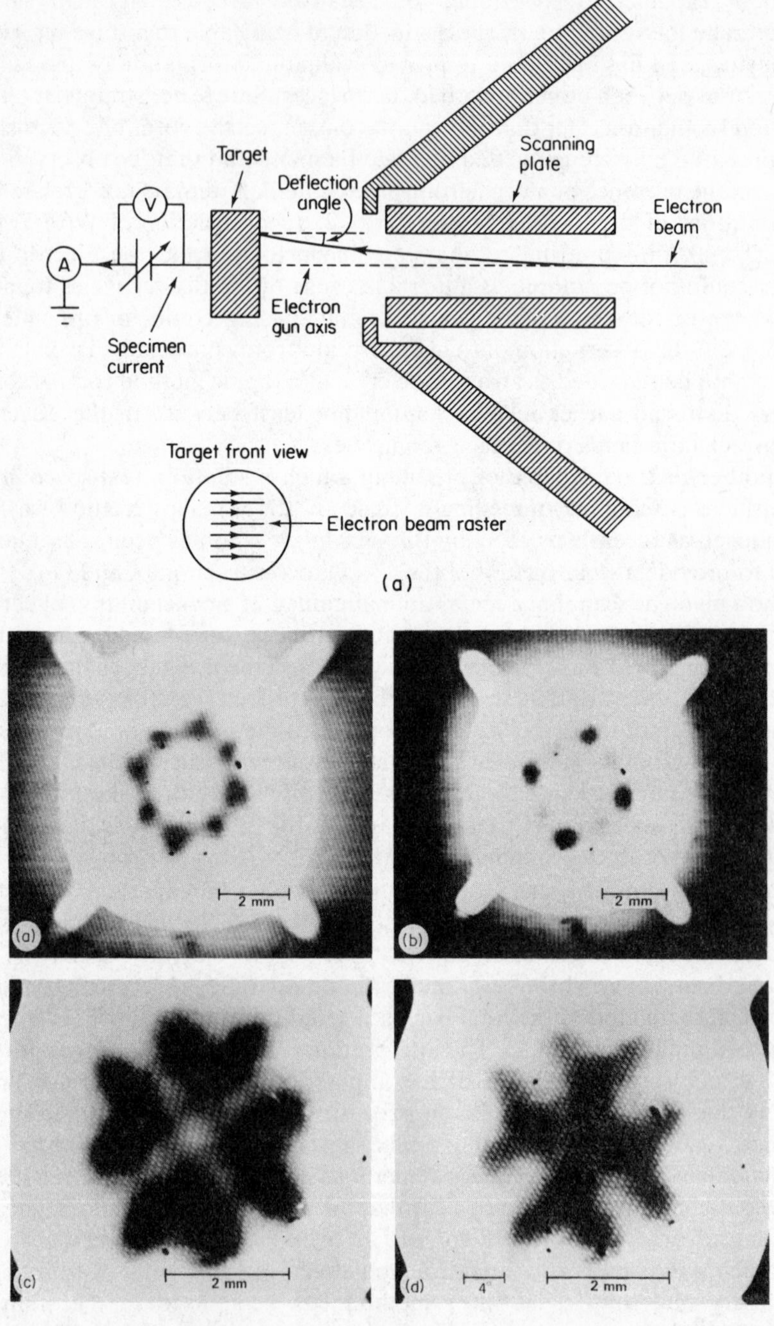

Figure 8.9. (a) Schematic diagram of an experimental apparatus suitable for carrying out CID (current image diffraction); (b) typical patterns observed. (Reproduced by permission of The American Physical Society from Nall *et al.*[375])

of these may be related in some way, directly or indirectly, to the results of specific crystallographic techniques. It is not possible to provide an exhaustive list of instances in which links have been sought between the results of various techniques, but it is hoped that the examples provided in this chapter offer a representative cross-section of the more important and closely related methods currently developed. In many cases, the nature of the investigation will itself determine the most appropriate methods to be used. If the problem is particularly complex—for example the (7×7) reconstruction of Si(111), a variety of different techniques may be appropriate, each making its own contribution to the complete picture.

8.7 Summary

There are many surface sensitive techniques currently in use, each with its own advantages and limitations. Some are capable of providing crystallographic data directly, others provide information from which crystallographic information may be deduced. Some are suited to the study of metal surfaces, others to semiconductor surfaces, others to adsorbed molecules, and so on. None, however, provide a fully comprehensive picture of the surface for all types of surface structure.

The study of surface atomic structures is at best a difficult process, limited by the ability to prepare and maintain reproducible surfaces that are not in turn altered or destroyed by the probe itself. Consequently, any opportunity to compare results obtained from independent techniques must be greeted as a valuable contribution to the ongoing process of learning that this work entails. LEED stands out as one of the most widely available and extensively researched of these crystallographic probes, and will undoubtedly continue to provide a major contribution to the study of surface structures in the foreseeable future. As alternative methods become established, comparisons will be made between results, and it will be important to recognize the strengths and weaknesses of each technique if our knowledge of surface structures is to expand reliably and unambiguously.

Questions

1. If the angle subtended by the LEED screen at the crystal is 120°, how many diffraction beams may be observed if the primary beam is incident normally (perpendicularly) on a Ni(001) surface up to 250 eV? Assume the inner potential is 10 V. How many equivalent beam sets can be observed? Calculate the energies of the Bragg peaks up to 250 eV. If an adsorbate forms a $p(2 \times 2)$ structure on the surfaces, how may equivalent beam sets will be visible at 100 eV?
2. Determine the shape of the diffraction pattern unit cell (the reciprocal lattice) produced by electron diffraction from the following surfaces: (i) bcc (2 1 0); (ii) hcp (1 0 $\bar{1}$ 2); (iii) fcc (2 1 1).
3. Describe the diffraction patterns given in Figure Q1 using either Wood's, or matrix, notation, as appropriate. Determine the real structure unit cell in each case.
4. Calculate the interlayer spacing normal to a: (i) bcc (111) face, for which $a = 2.87$ Å; (ii) fcc (110) face, for which $a = 3.62$ Å; (iii) hcp (1 1 $\bar{2}$ 0) face for which $a = 2.98$ Å and $c = 5.63$ Å; (iv) hcp (1 0 $\bar{1}$ 2) face for which $a = 2.98$ Å and $c = 4.45$ Å.
5. Suggest a surface structure that could produce the $c'(12 \times 8)$–$p2gg$ diffraction pattern given in Figure 1.44.
6. If the component of the earth's magnetic field normal to an electron beam trajectory is 0.4 gauss, and the distance of travel from the gun exit to the sample is 7 cm, how great will be the deflection off-axis if the electron energy is 100 eV? If the magnetic field is imperfectly cancelled, leaving a residual field of 0.05 gauss, what would be the deflection at 25 eV?
7. The maximum muffin-tin sphere radius is obtained when adjacent spheres are touching. Determine the ratio of the Wigner–Seitz volume to the maximum muffin-tin sphere volume for a (i) bcc (ii) fcc (iii) hcp crystal.
8. The Fe(12 1 0) surface has the form of (100) terraces with periodic step faces. In a LEED experiment, in which the primary beam incident with a polar angle 5° and azimuthal angle 0°, the (0, 0) beam was found to vary between a single spot at about 60 eV primary beam energy, and a split spot at about 85 eV. Are the step heights monoatomic or greater? Calculate the energies at which the (0, 0) beam will exhibit splitting with maximum angular disperson, in the range 50–300 eV.
9. The compound $BaTiO_3$ has a cubic perovskite structure. Parallel to an ideal (100) surface, this comprises alternative layers of TiO_2 And BaO

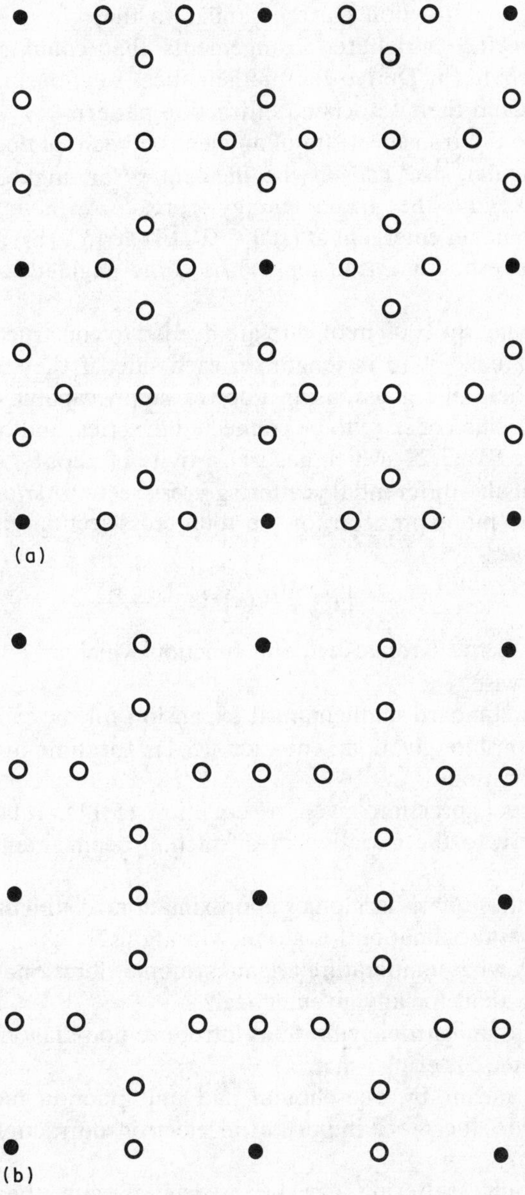

Figure Q1. Diffraction patterns formed by adsorbed species on certain square lattice surfaces: (●) substrate beams; (○) adsorbate-related beams (see Figure Q2(a)). It has been found that the resulting LEED patterns comprise the superposition of domains aligned at 90° to each other, each exhibiting a diffraction pattern similar to that given in Figure Q2(b). Suggest surface structure arrangements that could give rise to the observed pattern. Derive the mathematical relationship between these structures and their associated diffraction patterns.

10. Determine the transfer width of an electron beam of energy 200 eV and source angular size 1.5°, if the incident polar angle is zero (normal incidence) and the beam energy spread $\Delta E/E = 5 \times 10^{-3}$ for a diffraction beam emergent at (i) $\theta = 0°$, (ii) at 45°, (iii) at 60°. What will be the corresponding transfer widths if the angular size is reduced to 0.75°?
11. Show that nearly 1000 m of wire are needed to construct a pair of square Helmholtz coils, 1 m in length on each side, if they are to provide a magnetic field of 1 gauss, using a power supply capable of delivering 0.5 A at 25 V. The coils are to be connected in series, and constructed using wire gauge SWG 23, which has a resistivity of about 0.058 Ω/m.
12. Given that the differential scattering cross-section $d\sigma/d\Omega(k, \theta) = |f(k, \theta)|^2$ derive the expression for the total cross-section given by equation (3.34). *Note:*

$$\int_{-1}^{+1} P_l(x)\, P_{l'}(x)\, dx = \frac{2}{2l+1}\, \delta_{ll'}$$

where $\delta_{ll'}$ is the Kronecker delta function which = 1 when $l = l'$ and = 0 otherwise.
13. Using the standard mathematical expansion for an exponential, derive the relationship given in equation (5.4), for time-averaged periodic atomic vibrations.
14. Derive the expression given in equation (5.11), relating the Debye temperature to the variation of diffraction beam intensity with crystal temperature.
15. How is the quasi-harmonic approximation distinguished from the harmonic approximation for atomic vibrations?
16. At roughly what temperature are anharmonic vibrational effects likely to become evident for any given crystal?
17. Why should anisotropic vibrations introduce non-diagonal elements into the ion core scattering *t*-matrix?
18. What are meant by one-phonon and multiphonon modes? Why are they likely to be more important in electron diffraction than in X-ray diffraction?
19. Zero-phonon scattering can be associated with the attenuation of diffraction beams as described by the Debye–Waller factor. What is the effect of one-phonon and multiphonon scattering?
20. Calculate the energy shift in Bragg peaks between 100 and 200 eV for Cu(100), when warming a crystal from liquid helium temperature to room temperature.

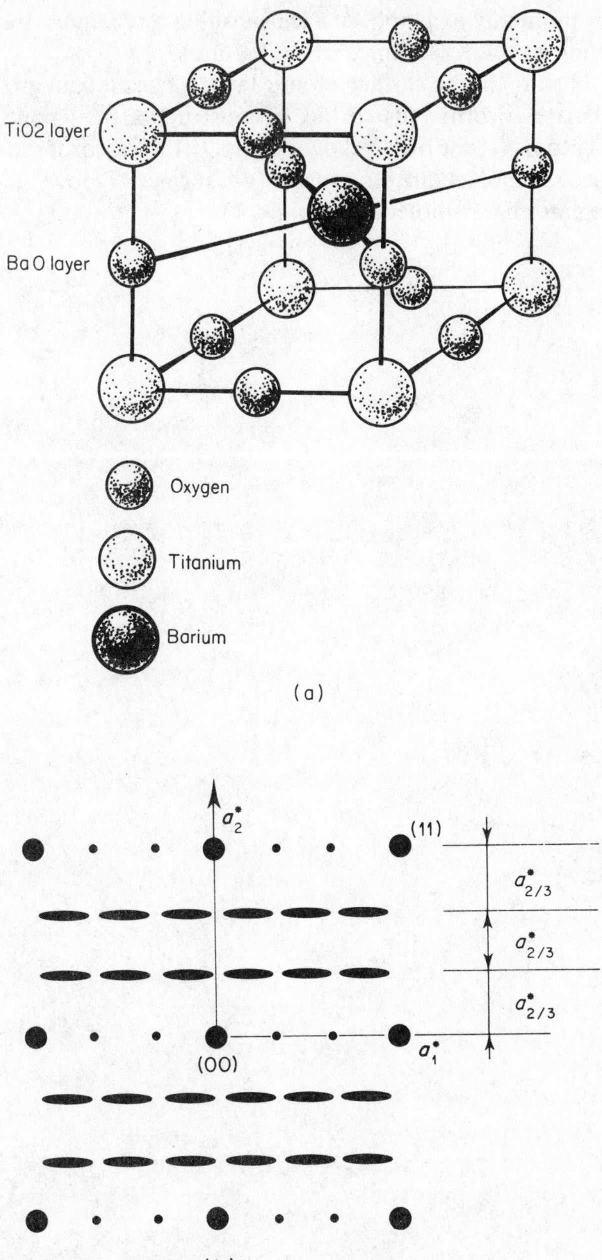

Figure Q2. (a) The cubic perovskite structure of $BaTiO_3$; (b) the $BaTiO_3$ $(3 \times 3n)$ diffraction pattern observed from the (001) face. (Reproduced by permission of North-Holland Physics Publishing from Aberdam, D., Bouchet, G. and Ducros, P. (1971) *Surface Sci.*, **27**, 559)

21. Which of the many available surface sensitive techniques are best suited to the study of the following surface features?
(a) Lateral shifts in the surface atomic layer; (b) epitaxial growth; (c) the surface barrier profile; (d) surface reconstruction; (e) bonding sites of adsorbed atoms (i) for ordered overlayers, (ii) for disordered overlayers; (f) adsorbed molecular structure; (g) surface Debye temperatures (including atomic or molecule vibrations).

Appendix 1

Element	a, c (Å)	Coefficient of thermal expansion (10^6 deg^{-1})	Debye temp. (K)	Crystal structure	Atomic weight	Work function (eV)
Be	2.283, 3.607	16.2	1200	hcp	9.0	4–6
C (diamond)	3.567	0.6	1860	diamond	12.0	5.0
(graphite)	2.461, 6.709	2.1		hex	12.0	5.0
Na	4.30	72.0	160	bcc	23.0	2.75
Mg	3.203, 5.202	27.5	405	hex	24.0	—
Al	4.043	23	420	fcc	27.0	4.28
Si	5.431	2.5	670	diamond	28.1	4.75
K	5.20 (−150 °C)	84	90	bcc	39.1	2.30
Ca	5.56	18.5	220	fcc	40.1	2.9
Ti	2.92, 4.67	9.2	430	hcp	47.9	4.53
V	3.04	6.15	380	fcc	51.0	4.3
α-Cr	2.87	6.25	585	bcc	52.0	4.5
Mn	8.913	27	450	bcc	54.9	4.1
α-Fe	2.866	15	467	bcc	55.9	4.5
α-Co	3.55	13.5	445	fcc	58.9	5.0
Ni	3.524	13.2	465	fcc	58.7	5.15
Cu	3.615	16.9	343	fcc	63.5	4.65
Zn	2.665, 4.946	30	305	hcp	65.4	3.63
Ga	4.51, 7.86	18.1	333	tetrag.	69.7	4.2
Ge	5.62	5.8	370	diamond	72.6	5.0
As	3.76	6.0	291	rhomb	74.9	—
Se	4.365, 4.957	37.9	89	hex–rh	79.0	—
Rb	5.62 (−76 °C)	90	55	bcc	85.4	2.2
Sr	6.05	21	129	fcc	87.6	2.59

Element	a, c (Å)	Coefficient of thermal expansion (10^6 deg^{-1})	Debye temp. (K)	Crystal structure	Atomic weight	Work function (eV)
Zr	3.23, 5.14	6.0	310	hcp	91.2	4.05
Nb	3.27	8.0	250	bcc	92.9	4.3
Mo	3.142	7.0	470	bcc	96.0	4.6
Ru	2.686, 4.272	9.9	600	hcp	101.1	4.71
Rh	3.820	6.6	480	fcc	102.9	4.98
Pd	3.859	9.5	275	fcc	106.4	5.12
Ag	4.086	18.6	225	fcc	107.9	4.26
Cd	2.98, 5.63	29	190	hex	112.4	4.08
Sn	5.831	21	190	tetrag.	118.7	4.42
Sb	4.310	9.2	204	rhomb	121.7	4.6
Te	4.45, 5.90	17.0	153	hex–rh	127.7	—
Cs	6.05 (−76 °C)	97.0	39.2	bcc	132.9	2.14
Ba	5.01	19	95	bcc	137.4	2.52
α-La	—	5.2	132	hex	138.9	3.5
Ta	3.272	7.5	245	fcc	181.0	4.25
W	3.155	5.5	405	bcc	183.9	4.55
Re	2.752, 4.448	6.7	450	hcp	186.2	4.72
Os	2.714, 4.32	6.1	500	hcp	190.2	5.93
Ir	3.823	6.5	420	fcc	192.2	5.6
Pt	3.923	9.5	233	fcc	195.1	5.65
Au	4.079	14.0	165	fcc	197.0	5.1
α-Tl	3.47, 5.52	28	89	hcp	204.4	—
Pb	4.950	28.3	95	fcc	207.2	4.25
Bi	4.546	13.3	117	rhomb	209.0	4.34

Sources: Handbook of the Physicochemical Properties of the Elements (1968), ed. G. V. Samsanov (IFI/Plenum, New York and Washington); *International Handbook of X-ray Crystallography* (1954), ed. K. Lonsdale (Kynoch Press, Birmingham, UK); 'Compilation of Work function measurements', by H. B. Michaelson, *J. Appl. Phys.*, 1977, **48**, 4729; *The Structure of Crystals* (1931), by R. W. G. Wyckoff (American Chemical Society, New York); *Lange's Hanbook of Chemistry* 12th edn (1979), ed. J. A. Dean (McGraw-Hill); 'Work function of metals' (1979), by J. Holzl and F. K. Schulte in *Solid State Physics*, Springer Tracts in Modern Physics, Vol. 85 (Springer-Verlag, Berlin, Heidelberg and New York)

General References

In a book of this kind, it has not been possible to cover comprehensively all aspects of surface studies. For additional background, or more extensive treatment of certain topics, the following books may be useful.

Chapter 1

An Atlas of Crystal Surfaces by J. F. Nicholas (1965, Gordon and Breach, London) provides a pictorial representation of many ideal single-crystal surfaces. An extensive review of current knowledge of semiconductor surface structures is given in Semiconductor surface structures', by A. Kahn, *Surface Science Reports,* 1983 **3**, (4/5).

Chapter 2

For much useful information on vacuum technology, the following books can be recommended: *Vacuum Manual*, by L. Holland, W. Steckelmacher and J. Yarwood (eds.) (1974, E. and F. N. Spon, London); *The Physical Basis of Ultrahigh Vacuum*, by P. A. Redhead, J. P. Hobson and E. V. Kornelses (1968, Chapman and Hall, London); *The Physical Principles of Ultrahigh Vacuum Systems and Equipment,* by N. W. Robinson, (1968, Chapman and Hall, London).

The following contain many details of electron-gun design: *Electron Beams, Lenses and Optics* (2 vols), by A. B. and J. C. J. El-Kareh (1970, Academic Press, London); *Electron Optics*, by P. Grivet, (1965, Pergamon, Oxford).

Chapter 3

There are many books on atomic structure and scattering theory, ranging from elementary to very advanced treatments. The following are just two examples, that cover aspects relevant to LEED: *Quantum Mechanics* (3rd edn.) by L. J. Schiff (McGraw-Hill, London and New York) provides a comprehensive text on quantum mechanics, leading into scattering theory; *Scattering Theory of Waves and Particles* by R. G. Newton (McGraw-Hill, London and New York) is an advanced text on scattering theory.

Chapter 4

Many reviews of LEED theory have been written, but few comprehensive texts. The following contain many of the basic mathematical and computational details of the more commonly used LEED programmes; *Low Energy Electron Diffraction*, by J. B. Pendry, (1974, Academic Press, London and New York) describes the mathematical basis of LEED theory using momentum representation, and provides essential background to the structure of the CAVLEED computer program package. In *Progress in Surface Science*, Vol. 7 (1975), S. Y. Tong provides a rigorous mathematical framework for LEED theory using either momentum representation of angular-momentum representation. *Surface Crystallography by LEED* by M. Van Hove, and S. Y. Tong (1979, Springer-Verlag, Berlin) is essentially a description of the combined-space method computer program package and its operation, including a complete program listing and test examples. However, a few minor errors have since been found in these programs, and it is advisable to contact the authors directly for a recent copy of the programs, or a list of corrections, rather than utilize the programs directly from this book.

Chapter 5

In Chapter 9 of *Surface Physics of Materials*, Vol II, J. M. Blakely (ed.) (1975, Academic Press, New York, San Francisco and London), M. G. Lagally provides a good review of surface vibrational behaviour, and certain aspects of its influence on LEED beams.

Chapter 6

In *Electron Spectroscopy for Surface Analysis*, Vol. 4 of 'Topics in Current Physics', (1977, Springer-Verlag, Berlin), M. Henzler describes in detail many aspects of the effects of surface defects on LEED beams. In *Physical and Chemical Properties of Stepped Surfaces* (Vol. 85 of Springer Tracts in Modern Physics, 1939, Springer-Verlag, Berlin), H. Wagner covers many aspects of the formation of surface steps and their effect on LEED beams.

Chapter 8

A number of surface analytical techniques are described in the following volumes: *Methods of Surface analysis, Vol 1: Methods and Phenomena*, A. W. Czanderna (ed.), (1975, Elsevier, New York); *Vibrational Spectroscopies for Adsorbed Species* A. T. Bell and M. L. Hair (eds.) (1980, American Chemical Soc., Washington, DC).

In addition the following provide a wide range of useful background material concerning the study of surface crystal structures: *Nature of the Surface Chemical Bond*, T. N. Rhodin and G. Ertl (eds.) (1979, North-Holland Amsterdam); *Chemistry in Two Dimensions*, by G. A. Somorjai (1981, Cornell University Press, Ithaca, New York).

References

1. Davisson, C. J., and Germer, L. H. (1927) *Phys. Rev.*, **30**, 705.
2. Thomson, G. P., and Reid, A. (1972) *Nature, Lond.*, **119**, 890.
3. Thomson, G. P. (1972) *Nature, Lond.*, **120**, 802.
4. Rivière, J. C. (1973) *Contemp. Physics*, **14**, 513.
5. Ehrenberg, W. (1934) *Phil. Mag.*, **18**, 878.
6. Auger, P. (1925) *J. Phys. Radium*, **6**, 205.
7. Lander, J. J. (1953) *Phys. Rev.*, **91**, 1382.
8. Harris, L. A. (1968) *J. Appl. Phy.*, **39**, 1419, 1428.
9. Weber, R. E., and Peria, W. T. (1967) *J. Appl. Phys.*, **38**, 4355.
10. Palmberg, P. W. (1968) *Appl. Phys. Lett.*, **13**, 183.
11. Clarke, L. J. (1981) *Surface Sci.*, **102**, 331.
12. Pantel, R., Bujor, M., and Bardolle, J. (1979) *Surface Sci.*, **83**, 228.
13. Shaw, C. G., Fain, S. C. and Chinn M. D. (1978) *Phys. Rev. Lett.*, **41**, 955.
14. Rundgren, J., and Salwen, A. (1976) *J. Phys. C.*, **9**, 3701.
15. Clarke, L. J. (1980) *Surface Sci.*, **91**, 131.
16. (a) Van Hove, M. A. and Tong, S. Y. (1979) *Surface Crystallography by LEED* (Springer-Verlag, Berlin); (b) Tong, S. Y. and Van Hove, M. A. (1977) *Phys. Rev.*, **B16**, 1459.
17. Phillips, F. C. (1946) *An Introduction to Crystallography* (Longman, Green, London, New York and Toronto).
18. Nicholas, J. F. (1965) *An Atlas of Crystal Surfaces* (Gordon and Breach, London).
19. Andersson, S., Pendry, J. B. and Echenique, P. M. (1977) *Surface Sci.*, **65**, 539.
20. (a) Gafner, G. and Feder, R. (1976) *Surface Sci.*, **57**, 37; (b) Shih, H. D., Jona, F., Jepsen, D. W., and Marcus, P. M. (1977) *Bull. Am. Phys. Soc.*, **22**, 357.
21. Morales de la Garza, L., and Clarke, L. J. (1981) *J. Phys. C.*, **14**, 5391.
22. Van Hove, M. A. and Tong, S. Y. (1976) *Surface Sci.*, **54**, 91.
23. Lee, B. W., Ignatiev, A., Tong, S. Y., and Van Hove, M. A. (1977) *J. Vac. Sci. Technol.*, **14**, 291.
24. Feder, R. (1973) *Phys. Stat. Solidi*, **58**, K137.
25. Legg, K. O., Jona, F., Jepsen, D. W., and Marcus, P. M. (1977) *J. Phys. C.*, **10**, 937.
26. Van Hove, M. A., and Tong, S. Y. (1976) *Surface Sci.*, **54**, 91.
27. Lee, B. W., Ignatiev, A., Tong, S. Y., and Van Hove, M. A. (1977) *J. Vac. Sci. Technol.*, **14**, 291.
28. Debe, M. K., King, D. A., and Marsh, F. S. (1977) *Surface Sci.*, **68**, 437.
29. Kirschner, J., and Feder, R. (1979) *Surface Sci.*, **79**, 176.
30. Read, M. N., and Russell, G. J. (1979) *Surface Sci.*, **88**, 95.
31. Heilmann, P., Heinz, K., and Muller, K. (1979) *Surface Sci.*, **89**, 84.
32. Clarke, L. J., and Morales de la Garza, L. (1980) *Surface Sci.*, **99**, 419.
33. Ignatiev, A., Jona, F., Shih, H. D., Jepsen, D. W., and Marcus, P. M. (1975) *Phys. Rev.*, **B11**, 4787.
34. Van Hove, M. A. (1979) in *Nature of the Surface Chemical Bond*, eds Rhodin, T. N., and Ertl, G. (North-Holland, Amsterdam).
35. Bridge, M. E., Comrie, C. M., and Lambert, R. M. (1977) *Surface Sci.*, **67**, 393.

36. Frost, D.C., Hengrasmee, S., Mitchell, K. A. R., Shepherd, F. R., and Watson, P. R. (1978) *Surface Sci.,* **76**, L585.
37. Gauthier, Y., Baudoing, R., and Clarke, L. J. (1982) *J. Phys. C.,* **15**, 3231.
38. (a) Laramore, G. E., and Duke, C. B. (1972) *Phys. Rev.,* **B5**, 267; (b) Martin, M. R., and Somorjai, G. A. (1973) *Phys. Rev.,* **B7**, 3607.
39. Fukuda, Y., Lancaster, G. M., Honda, F., and Rabalais, J. W. (1978) *Phys. Rev.,* **B18**, 6191.
40. Prior, K A., Schwaha, K., and Lambert, R. M. (1978) *Surface Sci.,* **77**, 193.
41. Lurie, P. G. and Wilson, J. M. (1977) *Surface Sci;* **65**, 453.
42. Lander, J. J., Gobeli, G. W., and Morrison, J. (1963) *J. Appl. Phys.,* **34**, 2298.
43. Haneman, D. (1961) *Phys. Rev.,* **121**, 1093.
44. (a) Jona, F., Shih, H. D., Jepsen, D. W., and Marcus, P. M. (1979) *J. Phys.* **C12**, L455; (b) Cardillo, M. J., and Becker, G. E. (1980) *Phys. Rev.,* **B21**, 1497.
45. Kahn, A. (1983) *Surface Sci. Reports,* **3**, No. 5.
46. Laramore, G. E. and Switendick, A. C. (1973) *Phys. Rev.,* **B7**, 3615.
47. Kinniburgh, C. (1975) *J. Phys. C.,* **8**, 2382 and 2696.
48. Kinniburgh, C. G. and Walker, J. A. (1977) *Surface Sci.,* **63**, 274.
49. Walker, J. A., Kinniburgh, C. G., and Matthew, J. A. D. (1977) *Surface Sci.,* **68**, 221.
50. Felton, R. P., Prutton, M., Matthew, J A. D., and Zinn, W. (1979) *Surface Sci.,* **79**, 117.
51. MacRae, A. U. (1966) *Surface Sci.*, **4**, 247.
52. Grant, J. T., and Haas, T. W. (1970) *J. Vac. Sci. Technol.*, **7**, 77.
53. Duke, C. B., Meyer, R. J., Paton, A., Yeh, J. L., Tsang, J. C., Kahn, A., and Mark, P. (1980) *J. Vac. Sci. Technol.,* **17**, 501.
54. Masud, N. (1982) *J. Phys. C.,* **15**, 3209.
55. Henrich, V. E., Zeiger, H. J. Solomon, E. I., and Gay, R. R. (1978) *Surface Sci.,* **74**, 682.
56. Wood, E. A. (1964) *J. Appl. Phys.,* **35**, 1306.
57. Battacharya, A. K., Clarke, I. J., and Morales de la Garza, L. (1980) *Indian J. Chem.,* **A19**, 680.
58. Peralta, L., Berthier, Y. and Oudar, J. (1976) *Surface Sci.,* **55**, 199.
59. Clarke, L. J. and Morales de la Garza, L. (1982) *Surface Sci.,* **12**, 32.
60. Wulff, G. (1901) *Z. Krist.,* **34**, 449.
61. Herring, G. (1951) *Phys. Rev.,* **82**, 87.
62. Gjostein, N. A. (1963) *Acta Met.*, **11**, 957; Gjostein, N. A. (1963) *Acta. Met.*, **11**, 969.
63. Baron, Blakely, J. M. and Somorjai, G. A. (1973) *Surface Sci*., **41**, 45.
64. Ferrante, J., and Barton, G. C. (1968) *NASA Tech. Note D-4735.*
65. Lambert, R. M., Linnett, J. W., and Schwarz, J. A. (1971) *Surface Sci.,* **26**, 572.
66. Clarke, L. J. (1979) Ph.D. thesis, Cambridge, UK.
67. Gruber, E. E., and Mollins, W. W. (1967) *J. Phys. Chem. Solids (GB)* **28**, 875.
68. Blakely, J. M., and Schoebel, R. L. (1971) *Surface Sci.,* **26**, 321.
69. Wynblatt, P. (1972) in *Interatomic Potentials and Simulation of Lattice Defects,* eds. Gehlen, Beeler and Jaffee, p. 633 (Plenum Press, New York).
70. Lang, B., Joyner, R. W., and Somorjai, G. A. (1972) *Surface Sci.,* **30**, 440.
71. Wagner, H. (1979) in *Physical and Chemical Properties of Stepped Surfaces,* Springer Tracts in Modern Physics, Vol. 85 (Springer-Verlag, Berlin).
72. Castner, D. G., and Somorjai, G. A. (1979) *Chemical Rev.,* **79**(3), 233.
73. Van Hove, M. A., and Somorjai, G. A. (1980) *Surface Sci.,* **92**, 489.
74. Marcus, P. M., and Jona, F. (1972) *Surface Sci.,* **31**, 355.
75. Rundgren, J. and Salwen, A. (1974) *Computer Phys. Commun.,* **7**, 369; **9**, 312.
76. *Internation Tables for X-Ray Crystallography,* ed. K. Lonsdale (Kynoch Press, Birmingham).
77. Onuferko, J. H., and Woodruff, D. P. (1979) *Surface Sci.,* **87**, 357.

78. Debe, M. K., and King, D. A. (1977) *Phys. Rev. Lett.,* **39**, 708; (1979) *Surface Sci.,* **81**, 193.
79. Tung, R. T., and Graham, W. R. (1982) *Surface Sci.,* **115**, 576.
80. Dahlgren, D., and Hemminger, J. C. (1981) *Surface Sci.,* **109**, L513.
81. Park, R. L., and Farnsworth, H. E. (1964) *Rev. Sci. Intstrum.,* **35**, 1592.
82. Berndt, W. (1980) *Verhandl DPG (VI)*, **15**, 718.
83. Stair, P. C., Kaminska, T. J., Kesmodel, L. L., and Somorjai, G. A. (1975) *Phys. Rev.,* **B11**, 623.
84. Moss, H. (1968) Narrow angle electron guns and cathode ray tubes', in *Advances in Electronics and Electron Physics*, ed. L. Marton, Supplement 3 (Academic Press, London).
85. DeBersuder, L. (1974) *Rev. Sci. Instrum.,* **45**, 1569.
86. Chutjian, A. (1979) *Rev. Sci. Instrum.,* **50**, 347.
87. Cowell, P. G. (1982) *J. Phys. E (GB),* **15**, 994.
88. Drahos, V., Delong, A., Kolarik, V., and Lenc, M. (1973) *J. Microscopie,* **18**, 135.
89. O'Neill, M. R., and Dunning, F. B. (1974) *Rev. Sci. Instrum;* **45**, 1611.
90. Heilmann, P., Lang, E., Heinz, R., and Muller, K. (1976) *Appl. Phys.,* **9**, 247.
91. Heilmann, P., Lang, E., Heinz, R. and Muller, K. (1980) in *Proc. Conf. on Determination of Surface Structures by LEED* (IBM Conf. Series) to be published.
92. Welkie, D. G., and Lagally, M. G. (1979) *Appl. Surf. Sci.,* **3**, 272.
93. Weeks, S. P., Rowe, J. E., Christman, S. B. and Chaban, E. E. (1979) *Rev. Sci. Instrum,* **50**, 1249.
94. Fujiwara, K., Hayakawa, K., and Miyake, S. (1966) *Japan J. App. Phys.*, **5**, 295.
95. Tucker, C. W. Jr (1964) *J. Appl. Phys.,* **35**, 1897.
96. Scheibner, E. J., Germer, L. H., and Hartman, C. D. (1960) *Rev. Sci. Instrum,* **32**, 112.
97. Delong, A., and Drahos, V. (1971) *Nature Phys. Sci.,* **230**, 196.
98. Laydevant L., Guittard, C., and Bernard, R. (1972) *Proceedings of Fifth European Congress on Electron Microscopy,* 662 (Inst. of Phys., Manchester).
99. Berger, C., Dupuy, J. C., Laydevant, L., and Bernard, R. (1977) *J. Appl. Phys.,* **48** 5027.
100. Clarke, L. J. (1981) *Phil. Mag.,* **A43,** 779.
101. Dietz, R. E., McRae, E. G., and Cambell, R. L. (1980) *Phys Rev. Lett.,* **45**, 1280.
102. Taylor, N. J. (1969) *Rev. Sci. Instrum.*, **40**, 792.
103. Price, G. L. (1980) *Rev. Sci. Instrum.,* **51**, 605.
104. Wendelken, J. F. and Propst, F. M. (1976) *Rev. Sci. Instrum.,* **47**, 1069.
105. Pendry, J. B. (1974) *Low Energy Electron Diffraction*, pp. 5–6 (Academic Press, London).
106. Park, R. L., Houston, J. E., and Schreiner, D. G. (1971) *Rev. Sci. Instrum.,* **42**, 60.
107. Comsa, G. (1979) *Surface Sci.,* **81**, 57.
108. Wang, G. C., and Lagally, M. G. (1979) *Surface Sci.,* **81**, 69.
109. Lu, T. M., and Lagally, M. G. (1980) *Surface Sci.,* **99**, 695.
110. Heinz, K. (1980) *Surface Sci.,* **99**, 440.
111. Firester, A. H. (1966) *Rev. Sci. Instrum.,* **37**, 1264.
112. Taylor, N. J. (1964) *Varian Products Division Bulletin VR-29*.
113. Cunningham, S. L., and Weinberg, W. H. (1978) *Rev. Sci. Instrum.,* **49**, 752.
114. Sobrero, A. C., and Weinberg, W. H. (1982) *Rev. Sci. Instrum.,* **53**, 1566.
115. Battacharya, A. K., Clarke, L. J., and Morales de la Garza, M. (1981) *J. Chem. Soc. Faraday Trans. I,* **77**, 2223.
116. Bénard, J. and Laurent, J.-F (1965) *J. Chim. Physique,* **53**, 593.
117. Perdereau, M. (1971) *Surface Sci.,* **24**, 239.

118. Berthier, Y., Oudar, J., and Huber, M. (1977) *Surface Sci.,* **65**, 361.
119. Matthews, J. D. (1971) *Surface Sci.,* **24**, 248.
120. Davisson, C. J., and Germer, L. H. (1929) *Phys. Rev.,* **33**, 760.
121. Joffe, A. E., and Arsenieva, A. N. (1929) *C. R. Acad. Sci. Paris*, **188**, 152.
122. Myers, F. E. and Cox, R. T. (1929) *Phys. Rev.,* **34**, 106.
123. Wolf, F. (1928) *Z. Phys.,* **52**, 314.
124. Mott, N. F. (1929) *Proc. Roy. Soc.*, **A124**, 425.
125. Mott, N. F. (1932) *Proc. Roy. Sci.,* **A135**, 429.
126. Shull, C. G., Chase, C. T., and Myers, F. E. (1943) *Phys. Rev.,* **63**, 29.
127. See Van Klinken, J. (1966) *Nucl. Phys.*, **75**, 161 and Kessler, J. (1976) in *Polarized Electrons* (Springer-Verlag, Berlin).
128. Kuyatt, C. E. (1975) *Phys. Rev.,* **B12**, 4851.
129. Feder, R. (1977) *Phys. Rev.,* **B15,** 1751.
130. Kirschner, J., and Feder, R. (1979) *Phys. Rev. Lett.,* **42**, 1008.
131. Palmberg, P. W., De Wames, R. E., and Vredevoe, L. A. (1968) *Phys. Rev. Lett.,* **21**, 682.
132. Bas, E. B., Banninger, U., and Muhlethaler, H. (1974) *Jap. J. Appl. Phys. suppl.* 2, *pt*2, 197.
133. Busch, G., Campagna, M., Cotti, P., and Siegmann, H. C. (1969) *Phys. Rev. Lett.,* **22**, 597.
134. Pierce, D. T., Meier, F., and Zurcher, P. (1975) *Phys. Rev. Lett.*, **51A**, 465; (1975) *Appl. Phys. Lett.*, **26**, 671.
135. O'Neill, M. R., Kalisvaart, M., Dunning, F. B., and Walters, G. K. (1975) *Phys. Rev. Lett.,* **34**, 1167.
136. Muller, N., Wolf, D., and Feder, R. (1978). *Inst. Phys. Conf. Series,* **41**, 281.
137. Aberdam, D., Baudoing, R., and Gaubert, C. (1979) in *Handbook of Surfaces and Inerfaces,* Vol. 2, ed. Dobrzynski, (Garland SLTPM Press, New York and London).
138. Feder, R., and Kirschner, J. (1981) *Surface Sci.,* **103**, 75.
139. Pierce, D. T., Cellotta, R. J., Wang, G.-C., Unertl, W. N., Galejs, A., Kuyatt, C. E., and Mielczarek, S. R. (1980) *Rev. Sci. Instrum.,* **51**, 478.
140. Herman, F. and Skillman, S. (1963) *Atomic Structure Calculations* (Prentice-Hll, Hemel Hempstead).
141. Clementi, E. (1965) *IBM Journal of Research and Development,* **9**, 2.
142. Clementi, E. and Roetti, C. (1974) *Atomic Data and Nuclear Data Tables*, **14**, 177.
143. Loucks, T. (1967) *Augmented Plane Wave Method* (Benjamin, New York).
144. Mattheiss, L. F. (1964) *Phys. Rev.,* **134**, 970.
145. Andersson, S., and Pendry, J. B. (1073) *J. Phys. C.,* **6**, 601.
146. Dirac, P. W. M. (1930) *Proc. Camb. Phil. Soc.,* **26**, 676.
147. Slater, J. C. (1951) *Phys. Rev.,* **81**, 385.
148. Gaspar, R. (1954) *Act. Phys. Acad. Sci. Hung.,* **3**, 1263.
149. Kohn, W. and Sham, L. J. (1965) *Phys. Rev.,* **140A**, 1133.
150. Liberman, D. A. (1968) *Phys. Rev.,* **171**, 1.
151. Slater, J. C., Wilson, T. M., and Wood, J. H. (1969) *Phys. Rev.,* **179**, 28.
152. (a) Schwarz, K. (1972) *Phys. Rev.,* **B5**, 2466; (b) Schwarz, K. (1974) *Theor. Chim. Acta (Berlin)*, **34**, 225.
153. Echenique, P. M. (1976) *J. Phys. C.,* **9**, 3193.
154. Hedin, L. and Lundqvist, B. I. (1971) *J. Phys. C.,* **4**, 2064.
155. Echenique, P. M. and Titterington, D. J. (1977) *J. Phys. C.,* **10**, 625.
156. Jennings, P. J. (1971) *Surface Sci.,* **25**, 531.
157. Jennings, P. J. and Read, M. N. (1975) *J. Phys. C.,* **8**, L285.
158. Feder, R. (1981) in *Surface Structures by LEED,* eds. Marcus, P. M. and Jona, F. (Plenu Press, New York).

159. Pendry, J. B. (1978) in 'Proc. 50th Anniversary of LEED'; *Inst. Phys. Conf. Ser.*, **41**, 205.
160. Titterington, D. J., private communication.
161. Schiff, L. J. (1968) *Quantum Mechanics*, 3rd edn. (McGraw-Hill, New York and London).
162. Levinson, N. (1949) *Kgl. Danske Videnskab. Selskab. Mat.-fys. Medd.*, **25**, 9.
163. Calogero, F. (1967) *Variable Phase Approach to Potential Scattering* (Academic Press, New York).
164. Holzl, J. and Schulte, F. W. (1979) 'The work function of metals' in *Solid Surface Physics*, Springer Tracts in Modern Physics, Vol. 85 (Springer-Verlag Berlin).
165. Jennings, P. J., and Thurgate, S. M. (1981) *Surface Sci.*, **104**, L210.
166. (a) Andersson, S., Kasemo, B., Pendry, J. B., and Van Hove, M. A (1973) *Phys. Rev. Lett.*, **31**, 595; (b) Tong, S. Y. Kesmodel, L., and Rhodin, T. N. (1971) *Phys. Rev. Lett.*, **26**, 711.
167. (a) Inkson, J. C. (1971) *Surface Sci.*, **28**, 69; (b) Price, G. L. Jennings, P. J., Best, P. E., and Cornish, J. C. L. (1979) *Surface Sci.*, **89**, 151.
168. Clarke, L. J., Baudoing, R. and Gauthier, Y. (1982) *J. Phys. C.*, **15**, 3249.
169. Titterington, D. J. and Kinniburgh, C. G. (1980) *Computer Phys. Commun.*, **20**, 237.
170. Beeby, J. L. (1968) *J. Phys. C.*, **1**, 82.
171. McRae, E. G. (1968) *Surface Sci.*, **11**, 479.
172. Jepsen, D. W. Marcus, P. M., and Jona, F. (1971) *Phys. Rev. Lett.*, **26**, 1365.
173. Tong, S. Y., and Rhodin, T. N. (1971) *Phys. Rev. Lett.*, **26**, 711.
174. Jona, F. (1970) *IBM J. Res. Develop.*, **14**, 444.
175. Kambe, K. (1967) *Z. Naturf.*, **22a**, 322, 433; (1968) **23a**, 1280.
176. Stoner, N., Van Hove, M. A., and Tong, S. Y (1977) in *Characterization of Metal and Polymer Surfaces*, ed. Lieng-Huang Lee (Academic Press, New York, San Francisco and London).
177. Ignatiev, A., Pendry, J. B., and Rhodin, T. N. (1971) *Phys. Rev. Lett.*, **26**, 189.
178. McRae, E. G. (1968) *Surface Sci.*, **11**, 492.
179. Strozier, J. A. Jr., and Jones, R. O. (1970) *Phys. Rev. Lett.*, **23**, 1163; (1970) **25**, 516.
180. Tong, S. Y. (1975) *Progress in Surface Sci.*, **7**, 1.
181. Jennings, P. J. (1974) *Surface Sci.*, **41**, 67.
182. Groupe d'Etude des Surfaces (1975) *Surface Sci.*, **48**, 509.
183. Pendry, J. B. (1971) *Phys. Rev. Lett.*, **27**, 856.
184. Zimmer, R. S. and Holland, B. W. (1975) *Surface Sci.*, **47**, 717.
185. Zimmer, R. S. and Holland, B. W. (1975) *J. Phys., C.*, **8**, 2395.
186. Adams, D. L. (1981) *J. Phys. C.*, **14**, 789.
187. Gauthier, Y., Baudoing, R., and Clarke, L. J., to be published.
188. Pendry, J. B., and Gard, P. 1975) *J. Phys., C.*, **8**, 2048.
189. Masud, N. and Pendry, J. B. (1976) *J. Phys. C.*, **10**, 1.
190. Masud, N., Kinniburgh, C. G., and Pendry, J. B. (1979) *J. Phys. C.*, **12**, 5263.
191. Prutton, M., Walker, J. A., Welton-Cooke, M. R., and Felton, R. C. (1979) *Surface Sci.*, **89**, 95.
192. Yang, W. S., and Jona, F. (1981) *Surface Sci.*, **109**, L505.
193. Stoner, N., Van Hove, M. A., Tong, S. Y., and Webb, M. B. (1978) *Phys. Rev. Lett.*, **40**, 243.
194. Van Hove, M. A., Koestner, R. J., Stair, P. C., Biberian, J. P., Kesmodel, L. L., Bartos, I., and Somorjai, G. A. (1981) *Surface Sci.*, **103**, 218.
195. Shaw, C. G., Fain, S.C. Jr., Chinn, M. D., and Toney, M. F. (1980) *Surface Sci.*, **97**, 128.
196. Levine, J. D., McFarlane, S. H., and Mark, P. (1977) *Phys. Rev.*, **B16**, 5415.

197. Feder, R. (1979) *Solid State Commun.*, **31**, 821.
198. Celotta, R. J., Pierce, D. T., Wang, G.-C., Bader, S. D., and Felcher, G. P. (1979) *Phys. Rev. Lett.*, **43**, 728.
199. Feder, R. and Moritz, W. (1978) *Surface Sci.*, **77**, 505.
200. Van Hove, M. A., and Somorjai, G. A. (1982) *Surface Sci.*, **114**, 171.
201. Pendry, J. B. (1981) in *Surface Structures by LEED,* eds. Marcus, P. M. and Jona, F. (Plenum Press, New York).
202. Andesson, S., and Pendry, J. B. (1980) *J. Phys. C.*, **13**, 3547.
203. Jagodzinski, H., Mortiz, W., and Wolf, D. (1978) *Surface Sci.*, **77**, 233, 249, 265, 283.
204. Bennett, P. A., and Webb, M. W. (1981) *Surface Sci.*, **104**, 74.
205. Jona, F. (1967) *Surface Sci.*, **8**, 478.
206. Hoffstein, V. (1974) *Computer Phys. Commun.*, **7**, 50.
207. Rundgren, J., and Salwen, A. (1975) *Computer Phys. Commun.*, **9**, 312.
208. Jepsen, D. W. (1981) *Phys. Rev.*, **B22**, 5701.
209. Shih, H.-D. (1981) in *Surface Structures by LEED* eds. Marcus P. M. and Jona, F. (Plenum Press, New York).
210. Ott, H. (1935) *Ann. d. Physik,* **23**, 169.
211. James, R. W. (1962) *The Optical Principles of the Diffraction of X-rays* (Bell, London).
212. Wallis, R. F., Clark, B. C., and Herman, R. (1969) in *The Physics and Chemistry of Solid Surfaces*, ed. G. A. Somorjai) **17**, 1 (Wiley, New York).
213. Dennis, R. L., and Webb, M. B. (1973) *J. Vac. Sci. Technol.*, **10**, 192.
214. Theeten, J. B., Domange, J. L., and Herault, J. P. (1973) *Surface Sci.*, **35**, 145.
215. Wilson, J. M. and Bastow, T. J. (1971) *Surface Sci.*, **26**, 461.
216. Gelatt, C. D., Lagally, M. G., and Webb, M. B. (1969) *Bull. Am. Phys. Soc.*, **14**, 793.
217. Woodruff, D. P., and Seah, M. P. (1970) *Phys. Stat.Solidi (a)*, **1**, 429.
218. Unertl, W., and Webb, M. B. (1973) *J. Vac. Sci. Technol.*, **11**, 193.
219. Dobryzinski, L. and Marududin, A. A. (1973) *Phys. Rev.*, **137**, 1267.
220. Kenner, V. E., and Allen, R. E. (1973) *Phys. Rev.*, **138**, 2916.
221. (a) Allen, R. E. and deWette, F. W. (1969) *Phys. Rev.*, **179**, 873; (b) Allen, R. E., DeWette, F. W., and Rahman, A. (1969) *Phys. Rev.*, **179**, 887.
222. Clarke, B. C., Herman, R., and Wallis, R. F. (1965) *Phys. Rev.*, **A139**, 860.
223. Allen, R. E. and deWette, F. W. (1969) *Phys. Rev.*, **188**, 1320.
224. Tong, S. Y., Rhodin, T. N., and Ignatiev, A. (1973) *Phys. Rev.*, **B8**, 906.
225. Wallis, R. F., Clark, B. C., and Herman, R. (1968) *Phys. Rev.*, **167**, 625.
226. Vail, J. (1967) *Can. J. Phys.*, **45**, 2661.
227. Bortolani, V., Nizzoli, F., and Santoro, G. (1978) *Inst. Phys. Conf. Series,* **39**, 326.
228. Legg, K. O. Jona, F., Jepsen, D. W., and Marcus, P. M. (1977) *J. Phys. C. Solid State Phys.*, **10**, 937.
229. Tabor, D., and Wilson, J. M. (1971) *J. Vac. Sci. Technol.*, **9**, 695.
230. Heilmann, P., Heinz, K., and Müller, K. (1979) *Surface Sci.*, **89**, 85.
231. Debe, M., King, D., and Marsh, F. (1977) *J. Phys.*, **C10**, L303.
232. Lindemann, F. A. (1910) *Phys. Z.*, **11**, 609.
233. Chatterjee, B. (1978) *Nature*, **275**, 203.
234. Wilson, J. M. (1973) Ph.D. thesis, Cambridge.
235. Aberdam, D. (1978) *Inst. Phys. Conf. Ser.*, **41**, 239.
236. Duke, C. B. and Laramore, G. E. (1970) *Phys. Rev.*, **B2**, 4783.
237. Holland, B. W. (1971) *Surface Sci.*, **28**, 258.
238. McKinney, J. T., Jones, E. R., and Webb, M. B. (1967) *Phys. Rev.*, **160**, 523.
239. Jones, E. R., McKinney, J. T., and Webb, M. B. (1966) *Phys. Rev.*, **151**, 476.
240. Wallis, R. F. (1964) *Surface Sci.*, **2**, 146.

241. MacRae, A. U. (1964) *Surface Sci.*, **2**, 522.
242. Mroz, S., and Mroz, A. (1981) *Surface Sci.*, **109**, 444.
243. Morabito, J. M., Steiger, R. F., and Somorjai, G. A. (1969) *Phys. Rev.*, **179**, 638.
244. Ulehla, M. and Davis, H. L. (1978) *J. Vac. Sci. Technol.*, **15**, 642.
245. Theeten, J. B., Dobrzynski, L. and Domange, J. L. (1973) *Surface Sci.*, **34**, 145.
246. Hjelmberg, H. (1979) *Surface Sci.*, **81**, 539.
247. Richardson, N. V., and Bradshaw, A. M. (1980) *Surface Sci.*, **88**, 255.
248. Andersson, S. (1977) *Solid State Commun.*, **21**, 75.
249. Cyvin, S. J. (1968) *Molecular Vibrations and Mean Square Vibrational Amplitudes* (Elsevier, Amsterdam).
250. Bader, S. D. (1980) *Surface Sci.*, **99**, 392.
251. Van Hove, M. H. and Tong, S. Y. (1979) *Surface Crystallography by LEED* (Springer-Verlag, Berlin).
252. Rieder, K. H., and Wilsch, H. (1983) *Surface Sci.*, **131**, 245.
253. Levi, A. C., and Suhl, H. G. (1979) *Surface Sci.*, **88**, 221.
254. Meyer, H. D. (1981) *Surface Sci.*, **104**, 117.
255. Mattera, L. Rocca, M., Salvo, C., Terrini, S., Tommasini, F., and Valbusa, U. (1983) *Surface Sci.*, **124**, 571.
256. Davies, J. A., Jackson, D. P., Mitchell, J. B., Norton, P. R., and Tapping, R. L. (1976) *Nucl. Instrum. Meth.*, **132**, 609.
257. Kesmodel, L. L., and Somorjai, G. A. (1977) *Surface Sci.*, **64**, 341.
258. Jackson, D. P., and Barrett, J. H. (1977) *Computer Phys. Commun.*, **13**, 157.
259. Davies, J. A., Jackson, D. P., Matsunami, N. Norton, P. R., and Andersen, J. U. (1978) *Surface Sci.*, **78**, 274.
260. Bogh, E. and Stensgaard, I. (1978) *Phys. Lett.*, **65A**, 357.
261. Van der Veen, J. F., Smeenk, R. G., Tromp, R. M., and Saris, F. W. (1979) *Surface Sci.*, **79**, 219.
262. Poelsma, B., Verheij, L. K. and Boers, A. L. (1983) *Surface Sci.*, **133**, 344.
263. Ertl, G. and Kuppers, J. (1970) *Surface Sci.*, **21**, 61.
264. Maurice, V., Legendre, J. J. and Huber, M. (1983) *Surface Sci.*, **129**, 301.
265. (a) Ellis, W. P. (1972) in *Optical Transforms*, ed. Lipson, H. (Academic Press, London); (b) Ellis, W. P., and Campelt, B. D. (1968) in *Proc. LEED Symposium, Am. Cryst. Assn.*, Tucson (Polycrystal Book Service, Pittsburg); (c) Fedata, D. G., Fischer, T. E., and Robertson, W. P. (1968) *J. Appl. Phys.*, **39**, 5658.
266. Felter, T. E., and Estrup, P. J. (1976) *Surface Sci.*, **54**, 179.
267. Riwan, R., Guillot, C., and Paigne, J. (1975) *Surface Sci.*, **54**, 179.
268. Jona, F., Legg, K. O., Shih, H. D., Jepsen, D. W., and Marcus, P. M. (1978) *Phys. Rev. Lett.*, **40**, 1466.
269. Wagner, H.(1979) in *Physical Chemical Properties of Stepped Surfaces*, Springer Tracts in Modern Physics, Vol. 85 (Springer-Verlag, Berlin).
270. Henzler, M. (1977) in *Electron Spectroscopy for Surface Analysis' Topics in Current Physics*, Vol. 4, ed. Ibach, H. (Springer-Verlag, Berlin).
271. Houston, J. E., and Park, R. L. (1970) *Surface Sci.*, **21**, 209; (1971) **26**, 269.
272. Henzler, M., and Wulfert, F. W. (1976) in *Proc. XIII Intern. Conf. Phys. Semiconductors*, Rome.
273. Laramore, G. E., Houston, J. E., and Park, R. L. (1973) *J. Vac. Sci. Technol.*, **10**,
274. Zanazzi, E., Jona, F., Jepson, D. W., and Marcus, P. M. (1977) *J. Phys. C: Solid State Phys.*, **10**, 375.
275. Clarke, L. J., Baudoing, R. and Gauthier, Y. (1983) *J. Phys. C.*, **15**, 3249.
276. Aberdam, D., Baudoing, R., and Gaubert, C. (1977) *Surface Sci.*, **62**, 567.
277. Henzler, M. (1973) *Surface Sci.*, **36**, 109.
278. Henzler, M. (1970) *Surface Sci.*, **19**, 159.

279. Grunze, M. Bozso, F., Ertl, G., and Weiss, M. (1978) *Applications of Surface Sci.,* **1**, 241.
280. Dowben, P. A., Grunze, M. and Jones, R. G. (1981) *Surface Sci.,* **109**, L519.
281. McKee, C. S., Perry, D. L., and Roberts, M. W. (1973) *Surface Sci.,* **39**, 176.
282. McKee, C. S., Roberts, M. W., and Williams, M. L. (1977) *Advances in Colloid and Interface Sci.,* **8**, 29.
283. Avery, N. R. (1974) *Surface Sci.,* **43**, 101.
284. Dowben, P. A. and Jones, R. G. (1981) *Surface Sci.,* **105**, 334.
285. Barker, R. A. and Estrup, P. J. (1981) *J. Chem. Phys.,* **74**, 1442.
286. Einstein, T. L. and Schreiffer, J. R. (1973) *Phys. Rev.,* **B7**, 3629.
287. Grunze, M. and Dowben, P. J. (1982) *Appl. Surface Sci.,* **10**, 209.
288. Andersson, S. (1973) *Collective Properties of Physical Systems,* p. 188 Nobel Symposia Series, Vol. 24.
289. Estrup, P. J. (1969) in *Structure and Chemistry of Solid Surfaces,* ed. Somorjai, G. A. (Wiley, New York).
290. Roelofs, L. D., Park, R. L., and Einstein, T. L. (1979) *J. Vac. Sci. Technol.,* **16**, 478.
291. Wang, G. C., Lu, T.-M., and Lagally, M. G. (1978) *J. Chem. Phys.,* **69**, 479.
292. Tracy, J. C. and Blakely, J. M. (1969) in *Structure and Chemistry of Solid Surfaces,* ed. Somorjai, G. A. (Wiley, New York).
293. Gerlach, R. L., and Rhodin, T. N. (1969) *Surface Sci.,* **17**, 32.
294. Ertl, G., and Plancher, M. (1975) *Surface Sci.*, **48**, 364.
295. McKee, C. S., Roberts, M. W. and Williams, M. L. (1977) *Advances in Colloid and Interface Science,* **8**, 29.
296. Yang, W. S., Jona F., and Marcus, P. M. (1983) *Phys. Rev.,* **B27**, 1394.
297. Lagally, M. G., Ngoc, T. C., and Webb, M. B., (1971) *Phys. Rev. Lett.,* **26**, 1557; (1973) *Surface Sci.,* **35**, 117.
298. Aberdam, D. and Baudoing, R. (1972) *Solid State Commun.,* **10**, 1199.
299. Burkstrand, J. M., Kleiman, G. G., and Arlinghaus, F. J. (1974) *Surface Sci.,* **46**, 43.
300. Quinto, D. T., and Robertson, W. D. (1973) *Surface Sci.*, **34**, 501.
301. Buchholz, J. C., Wang, G. C., and Lagally, M. G. (1975) *Surface Sci.,* **49**, 508.
302. McDonnel, L., Woodruff, D. P., and Mitchell, K. A. R. (1974) *Surface Sci.,* **45**, 1.
303. Alff, M., and Moritz, W. (1979) *Surface Sci.,* **80**, 24.
304. Maglietta, M., Zanazzi, E., Jona, F., Jepson, D. W., and Marcus, P. M. (1977) *J. Phys. C.*, **10**, 3287.
305. Pendry, J. B. (1972) *J. Phys. C.,* **5**, 2567.
306. Miller, D. J., and Haneman, D. (1981) *Surface Sci.,* **104**, L237.
307. Stevans, M. A., and Russell, G. J. (1981) *Surface Sci.,* **104**, 356.
308. Davis, H. L., and Noonan, J. R. (1982) *Surface Sci.,* **115**, L75.
309. Williams, A. R., and Morgan, J. van W. (1974) *J. Phys. C.,* **7**, 37.
310. Azaroff, L. V. (1968) in *Elements of X-ray Crystallography* (McGraw-Hill, New York).
311. Van Hove, M. A., Tong, S. Y., and Elconin, M. H. (1977) *Surface Sci.,* **64**, 85.
312. Zanazzi, E., and Jona, F. (1977) *Surface Sci.,* **62**, 61.
313. Pendry, J. B. (1980) *J. Phys. C.,* **13**, 937.
314. Philip, J. and Rundgren, (1980) in *Proc. Conf. on Determination of Surface Structures by LEED,* ed. Marcus, P. M. (IBM Conference Series).
315. Gnedenko, B. V., and Kolmogorov, A. N. (1954) *Limit Distribution for Sums of Independent Random Variables* (Addison-Wesley, Cambridge, Mass.).
316. Kelley, J. L. (1955) *General Topology* (Van Nostrand, New York).
317. Feder, R., and Kirschner, J. (1978) *Phys. Stat. Sol (a),* **45**, K117.

318. Kirschner, J., and Feder, R. (1979) *Surface Sci.,* **79**, 176.
319. Masud, N., Kinniburgh, C. G., and Pendry, J. B. (1977) *J. Phys. C: Solid State Phys.,* **10**, 1.
320. Clarke, L. J. (1979) *Vacuum,* **29**, 405.
321. Clarke, L. J. (1979) *Surface Sci.,* **80**, 32.
322. Gauthier, Y., Baudoing, R., Clarke, L. J., and Gaubert, C. G. (1982) *J. Phys. C.*, **15**, 3249.
323. Gauthier, Y., Baudoing, R., Joly, J., Gaubert, C. G., and Rundgren, J. (1984) *J. Phys. C.*, to be published.
324. Watson, P. R., Shepherd, F. R., Frost, D. C., and Mitchell, K. A. R. (1978) *Surface Sci.,* **72**, 562.
325. Heilmann, P., Heinz, K. and Muller, K. (1979) *Surface Sci.,* **89**, 84.
326. Ignatiev, A., Jona, F., Debe, M., Johnson, D. C., White, S. J., and Woodruff, D. P. (1976) *J. Phys. C.,* **10**, 1109.
327. Barker, R. A., Horlacher, A. M., and Estrup, P. J. (1982) *J. Vac. Sci. Technol.*, **20**, 536.
328. Legg, K. O., Jona, F., Jepsen, D. W., and Marcus, P. M. (1977) *Phys. Rev.*, **B16**, 5271.
329. Debe, M. K., and King, D. A. (1982) *J. Phys. C.*, **15**, 2257.
330. Clarke, L. J., unpublished.
331. Barker, R. A., Estrup, P. J., Jona, F., and Marcus, P. M. (1978) *Solid State Commun.*, **25**, 375.
332. Lichtman, D. (1975) in *Methods and Phenomena*, Ch. 2, ed. Czanderna, A. W., Methods of Surface Analysis, Vol. 1 (Elsevier, New York).
333. Rosenberg, I. J., Weiss, A. H., and Canter, K. F. (1980) *Phys. Rev. Lett.,* **44**, 1139.
334. Weiss, A. H., Rosenberg, I. J., Canter, K. F., Duke, C. B., and Paton, A. (1983) *Phys. Rev.,* **B27**, 867.
315. Masud, N., Baudoing, R., Aberdam, D., and Gaubert, C. (1983) *Surface Sci.,* **133**, 580.
336. Ichikawa, T., and Ino, S. (1980) *Surface Sci.,* **97**, 489.
337. Liebsch, A. (1974) *Phys. Rev. Lett.,* **32**, 1203.
338. Johnson, P. D., Woodruff, D. P., Farrell, H. H., Smith, N. V., and Traum, M. M. (1983) *Surface Sci.,* **129**, 366.
339. Fadley, C. S., Kono S., Peterson, L.-G., Goldberg, S. M., Hall, N. F. T. Lloyd, J. T., and Hussain, Z. (1979) *Surface Sci.,* **89**, 52.
340. Hopkinson, J. F. L., Pendry, J. B., and Titterington, D. J. (1980) *Comput. Phys. Commun.,* **19**, 69.
341. Baudoing, R., Blanc, E., Gaubert, C., Gauthier, Y., and Gnuchev, N. (1983) *Surface Sci.,* **128**, 22.
342. Peterson, L.-G., Kono, S., Hall N. F. T., Goldberg, S., Lloyd, J. T., Fadley, C. S., and Pendry, J. R. (1980) *Mater. Sci. and Eng.,* **42**, 111.
343. Lytle, F. W., Via, G. H., and Sinfelt, J. H. (1977) *J. Chem. Phys.,* **67**, 3831.
344. Citrin, P. H., Eisenberger, P., and Hewitt, R. C. (1978) *Phys. Rev. Lett.,* **41**, 309.
345. Bunker, B. A., and Stern, E. A. (1983) *Phys. Rev.,* **B27**, 1017.
346. Durham, P. J., Pendry, J. B., and Norman, D. (1982) *J. Vac. Sci. Technol.,* **20**, 665.
347. Durham, P. J., Pendry, J. B., and Hodges, C. H. (1981) *Solid State Commun.,* **38**, 159.
348. Ibach, H., Horn, K., Dorn, R., and Luth, H. (1973) *Surface Sci.,* **38**, 433.
349. Dubois, L. H. and Somorjai, G. A. (1980) in *Vibrational Spectroscopies for Adsorbed Species* Ch. 9, eds Bell, A. T. and Hair M. L. (American Chem Soc., Washington DC).

350. Thiel, P. A., and Weinberg, W. H. (1980) in *Vibrational Spectroscopies for Adsorbed Species* Ch. 10, eds Bell, A. T. and Hair, M. L. (American Chem. Soc., Washington DC).
351. Andersson, S. (1979) *Surface Sci.,* **79**, 385.
352. Eischens, R. P., Pliskin, W. A., and Francis, S. A. (1954) *J. Chem. Phys.*, **22**, 1786.
353. Greenler, R. G., Rahn, R. R., and Schwarz, J. P. (1971) *J. Catal.,* **23**, 42.
354. Tompkins, H. G. (1975) in *Methods of Surface Analysis,* ed. Czanderna, Methods and Phenomena Vol. 1 (Elsevier, New York).
355. Fedyk, J. D., and Dignam, M. J. (1980) in *Vibrational Spectroscopies for Adsorbed Species* Ch. 5 eds Bell, A. T., and Hair, M. L. (American Chem. Soc., Washington DC).
356. Norton, P. R., Davies, J. A., Jackson, D. P., and Matsunami, N. (1979) *Surface Sci.,* **85**, 269.
357. Cheung, N. W., Feldman, L. C., Silverman, P. J., and Stensgaard, I. (1979) *Appl. Phys. Lett.,* **35**, 859.
358. Turkenburg, W. C., Soszka, W., Saris, F. W., Kersten, H. H., and Colenbrander, B. G. (1976) *Nucl. Instrum and Methods,* **132**, 587.
359. Van der Veen, J. F., Tromp, R. M., Smeenk, R. G., and Saris, F. W. (1980) *Nucl. Instrum and Methods,* **171**, 143.
360. Stern, O. (1929) *Die Naturwissenschaften,* **17**, 291.
361. (a) Rieder, K. H. and Engel, T. (1980) *Phys. Rev. Lett.,* **45**, 824.
(b) Rieder, K. H. (1983) *Surface Sci.,* **128**, 325.
362. Estel, J., Hoinkes, H., Kaarmann, H., Nahr, H., and Wilsch, H. (1967) *Surface Sci.*, **54**, 393.
363. Bush, P. S., and Raff, L. M. (1979) *J. Chem. Phys.,* **70**, 5026.
364. Tommasini, F. (1983) *Surface Sci.,* **125**, 188.
365. Madey, T. E., Czyzewski, J. J., and Yates, J. T. Jr. (1975) *Surface Sci.,* **49**, 465.
366. Niehus, H. (1979) *Surface Sci.,* **80**, 245.
367. Preuss, E. (1980) *Surface Sci.*, **94**, 249.
368. Holland, S. P., Garrison, B. J., and Winograd, N. (1979) *Phys. Rev. Lett.,* **43**, 220.
369. Robinson, I. K. (1983) *Phys. Rev. Lett.,* **50**, 1145.
370. (a)Estrup, P. J., Roelofs, L. D., and Ying, S. C. (1982) *Surface Sci.,* **123**, L703; (b) Melmed, A. J., and Graham, W. R. (1982) *Surface Sci.,* **123**, L706.
371. (a) Muller, E. W. (1960) *Advan. Electron. Phys.,* **13**, 83; (b) Muller, E. W., and Tsong, T. T. (1969) *Field Ion Microscopy* (Elsevier, New York).
372. McMullen, E. R., Perdew, J. P., and Rose, J. H. (1982) *Solid State Commun.,* **44**, 945.
373. Binnig, G., Rohrer, H., Gerber, Ch., and Weibel, E. (1982) *Phys. Rev. Lett.,* **49**, 57; *Appl. Phys. Lett.,* **42**, 178.
374. Binnig, G., Rohrer, H., Gerber, Ch., and Weibel, E. (1983) *Phys. Rev. Lett.,* **50**, 120.
375. Nall, B. J., Jette, A. N., and Bargeron, C. B. (1982) *Phys. Rev. Lett.,* **48**, 882; Jette, A. N., Nall, B. H., and Bargeron, C. B. (1983) *Phys. Rev.,* **B27**, 708.
376. Negre, M., Mischler, J., Benazeth, N., Noguera, C., and Spanjaard, D. (1978) *Surface Sci.,* **78**, 174.

Index

Note this index contains references to all the main features included in this book, but terms that appear frequently (such as unit cell, wave vector, etc.) are only given the first page number on which they appear, since they subsequently appear frequently as part of the natural terminology of LEED. Bold type indicates the main page references to a particular topic; italics indicate figures only.